INTRODUCTION TO DYNAMICS

INTRODUCTION TO DYNAMICS

BRUCE H. KARNOPP
The University of Michigan

**ADDISON-WESLEY
PUBLISHING COMPANY**

Reading, Massachusetts
Menlo Park, California · London · Don Mills, Ontario

ISBN 0-201-03614-2
BCDEFGHIJK-MA-89876543210

To my sons:

John Forrest
James Douglas

PREFACE

With any subject matter as classical as elementary mechanics, it is difficult to conceive of a new book which is both unique and utilitarian. The present volume is no exception. Where it attempts to differ from most other books on this subject is in organization, choice of topics, and level of presentation.

Each chapter contains a large number of examples. The student can examine as many of these as suits his own needs. A collection of important equations is given in a summary at the end of each chapter. And appendixes on mathematics and experimental and theoretical methods for determining moments and products of inertia are given at the end of the book. Many homework problems are given for each section of the book, and answers have been provided for all odd-numbered problems, as well as for any even-numbered problem on which a subsequent odd-numbered problem might depend.

Chapter 1 includes a thorough discussion of the English–gravitational, metric–absolute, and metric–gravitational unit systems. It is clear that today's engineering students will have to deal with each of these unit systems in the years to come.

Chapter 2 is a brief presentation of the statics of rigid bodies. The discussion centers on the forces encountered in statics, the free-body diagram, and conditions of equilibrium. This chapter can be omitted if students have had prior exposure to statics. On the other hand, the book can be used as the basis for a combined statics–dynamics course.

Chapters 3, 4, 7, and 8 contain the classical material regarding the kinematics and kinetics of a single particle, a collection of particles, and a collection of rigid bodies.

Chapters 5 and 6 discuss topics which may be deleted from a dynamics course. Chapter 5 concerns central force motion and, in particular, satellite motion. Chapter 6 is a thorough presentation of elementary vibration theory. These topics are positioned early in the book for two reasons: (1) Chapters 3 and 4 provide all necessary kinematics and kinetics to describe these problem areas; (2) central force motion and vibration theory are clearly relevant engineering problems. In particular, nearly all engineers encounter vibration problems, in one form or another, during their professional careers. Thus Chapters 5 and 6 can provide a very strong motivation for the study of dynamics.

In many engineering schools, as at the University of Michigan, vibration theory is made available to the students in a separate course. However, the author recalls vividly the admonition (and *admonition* may be too weak a word!) of his Latin teacher, *"repetitio est mater discendi."* A student who in a dynamics course has attained an incomplete understanding of such ideas as phase, natural frequency, and driving frequency will have an increased likelihood of complete understanding when he encounters these concepts the second time around in a vibrations course.

A course in differential equations is *not* a prerequisite for the understanding of any portion of the book. In particular, the only mathematics necessary for the understanding of Chapters 5 and 6 is integral calculus. A complete set of vibration solutions is given at the end of Chapter 6.

The original plan for this book called for coverage of such topics as three-dimensional kinematics and kinetics of rigid bodies and Lagrange equations. However, to include these topics in a volume of reasonable size would have required a significant condensation of the entire book. Since the course at the University of Michigan on which this book is based occurs in the second year, it was decided that condensation would be unwise and that these topics should be saved for subsequent presentation.

In acknowledging the help of others in writing a book, an author must strike a delicate balance: he must list enough names to indicate some personal humility and respect, but not so many as to create the impression that the book was pirated entirely from others. The author has had a great deal of help in various phases of the book from students and colleagues at the University of Michigan. Professors James Daily and Sam Clark gave encouragement and direction to the project. Professor Roger Low gave the manuscript a very thorough review, making many valuable suggestions.

Other reviews of the manuscript were obtained from Professors W. Feng (Carnegie-Mellon), B. C. McInnis (Houston), R. C. Rosenberg (Michigan State), and D. A. Smith (Wyoming). Help on various aspects of the book was provided by Dr. T. Tielking, Dr. J. Bernard, Lt. R. S. Duncan, and R. Cheng. The manuscript was typed and retyped and retyped . . . with precision and reasonably good humor by Mrs. Shirley (Walton) Bretting and Mrs. Lisabeth Graves Fishinger. It is only Addison-Wesley house policy which prevent me from acknowledging the great help of many individuals of that organization.

And finally I thank my wife, Paula, for her patience, which was extensive (though not inexhaustible), and for her instrumental help in proofreading the galleys and page proofs.

It is inevitable that some errors, typographical and otherwise, will appear in the final book. Ultimately the author must take responsibility for each error, and he would appreciate learning of any error which is unearthed.

Ann Arbor, Michigan   B.H.K.
February 1974

CONTENTS

Chapter 1   The Principles of Newtonian Mechanics

1. Introduction   1
2. The Laws of Motion   2
3. Dimensional Analysis and Unit Systems   4
4. Conclusion   12

Chapter 2   Introduction to Statics

1. Introduction   18
2. Forces   18
3. Equilibrium of a Particle   20
4. Equilibrium of a Rigid Body   25
5. Couples   32
6. Transformation of Force Systems; Equivalent Force Systems   35
7. Forces in Statics   43
8. The Free-Body Diagram   48
9. Conclusion   61
10. Summary   61

Chapter 3   Kinematics of a Particle

1. Introduction   92
2. Natural Coordinates   93
3. Cartesian Coordinates   99
4. Cylindrical Coordinates   106
5. Spherical Coordinates   111
6. Summary   114

Chapter 4   Dynamics of a Particle

1. Introduction   125
2. $\mathbf{F} = \dot{\mathbf{p}}$ 126
3. $\mathbf{M} = \dot{\mathbf{H}}$ 129
4. The Work–Energy Relation   133
5. Conservative Forces   141
6. Forces in Dynamics   148
7. Examples   157
8. The Free-Body Diagram (Again)   177
9. Summary   179

Chapter 5   Central Force Motion

1. Introduction   220
2. General Properties of Central Force Motion   221
3. The Gravitational Field   231
4. The Geometry of a Conic Section   236
5. Trajectories in the Gravitational Field   241
6. Kepler's Laws   243
7. Solution of Central-Force Problems   245
8. Summary   247

Chapter 6   Elementary Vibration Theory

1. Introduction   263
2. Examples of Simple Vibration Problems   265
3. Free Motion of an Undamped System   268
4. Periodic Functions of Time   279
5. Harmonically Forced Motion of an Undamped System   283
6. Free Motion of a Damped System   291
7. Harmonically Forced Motion of a Damped System   299
8. Arbitrarily Forced Motion of a Damped System   309
9. Introduction to Vibration Isolation   313
10. Summary   318

Chapter 7   Dynamics of a System of Particles

1. Introduction   351
2. Preliminaries   351
3. Motion of the Center of Mass   353
4. Motion About the Center of Mass, $\mathbf{M} = \dot{\mathbf{H}}$ 358
5. The Work–Energy Relation   366
6. Transition to the Rigid Body   375
7. Collision Problems   380
8. Systems Which Accumulate and Disburse Mass   396
9. Summary   403

Chapter 8   Rigid Bodies in Plane Motion

1. Introduction   430
2. Kinematical Preliminaries: Relative Motion   431

3. The Instant Center   436
4. Acceleration of a Point of a Rigid Body   442
5. Equations for the Dynamics of a Rigid Body in Plane Motion   445
6. Moment of Inertia   457
7. The Work–Energy Relation   465
8. Vibration Problems Involving Rigid Bodies   477
9. Summary   479

Appendix 1   **Mathematics**

     A. Vector Algebra   523
     B. Vector Calculus   529
     C. Miscellaneous Elementary Mathematics   534
     D. Matrices   537

Appendix 2   **Moments of Inertia**

     A. Moment of Inertia About a Line   542
     B. Moments and Products of Inertia of Three-Dimensional
        Bodies   547

Appendix 3   **The Work–Energy Relation:**
               Rigid-Body Motions   559

               Bibliography   563

               Answers to Selected Problems   567

               Index   581

1

THE PRINCIPLES OF NEWTONIAN MECHANICS

So act, that the rule on which thou actest would admit of being adopted as a law by all rational beings.

Metaphysics of Ethics
Immanuel Kant

But when he begins to deduce from this precept any of the actual duties of morality, he fails, almost grotesquely, to show that there would be any contradiction, any logical (not to say physical) impossibility, in the adoption by all rational beings of the most outrageously immoral rules of conduct. All he shows is that the *consequences* of their universal adoption would be such as no one would choose to incur.

Utilitarianism
John Stuart Mill

1. INTRODUCTION

The main topic of this book is Newtonian dynamics. Our primary aim will be to show how one can formulate and solve problems involving the motion of particles, systems of particles, and, in particular, rigid bodies.

An extraordinarily wide range of problems can be formulated with acceptable accuracy by means of Newtonian dynamics. Some typical examples are the dynamics of an automobile, the dynamics of machinery, the motion of a satellite, and the vibrations of a ship.

Most mechanical problems of interest to the engineer can properly be discussed within the Newtonian context. However, there are problems which are not ordinarily discussed in this context. These include the motion of atomic particles at speeds close to the speed of light and the determination of the motion of the planet Mercury. This latter set of problems is generally discussed in terms of the theory of relativity.

In this chapter we discuss the form of Newton's laws as they will be employed in this book. The remainder of the chapter deals with the theory of dimensions and unit systems. In particular, the English and metric unit systems are presented.

2. THE LAWS OF MOTION

The basis of Newtonian dynamics consists of three laws.

Newton's Laws

I. A particle remains at rest or in uniform motion along a straight line until acted upon by a force.

II. The rate of change of momentum of a particle is proportional to the force on the particle and has the same direction as the force.

III. If one particle exerts a force on a second particle, the second particle exerts a force on the first which is equal and opposite to the first force. Both forces lie on the line between the particles.*

To be meaningful, these laws require further explanation. Let us first try to define the terms involved.

Particle. A point in space with the property of mass.

Mass. A measure of the inertia of a body. In Newtonian dynamics, the mass of a particle is constant. In relativistic mechanics, the mass of a particle can change. Mass is a scalar quantity.

Force. The quantity which causes the nonuniform motion of a particle. A force can have magnitude, sense, and direction. Thus force has all the properties of a vector. The "force" described in the second law is the vector sum of all the forces on the particle.

Coordinate system. Newton's laws are valid in Newtonian (or inertial) frames of reference. These are either fixed in space or moving uniformly with respect to a coordinate system which is fixed in space. Often we write equations in a noninertial reference frame as if the frame were inertial (for example, a frame fixed to the earth). This is not strictly valid. In Chapter 8 we discuss this procedure.

---

* The form we have given for the Third Law is called the "strong" form. If the final sentence is deleted from our statement, we are left with the "weak" form of the law. See K. R. Symon, *Mechanics*, Addison-Wesley (1960), page 157.

Momentum. The momentum of a particle (sometimes called its linear momentum to distinguish it from its angular momentum or moment of momentum) is defined in terms of its mass and velocity:

$\mathbf{p} = m\mathbf{v}.$

We have not attempted here to answer any of the philosophical questions which arise from Newton's laws. There is much ambiguity in the concepts considered above. Perhaps the most difficult concept to appreciate is the inertial frame of reference. For if we are forced to use a coordinate system attached to the "fixed stars," as is commonly suggested, we are in somewhat of a bind. With respect to what reference are the fixed stars fixed?

In actual fact, the "truth" of Newton's laws consists only in their great utility in predicting the motions of physical systems. This, of course, does not mean that the Newtonian view of nature is correct. But it does mean that even though Newton's laws may not be based on the most accurate model of nature, they do give us a means of accurately predicting the motions of a wide range of physical systems.

The difficulty is much the same as in astronomy. If we wish to know only where to find a heavenly body at some time, the tables of Ptolemy* do a remarkably good job. However, Ptolemy was under the impression that the earth was the center of the universe. But the fact that Ptolemy had the wrong idea about the universe does not make his tables less useful.

There are engineering problems in which Newtonian mechanics does not hold with sufficient accuracy. Here relativistic effects must be considered. The correction terms which arise are of the form $\sqrt{1 - (v/c)^2}$, where v is the speed of the particle and c is the speed of light.†

Mathematically we can express the second and third laws as follows.

II. $\mathbf{F} = \dot{\mathbf{p}} = m\mathbf{a},$ \hfill (1.1)

$$\mathbf{v} = \frac{d\mathbf{r}}{dt}, \qquad \mathbf{a} = \frac{d^2\mathbf{r}}{dt^2},$$

where $\mathbf{r}$ is the vector from a fixed point O to the particle m and $\mathbf{F} = \mathbf{F}_1 + \mathbf{F}_2 + \cdots + \mathbf{F}_n$ is the sum of the forces on the particle (see Fig. 1.1).

III. Let $\mathbf{f}_{ij}$ be the force on m_i due to m_j. Then

$$\mathbf{f}_{ij} = -\mathbf{f}_{ji} \hfill (1.2)$$

and $\mathbf{f}_{ij}$ and $\mathbf{f}_{ji}$ lie on the line between m_i and m_j (Fig. 1.2).

---

* Ptolemy was an Egyptian mathematician and astronomer of the second century A.D.

† Speed of light = 186,000 miles/sec = 3.0×10^{10} cm/sec.

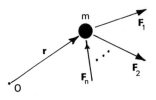

Figure 1.1

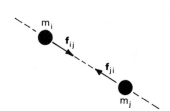

Figure 1.2

3. DIMENSIONAL ANALYSIS AND UNIT SYSTEMS

In order to write and evaluate the equations of statics and dynamics, we must
ultimately express all quantities in some system of units. There are four basic
quantities: force, mass, length, and time. We have one relation (1.1) between
the quantities. Thus we can assign units to three of the four quantities, and the
fourth will then be derived in terms of (1.1).

To facilitate our discussion of unit systems, let us begin with the introduction of the more fundamental concept of dimension. Suppose we are
measuring the distance between two points. We say that the *dimension* of distance is *length*. The dimension of area is length squared. Speed is defined as
the quotient: distance ÷ time. Thus the dimension of speed is the ratio of
length to time.

In dynamics, there are four dimensional quantities: force, mass, length,
time. Let us denote the dimensions of these quantities by

Force: F Length: L
Mass: M Time: T

Then the quantities below have the following dimensions.

Position: L Angle: 1
Velocity: L/T Area: L^2
Acceleration: L/T^2 Volume: L^3

A quantity with dimension 1 is called a *dimensionless* quantity. For example,
angles are dimensionless. To see this, consider a sector of a circle (Fig. 1.3).

Figure 1.3

The arc length s is given by

$$s = R\theta \quad \text{or} \quad \theta = s/R.$$

Since the dimensions of s and R are both L, the dimension of the angle θ is 1.

Suppose we take L and T to be basic dimensions. For the third basic dimension, we can take either F or M.

Gravitational Dimensions

The basic dimensions are F, L, and T. When we write (1.1) dimensionally, we have

$$F = ML/T^2. \tag{1.3}$$

Thus the derived dimension of mass in the gravitational system is

$$M = FT^2/L. \tag{1.4}$$

Absolute Dimensions

The basic dimensions are M, L, and T. Force is the derived dimension. Thus (1.3) gives the derived dimension of force as

$$F = ML/T^2. \tag{1.5}$$

Thus we can write all physical quantities in terms of their gravitational or absolute dimensions. Engineers seem to prefer the gravitational system, with its emphasis on force, while physicists seem to prefer the absolute system.

We can now give dimensional expressions for the quantities which we will encounter in this book.

Quantity	Gravitational dimension	Absolute dimension
Force	F	ML/T^2
Mass	FT^2/L	M
Length	L	L
Time	T	T
Position	L	L
Velocity	L/T	L/T
Acceleration	L/T^2	L/T^2 (cont.)

Quantity	Gravitational dimension	Absolute dimension
Angle	1	1
Angular velocity	$1/T$	$1/T$
Angular acceleration	$1/T^2$	$1/T^2$
Moment	FL	ML^2/T^2
Impulse	FT	ML/T
Angular impulse	FLT	ML^2/T
Momentum	FT	ML/T
Moment of momentum	FLT	ML^2/T
Areal velocity	L^2/T	L^2/T
Work (energy)	FL	ML^2/T^2
Frequency	$1/T$	$1/T$
Period	T	T
Moment of inertia	FT^2L	ML^2

The above list of quantities is useful in two ways. First we can use the table to check the dimensionality of any equation we have derived. That is, if we have a certain relation, each term must, at least, have the same dimensionality as every other term. Suppose, for example, we had derived the following equation for a force N:

$$N^2 = 23mv^2/h + F_0^2 \sin^2 \omega t. \tag{a}$$

For this to be valid dimensionally, each term must have the same dimensions. Thus we must have each of the following terms dimensionally equal:

$$\dim (N^2), \qquad \dim (23 \ mv^2/h), \qquad \dim \ (F_0^2 \sin^2 \omega t).$$

The first term has dimension F^2. The second has dimension $ML^2/T^2L = ML/T^2 = F$. Thus the second term has dimension F and not F^2, as is required (assuming that h is a length and 23 is dimensionless). The dimension of the third term is F^2, since $\dim (\sin \theta) = 1$. (The sine is the ratio of two sides of a triangle.) Thus we see that expression (a) can never be valid.

The expressions

$$N^2 = (mv^2/h)^2 + F_0^2 \sin^2 \omega t \tag{b}$$

or

$$N = (mv^2/h) + F_0 \sin \omega t \tag{c}$$

are both valid dimensionally. This does not mean that either (b) or (c) is true; only that (b) or (c) could be true, whereas (a) could not.

In performing dimensional calculations, we often encounter derivatives and integrals. These offer little trouble. For example,

$$\frac{dx}{dt} = \text{ratio of length to time} : L/T,$$

$$\frac{dy}{dx} = \text{ratio of two lengths} : L/L = 1.$$

Second derivatives are simple:

$$\text{dim} \left[\frac{d^2x}{dt^2} \right] = \text{dim} \left[\frac{d}{dt} \left(\frac{dx}{dt} \right) \right] = \frac{1}{T} \left(\frac{L}{T} \right) = \frac{L}{T^2},$$

$$\text{dim} \left[\frac{d^2y}{dx^2} \right] = \text{dim} \left[\frac{d}{dx} \left(\frac{dy}{dx} \right) \right] = \frac{1}{L} (1) = \frac{1}{L}.$$

In fact, part of the usefulness of the Leibniz notation for derivatives (dy/dx, d^2y/dx^2, etc.) is its utility in dimensional calculations. For example,

$$\text{dim} \left[\frac{d^nA}{dB^n} \right] = \frac{\text{dim} (A)}{[\text{dim} (B)]^n}.$$

Integrals are similar. We can think of the integral in approximate terms. For example,

$$\text{dim} \left[\int F \, dt \right] = \text{dim} \left[\sum_i F_i \, \Delta t_i \right] = FT \qquad \text{or} \qquad ML/T,$$

$$\text{dim} \left[\int \frac{dy}{dx} \, dx \right] = 1 \cdot L = L.$$

Thus we can think of the differential at the end of an integral as contributing its dimension to the quantity.

The second use for the dimension table comes when we derive a set of physical units. There are several unit systems. These depend on two factors: (i) size scale and (ii) history. When we speak of size scale, we mean the magnitude of the quantities involved. One could express the distance from the sun to the earth in angstrom units, but it would be as inadvisable as expressing atomic distances in light years. Thus we should pick units which roughly correspond to the magnitudes expected.

Units are also decided on for historical reasons. In the United States and Canada, engineers have used the British gravitational units, whereas in the rest of the world (now including Britain), engineers and scientists use metric units.*

---

* The fact that people in North America use the British system of units while everyone else uses metric units reminds one of the Oxford Atlas, which splits the geography of the world into "The British Isles and Dominions" and "The Rest of the World."

It now appears that North America will soon join the rest of the world in the use of the metric system.

The British absolute units are rarely used, and seem to exist now only as a source of confusion. An American engineer who uses the metric system is likely to employ the absolute units because he probably learned these units in a physics course. A German engineer would probably use the metric gravitational system.

FPS		British gravitational units Derived unit
Force	Pound (lb)	1 slug = that mass which has an acceleration of
Mass	Slug (slug)	1 ft/sec² due to a force of 1 lb
Length	Foot (ft)	1 slug = 1 lb-sec²/ft
Time	Second (sec)	
a) CGS		Metric absolute units Derived unit
Force	Dyne (dy)	1 dyne = the force required to give a 1-g mass
Mass	Gram (g)	an acceleration of 1 cm/sec²
Length	Centimeter (cm)	1 dyne = 1 g-cm/sec²
Time	Second (sec)	
b) MKS		Derived unit
Force	Newton (newton)	1 newton = the force required to give a 1-kg
Mass	Kilogram (kg)	mass an acceleration of 1 m/sec²
Length	Meter (m)	1 newton = 1 kg-m/sec²
Time	Second (sec)	

To begin to evaluate any equation involving dynamics numerically, we must begin in one of the above unit systems. We can, of course, later transform the derived quantities (for example, we can translate ft/sec into in./sec). In the British system, the only problem comes in writing mass units. We commonly hear people say, "the body weighs W pounds." To get the number of slugs, we must translate this statement to read "the force of gravity on the mass is W pounds." Thus we have

$$W \text{ (lb)} = m \text{ (slugs)} \, g \text{ (ft/sec}^2\text{)}.$$

The acceleration of gravity g is about 32.2 ft/sec². Thus

$$m \text{ (slugs)} = \left(\frac{W}{32.2}\right) \text{ lb-sec}^2/\text{ft}.$$

Thus a body which weighs 32.2 lb has a mass of 1 slug.

In the metric system, there is a similar difficulty. When someone speaks of a "body which weighs a kilogram," he is expressing himself in the metric gravitational system.

Metric gravitational units

a) CGS	Derived unit
Force Gram force (g)	mass unit = g-sec²/cm
Mass *	(N.B.:† g = gram force)
Length Centimeter (cm)	
Time Second (sec)	

b) MKS	Derived unit
Force Kilogram force (kg)	metric slug = kg-sec²/m
Mass Metric slug (metric slug)	(N.B.: kg = kilogram force)
Length Meter (m)	
Time Second (sec)	

* No name is commonly assigned to this unit.
† *Nota bene* = Latin for "note well" = pay attention!

If we say that a body *weighs* X kilograms in the gravitational system, then in the absolute system we say that the *mass* of the body is X kilograms. The weight in the absolute system is then converted into newtons:

weight   =   mass · acceleration
(newtons)   (kilograms)   (m/sec²)

or

weight (newtons) = mass (kilograms) 9.81 (m/sec²).

In the gravitational system, the mass is adjusted as follows:

weight   =   mass · acceleration
(kilograms)   (metric slug)   (m/sec²)

or

mass (metric slugs) = [weight (kilograms)/9.81 m/sec²].

Similarly, if we speak of a weight of X grams, we translate this to mean weight (dynes) = mass (g) 981 (cm/sec²).

Thus

mass (*)   = [weight (g)/981 (cm/sec²)].

Finally, just for completeness, we include the British absolute system, although it is rarely used, and will not be used in this text.

		British absolute units Derived unit
Force	Poundal (pdl)	1 poundal = force required to give 1 lbm an
Mass	Pound mass (lbm)	acceleration of 1 ft/sec²
Length	Foot (ft)	1 pdl = (1/32.2) lb
Time	Second (sec)	

In the chart on the opposite page we evaluate in the FPS-gravitational and the CGS and MKS absolute-unit systems the physical quantities of the chart on pages 5–6.

In many problems, we will need to convert quantities from one system to another. Thus we have a conversion table.

Force units
1 lb = 4.44 × 10⁵ dy = 4.44 newtons
1 dy = 10⁻⁵ newton = 2.25 × 10⁻⁶ lb
1 newton = 2.25 × 10⁻¹ lb = 10⁵ dy

Mass units
1 slug = 1.46 × 10⁴ g = 1.46 × 10 kg = 1.49 metric slug
1 g = 10⁻³ kg = 6.85 × 10⁻⁵ slug
1 kg = 6.85 × 10⁻² slug = 10³ g
1 metric slug = 6.72 × 10⁻¹ slug

Length units
1 ft = 3.05 × 10 cm = 3.05 × 10⁻¹ m
1 cm = 3.28 × 10⁻² ft = 10⁻² m
1 m = 10² cm = 3.28 ft
1 in. = 2.54 cm = 2.54 × 10⁻² m

---

* No special name is assigned to this unit.

Quantity	FPS grav.	CGS abs.	MKS abs.
Force	lb	dy	newton
Mass	slug	g	kg
Length	ft	cm	m
Time	← sec →		
Position	ft	cm	m
Velocity	ft/sec	cm/sec	m/sec
Acceleration	ft/sec²	cm/sec²	m/sec²
Angle	← 1 →		
Angular velocity	← 1/sec →		
Angular acceleration	← 1/sec² →		
Moment	lb-ft	dy-cm*	newton-m*
Impulse	lb-sec	dy-sec*	newton-sec*
Angular impulse	lb-ft-sec	dy-cm-sec*	newton-m-sec*
Momentum	lb-sec	g-cm/sec	kg-m/sec
Moment of momentum	lb-ft-sec	g-cm²/sec	kg-m²/sec
Areal velocity	ft²/sec	cm²/sec	m²/sec
Work (energy)	lb-ft	erg†	joule†
Frequency	← 1/sec or cycles/sec = hertz →		
Period	← sec. →		
Moment of inertia	slug-ft²	g-cm²	kg-m²

* Strictly speaking, this quantity should be written in terms of the mass unit rather than the derived-force unit. However, the form shown is more common.

† 1 erg = 1 g-cm²/sec² = 1 dyne-cm.

1 joule = 1 kg-m²/sec² = 1 newton-m.

Energy units

1 ft-lb = 1.36×10^7 erg = 1.36 joule

1 erg = 10^{-7} joule = 7.38×10^{-8} ft-lb

1 joule = 7.38×10^{-1} ft-lb = 10^7 erg

1 in.-lb = 1.13×10^6 erg = 1.13×10^{-1} joule

Example 1

Derive the conversion factor between ft-lb, joules, and ergs.

Solution

$$1 \text{ joule} = 1 \text{ newton-m}$$
$$= (2.25 \times 10^{-1} \text{ lb})(3.28 \text{ ft})$$
$$= 7.38 \times 10^{-1} \text{ lb-ft}$$

and

$$1 \text{ newton-m} = (10^5 \text{ dy})(10^2 \text{ cm})$$
$$= 10^7 \text{ dy cm} = 10^7 \text{ erg.}$$

....

Example 2

A block 1 in. $\times$ 2 in. $\times$ 3 in. which has a density of 100 lb/ft³ has a total force of 2 newtons on it. Find its acceleration.

Solution

The units here are all mixed up. Before we can evaluate any equation, we must put all quantities into a single unit system. Let us use the British gravitational system. For the force, we have

$$F = 2 \text{ newtons} = 2(2.25 \times 10^{-1} \text{ lb}) = 4.50 \times 10^{-1} \text{ lb.}$$

The block has a volume V:

$$V = (1 \cdot 2 \cdot 3) \text{ in.}^3 = 6(\tfrac{1}{12} \text{ ft})^3 = 3.47 \times 10^{-3} \text{ (ft)}^3.$$

Thus the block weighs

$$(3.47 \times 10^{-3} \text{ ft}^3)(100 \text{ lb/ft}^3) = 3.47 \times 10^{-1} \text{ lb.}$$

And the mass of the block is $M = 1.08 \times 10^{-2}$ slug. So when we write $\mathbf{F} = m\mathbf{a}$, we get

$$|\mathbf{a}| = |\mathbf{F}|/m = \frac{4.50 \times 10^{-1} \text{ lb}}{1.08 \times 10^{-2} \text{ slug}} = 41.7 \text{ lb/slug.}$$

To see that this is in fact 41.7 ft/sec², recall that

$$1 \text{ slug} = 1 \text{ lb-sec}^2/\text{ft.}$$

Thus

$$41.7 \text{ lb/slug} = 41.7 \frac{\text{lb}}{\text{lb-sec}^2/\text{ft}} = 41.7 \text{ ft/sec}^2.$$

....

4. CONCLUSION

Throughout this book, we will be writing various equations to describe the behavior of physical systems. In actual engineering practice, these equations alone are seldom sufficient to handle a particular problem. The problem is complete only when numerical values have been fed into the equations. It is very important that you not only learn how to write proper equations but also that you have a good idea of the order of magnitude of the quantities involved.

On the other hand, we strongly caution you not to plug in numerical values until the end of the derivation of the system equations. For if the numerical values are inserted too soon, it is very difficult to detect the point at which errors have been made.

PROBLEMS

Section 1.2

The following problems involve the algebra and calculus of vectors. If you have trouble with the operations indicated, refer to Appendix 1, Sections A and B.

Problem 1.1

Consider the vectors:

$$\mathbf{v}_1 = \mathbf{i} + \mathbf{j}, \qquad \mathbf{v}_2 = \mathbf{i} - \mathbf{j}.$$

a) Compute $10\mathbf{v}_1$, $5\mathbf{v}_2$.
b) Compute $(10\mathbf{v}_1 - 5\mathbf{v}_2)$.
c) Compute $\mathbf{v}_1 \cdot \mathbf{v}_2$.
d) Compute $\mathbf{v}_1 \times \mathbf{v}_2$.

Problem 1.2

Consider the vector $\mathbf{v} = 2\mathbf{i} + \mathbf{j} - 2\mathbf{k}$.

a) Find the length of the vector $\mathbf{v}$.
b) Find a unit vector in the direction of $\mathbf{v}$.

Problem 1.3

Let $\mathbf{v}_1 = 2\mathbf{i} + 2\mathbf{j} + \mathbf{k}$ and $\mathbf{v}_2 = 2\mathbf{i} + 10\mathbf{j} - 11\mathbf{k}$.

a) Find the projection of $\mathbf{v}_1$ on $\mathbf{v}_2$ (see Fig. 1.4).
b) Find the projection of $\mathbf{v}_2$ on $\mathbf{v}_1$.

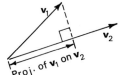

Figure 1.4

Problem 1.4

A parallelogram has edge vectors $\mathbf{v}_1$ and $\mathbf{v}_2$. Show that the area of the parallelogram (Fig. 1.5) can be expressed as

$$A = |\mathbf{v}_1 \times \mathbf{v}_2|.$$

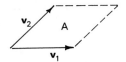

Figure 1.5

Problem 1.5

Find the area of the parallelogram with edge vectors

$$v_1 = 2i + 2j + k \text{ in.,}$$
$$v_2 = 2i + 10j - 11k \text{ in.}$$

Problem 1.6

Without evaluating each quantity in detail, show that the following quantities each equal zero.

a) $[(v_1 \times v_2) \cdot (10v_1)]$

b) $[(v_1 \times v_2) \times (v_1 \times v_2)]$

c) $[(10i - 5j) \cdot (i \times 5j)]$

d) $[(i + j) \times (i - j)] \times k$

Problem 1.7

Show that the volume of a parallelepiped (Fig. 1.6) with edge vectors v_1, v_2, v_3 can be expressed as

$$V = |(v_1 \times v_2) \cdot v_3|.$$

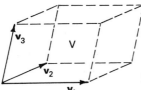

Figure 1.6

Problem 1.8

The edge vectors of a parallelepiped are

$$v_1 = 10i + j \text{ in.,} \qquad v_2 = 2i - 10j \text{ in.,} \qquad v_3 = 12k \text{ in.}$$

Find the volume of the parallelepiped. (*Hint:* See Problem 1.7)

Problem 1.9

The vector **v** has constant length, but its orientation changes with time. Show that

$$\frac{dv}{dt} \perp v.$$

Problem 1.10

Consider the vector

$$v = i \cos (10t) + j \sin (10t).$$

a) Show that **v** has constant length.

b) Compute $d\mathbf{v}/dt$.

c) Verify that $d\mathbf{v}/dt \perp \mathbf{v}$.

Section 1.3

Problem 1.11

Suppose we take, as fundamental dimensions, (a) F, M, L, (b) F, M, T. In each case, write

   i) the dimension of the remaining quantity,

  ii) the dimension of acceleration,

 iii) the dimension of energy.

Problem 1.12

Determine whether the following equation is dimensionally homogeneous:

$$\frac{d}{dt} \int_0^x F \, dx = \frac{1}{2} \frac{dm}{dt} v^2 + mva,$$

where F is a force, x is a distance, v is speed, a is acceleration, m is mass, and t is time.

Problem 1.13

Given that F is a force, x is a displacement, θ is an angle, and v is a speed, determine the dimensions of the quantities I and k in order that the following equation is dimensionally homogeneous:

$$\int_0^x F \, dx = \tfrac{1}{2} I \left(\frac{d\theta}{dt} \right)^2 + \tfrac{1}{2} k v^2.$$

Problem 1.14

A student has "derived" a new law of dynamics. He claims that the moment on a body equals the change in its kinetic energy. Suppose that

$$\mathbf{M} = \mathbf{r} \times \mathbf{F} \qquad \text{and} \qquad T = \tfrac{1}{2} m v^2,$$

where r is length, F is force, m is mass, and v is speed. Determine whether the equation

$$\mathbf{M} = \Delta T$$

is possible from a dimensional point of view.

Problem 1.15

The number of people p in a community as a function of time is given by the equation

$$p = k_0 + k_1 t \sin \left(\frac{2\pi t}{\tau} \right) + k_2 e^{k_3 t}.$$

The dimension of p is $[p] = P$ (people). Find the proper dimensions for the constants (i) k_0, (ii) k_1, (iii) τ, (iv) k_2, (v) k_3. In this equation, the variable t equals time.

Problem 1.16

A constant which is important for the study of satellite motion* about the earth is

$$(\gamma m_e) = 1.255 \times 10^{12} \text{ mi}^3/\text{hr}^2.$$

Express this constant in (a) ft^3/sec^2, (b) m^3/sec^2(m = meter)

Problem 1.17

A German-American wishes to express a quantity of energy equal to 20 ft-lb in (a) in.-newton, (b) cm-oz. In each case, find the equivalent quantity.

Problem 1.18

A body weighs 386 lb. Find the mass of the body expressed in (a) slugs, (b) kg (mass), (c) metric slugs, (d) g (mass).

Problem 1.19

In a swimming meet, a certain professor swims the 200-yd breast-stroke event in 1 min 30.6 sec.

   a) What is his average speed in miles/hr?

   b) If he maintains the average speed computed in (a), how long would it take him to swim (i) 100 meters, (ii) 1 mile?

Problem 1.20

Express the acceleration of gravity g in miles/millisecond^2.

Problem 1.21

The linear momentum of a certain body mv is given as 10 slug-ft/sec. Convert this quantity to CGS (absolute) units.

Problem 1.22

A car which weighs 2000 lb is traveling 60 mi/hr. Express the momentum of the car mv in terms of slug-ft/sec.

Problem 1.23

A jar of peanut butter weighing 12 oz has an acceleration of 1.0 in./sec^2 along a straight line. Find the total force acting on the jar, and express the force in ounces.

---

* We shall encounter these problems and use this constant in Chapter 5.

Problem 1.24

Suppose we wish to write the equations of dynamics with

time expressed in seconds,
force expressed in pounds,
length expressed in inches.

a) What mass unit should be used in order to write $\mathbf{F} = m\mathbf{a}$?

b) What value of the constant g should be employed?

c) An object weighs 400 lb; how many mass units would it have?

Problem 1.25

Suppose that, in writing the equations of dynamics, we express

force in pounds
mass in slugs
length in feet
time in milliseconds

Show that Newton's second law would have to be written $\mathbf{F} = 10^6 \, m\mathbf{a}$.

2

INTRODUCTION TO STATICS

I've got to admit it's getting better—
a little better all the time
(can't get any worse! . . .).

Getting Better
John Lennon, Paul McCartney

1. INTRODUCTION*

Statics is the study of systems that do not change with time. The fundamental notion of statics is to define the conditions under which a system remains in a given state. There are static fluids, static (electric) circuits, static structures, etc. If, for example, an apartment building ceased to be essentially static, its occupants would hardly have cause to applaud.

Thus statics is an important engineering subject on its own. However, the most important aspect of such a study concerns design considerations. What loads are to be expected in the various structure members? What design should those members have in order to safely withstand those loads? These are typical questions in statics. We are concerned here with statics as it relates to the main subject of this book: Newtonian dynamics.

The first step of analysis in Newtonian dynamics is to construct a free-body diagram. This usually means isolating each particle or rigid body from its immediate environment in order to examine the nature of the forces acting on it. Up to this point, statics and dynamics are identical. This is perhaps the most important reason for discussing statics here.

Pay close attention to the concept of the free-body diagram, for it is the heart and soul of the Newtonian approach to statics and dynamics. And if you go into the engineering world without being able to solve either statics or dynamics problems, you will be at a severe disadvantage.

2. FORCES

We use the term *force* frequently in everyday speech: He (she) was forcibly ejected from the bar; *La Forza del Destino;* the group forced a merger of the

---

* Many students using this book may have had a prior course in statics or an introduction to solid mechanics. If so, this chapter will serve as a review, or it can be skipped.

18

two companies. In each of the these phrases, the word *force* is used in a different manner. Yet there is one common denominator. The force is never defined—only its effect.

Traditionally, books on mechanics have defined mechanical forces as "pushes and pulls." To get more mathematical, one often quotes Newton's second law of motion,

F ∝ **a**,

which relates the proportionality between the force on a particle and the resulting acceleration. This statement leaves much to be desired, in that force is now defined only as that which produces an acceleration.

There are things which produce forces, e.g., springs, unbalanced masses, shock absorbers, dashpots. In each of these cases, however, the concept of force arises only as an intermediate variable related to displacement or velocity.

Modern researches into the fundamental nature of mechanics have decided that in a logical sense mechanics is very similar to geometry. To ask for a physical definition of a force is the same as asking for a physical definition of a straight line. The only definitions which make sense are those written in terms of the mathematical system within which we are operating.

In light of the above discussion, it is not surprising that there might have been attempts to define all mechanics without the concept of force. On the other hand, to the engineer, force is a very meaningful concept. Engineers can relate the resultant force in a structural member to its ultimate strength. From this point of view, it is also not surprising that there have been attempts to define mechanics, statics, and dynamics entirely in terms of forces.

The fact of the matter is that forces do play an important role in mechanics. From the point of view of statics, the important physical forces which arise are described in courses in strength of materials. Other physical forces that are important in dynamics will be described in Chapter 4 (specifically Section 6).

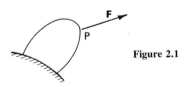

Figure 2.1

We say that a force **F** is applied to a point of a body (Fig. 2.1). The force has both a direction in space, a positive or negative sense along that direction, and a length or magnitude. The "force" described here has all the properties of a vector that is fixed at one point, its origin.

The body in Fig. 2.1 is "tied down" over a portion of its surface. The point *P* might move, even though a portion of the body is restrained from motion. Such motion is called *deformation,* since there is a relative motion between various parts of the body. That is, the distance of one point from another might vary when **F** is applied. Deformations are the subject of courses in strength of materials and elasticity. In this chapter, we shall be concerned primarily with *rigid bodies.* In particular, we are concerned with the conditions under which a rigid body, or system of rigid bodies, is in equilibrium. By saying that a rigid body is in equilibrium, we mean that each point of the body is fixed with respect to some one inertial reference frame. We shall examine this concept of equilibrium for both particles and rigid bodies.

In order to discuss equilibrium, we must know a little about dynamics. Again we quote Newton's second law:

$$\mathbf{F} \propto \mathbf{a}.$$

This law can be written in any inertial reference frame. If this is true, any two inertial reference frames can differ from each other only by a constant velocity (translation).

If a particle is to be in equilibrium, it must be at rest with respect to some inertial reference frame. And this in turn means that the sum of the forces on the particle must be zero:

$$\mathbf{F} = 0.$$

Conversely, if the sum of the forces on a particle is zero, it can have, at most, a constant velocity (since $\mathbf{a} = 0$). And this means in turn that the particle is at rest with respect to some inertial reference frame.* Thus we have:

> A necessary and sufficient condition for a particle to be in equilibrium is that the sum of the forces on the particle be zero.

3. EQUILIBRIUM OF A PARTICLE

Consider the particle in Fig. 2.2.

There are *n* forces $\mathbf{F}_1, \mathbf{F}_2, \ldots, \mathbf{F}_n$ which act on the particle. Under what condition is the particle in equilibrium? The sum of the forces on the particle is zero. Thus

$$\mathbf{F}_1 + \mathbf{F}_2 + \cdots + \mathbf{F}_n = 0. \tag{2.1}$$

---

* Suppose we have an inertial reference frame and a second frame which has a constant velocity with respect to the first frame. Then the second frame is also an inertial frame.

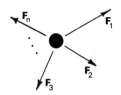

Figure 2.2

Condition (2.1) is both necessary and sufficient for the equilibrium of the particle.

The vector sum (2.1) is called the *force resultant* of a set of forces. It is denoted by $\mathbf{F}$:

$$\mathbf{F} = \mathbf{F}_1 + \mathbf{F}_2 + \cdots + \mathbf{F}_n = \sum_{i=1}^{n} \mathbf{F}_i. \tag{2.2}$$

> A necessary and sufficient condition for a particle to be in equilibrium is that the force resultant acting on the particle be zero.

We will also wish to speak of the moment of a force. Consider the force in Fig. 2.3.

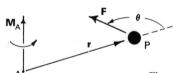

Figure 2.3

The force $\mathbf{F}$ is applied to the point P. The moment of $\mathbf{F}$ about a point A is then defined in terms of a cross product,*

$$\mathbf{M}_A = \mathbf{r} \times \mathbf{F}, \tag{2.3}$$

where $\mathbf{r}$ is the vector from A to P. This definition gives $\mathbf{M}_A$ as a vector perpendicular to the plane of $\mathbf{r}$ and $\mathbf{F}$, with the sense given by the right-hand rule. The moment vector is denoted by an ordinary vector with an associated curved vector, as shown in Fig. 2.3.

The above definition has two interpretations. Consider x and y axes in the plane of $\mathbf{r}$ and $\mathbf{F}$, with $\mathbf{r}$ in the x direction (Fig. 2.4). Then we can evaluate

$$\mathbf{M}_A = \mathbf{r} \times \mathbf{F} = \begin{vmatrix} \mathbf{i} & \mathbf{j} & \mathbf{k} \\ x & 0 & 0 \\ F_x & F_y & 0 \end{vmatrix} = xF_y\mathbf{k}.$$

---

* A review of the mathematics required for this book is given in Appendix 1.

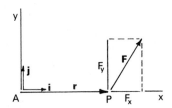

Figure 2.4

Thus the moment about A is given by the moment of the component of $\mathbf{F}$ perpendicular to $\mathbf{r}$. The moment is given by F_y multiplied by r as a moment arm. Now suppose we align x along $\mathbf{F}$ (Fig. 2.5).

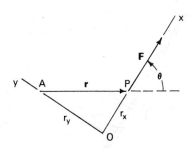

Figure 2.5

Now $\mathbf{F} = F\mathbf{i}$ and $\mathbf{r} = r_x\mathbf{i} - r_y\mathbf{j}$:

$$\mathbf{M}_A = \mathbf{r} \times \mathbf{F} = \begin{vmatrix} \mathbf{i} & \mathbf{j} & \mathbf{k} \\ r_x & -r_y & 0 \\ F & 0 & 0 \end{vmatrix} = Fr_y\mathbf{k}.$$

Thus the force $\mathbf{F}$ (the whole force) now acts though a moment arm $r_y = r\sin\theta$, where θ is the smaller of the two angles between $\mathbf{r}$ and $\mathbf{F}$.

These two interpretations result from

$$|\mathbf{M}_A| = |\mathbf{r}|(|\mathbf{F}|\sin\theta) = (|\mathbf{r}|\sin\theta)|\mathbf{F}|.$$

Returning to the particle, we can now consider the moments of the forces on the particle (Fig. 2.6).

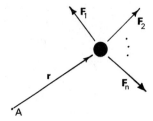

Figure 2.6

We now have n moment vectors:

$$\mathbf{M}_{A1} = \mathbf{r} \times \mathbf{F}_1, \qquad \mathbf{M}_{A2} = \mathbf{r} \times \mathbf{F}_2, \qquad \ldots, \qquad \mathbf{M}_{An} = \mathbf{r} \times \mathbf{F}_n.$$

We can sum up these vectors to get $\mathbf{M}_A$, the *moment resultant* about A of the forces on the particle:

$$\mathbf{M}_A = \sum_{i=1}^{n} \mathbf{r} \times \mathbf{F}_i = \mathbf{r} \times \sum_{i=1}^{n} \mathbf{F}_i = \mathbf{r} \times \mathbf{F}.$$

Thus, in the case of a single particle, the moment resultant and the force resultant are related:

$$\mathbf{M}_A = \mathbf{r} \times \mathbf{F}. \tag{2.4}$$

(This is not the case for a rigid body.) Thus if the particle is in equilibrium, $\mathbf{F} = 0$, and hence $\mathbf{M}_A = 0$. On the other hand, if $\mathbf{M}_A = 0$, there is no guarantee that $\mathbf{F} = 0$, and thus there is no guarantee that the particle is in equilibrium. Figure 2.7 gives two cases of this.

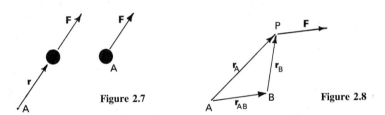

Figure 2.7   Figure 2.8

Thus the statement that $\mathbf{M}_A = 0$ is a "necessary" condition for the equilibrium, but not a "sufficient" condition.

Note that in writing $\mathbf{M}_A$ we purposely used the subscript A. We do this not (only) to give secretaries, typesetters, and notetakers fits. The trouble is that the moment of a force varies from point to point. Thus if we have two points, A and B (Fig. 2.8), then

$$\mathbf{M}_A = \mathbf{r}_A \times \mathbf{F}, \qquad \mathbf{M}_B = \mathbf{r}_B \times \mathbf{F}.$$

Or if we write

$$\mathbf{r}_A = \mathbf{r}_{AB} + \mathbf{r}_B,$$

we have

$$\mathbf{M}_A = \mathbf{r}_A \times \mathbf{F} = (\mathbf{r}_{AB} + \mathbf{r}_B) \times \mathbf{F}$$
$$= \mathbf{r}_{AB} \times \mathbf{F} + \mathbf{r}_B \times \mathbf{F} = \mathbf{r}_{AB} \times \mathbf{F} + \mathbf{M}_B.$$

Thus

$$\mathbf{M}_A = \mathbf{r}_{AB} \times \mathbf{F} + \mathbf{M}_B. \tag{2.5}$$

Thus if we know the force resultant, **F**, and the moment resultant, $\mathbf{M}_A$, about a point A, (2.5) gives us a means of finding the moment resultant about any other point, B.

From the point of view of the equilibrium of a particle, it is clear that the simplest equation says that the force resultant is zero, $\mathbf{F} = 0$. Moment equations become important when we are dealing with the equilibrium of rigid bodies.

Example 1

A particle rests on the sphere shown in Fig. 2.9. The weight of the particle is W (lb). What are the forces exerted on the particle by the sphere?

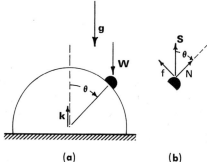

Figure 2.9

(a)   (b)

Solution

Since the particle is in equilibrium, we know that the weight must be counteracted by some equal and opposite force **S** exerted by the sphere:

$$\mathbf{S} = -(-W\mathbf{k}) = W\mathbf{k}.$$

It is interesting to note that we can divide **S** into two components, one normal to the sphere, N, and one tangential to the sphere surface, f. In Fig. 2.9(b), we see that

$$N = S \cos \theta, \qquad f = S \sin \theta,$$

and thus equilibrium requires that

$$N = W \cos \theta, \qquad f = W \sin \theta,$$

where θ defines the position of the particle on the sphere.

Thus far we have been discussing only (very trivial) mathematics. There is another equally important, and in a sense prior, question. By what physical mechanisms do the forces which constitute **S** arise? That is, are N and f possible and predictable? We shall discuss the matter of the physical origin of the forces in statics in Section 7.

···

Example 2

A particle which weighs $W = 10$ lb rests on a smooth, 30° incline restrained by a weight-less string (Fig. 2.10a). What is the tension in the string?

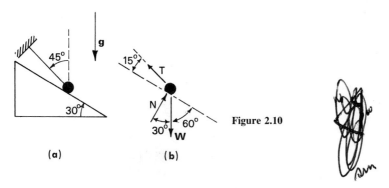

Figure 2.10

(a)   (b)

Solution

We begin the solution by examining the forces on the particle (Fig. 2.10b). The weight acts vertically downward. The incline is smooth, and thus there is no friction force. The only force which the incline can exert on the particle must be normal to the plane. This force is denoted by N. The string tension acts in the direction of the string. It is a tensile force; you can't push on a string.

Since the particle is at rest, the three forces must add (vectorially) to zero. We write the sum of the force components normal to the incline, F_n, and the sum of the force components tangential to the incline, F_t:

$$F_n = 0: \quad N - W \sin 60° + T \sin 15° = 0,$$

$$F_t = 0: \quad W \cos 60° - T \cos 15° = 0.$$

Recall that the weight W is given. Thus we have two equations for two unknowns, N and T. If we wish to find T, we can get it from the second equation:

$$T = W \left(\frac{\cos 60°}{\cos 15°}\right) = 10 \text{ lb} \left(\frac{0.500}{0.966}\right) \quad \text{Thus} \quad T = 5.17 \text{ lb.}$$

Note that once we have determined T, we can return to the first equilibrium equation, $F_n = 0$, to determine N.

··

4. EQUILIBRIUM OF A RIGID BODY

A rigid body is defined to be a body in which the distance between any two points of the body remains constant. This is an idealization, obviously, but it is an extremely useful concept in statics, and especially in dynamics.

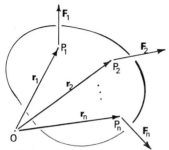

Figure 2.11

Suppose that a rigid body is subjected to forces F_1, F_2, . . . , F_n* with points of application, P_1, P_2, . . . , P_n. Figure 2.11 shows this rigid body and the forces F_1, F_2, . . . , F_n.

We wish to determine whether the rigid body is in equilibrium. It seems reasonable to say that a rigid body is in equilibrium if every point of it is at rest with respect to some (one) inertial reference frame. Again, to determine conditions which assure equilibrium for a rigid body, we refer to dynamics. In fact, in books on intermediate dynamics it is shown that the following conditions are necessary and sufficient for the equilibrium of a rigid body:

   i) $F = 0$ for all time $t \geq t_0$.

   ii) $M_O = 0$ for all time $t \geq t_0$.   (2.6)

   iii) The body is at rest with respect to some inertial reference frame at $t = t_0$.

In Eq. (2.6), we have (referring to Fig. 2.11)

$$F = F_1 + F_2 + \cdots + F_n \ldots \text{ force resultant}$$

$$M_O = r_1 \times F_1 + r_2 \times F_2 + \cdots + r_n \times F_n \ldots \text{ moment resultant}$$
$$\text{about } O.$$

The three parts of (2.6) require some explanation. First, $F = 0$ means that we take all the forces on the body, add them vectorially without regard to the points of application, and check to see that the vector sum is zero.

The second part of (2.6) says that we compute the moment of each force about some point O. These moments are then added vectorially to give the moment resultant about O. Again we must check to see that this moment resultant is zero. It is important to note that any point O could be used in this computation. Below we show why this is true.

---

   * For the time being, we shall consider a rigid body subject to applied forces only. In Section 5 we shall introduce the concept of a couple. In general, a rigid body is subject to both forces and couples.

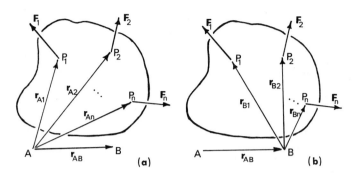

Figure 2.12

The third part of (2.6) prevents us from saying that a rigid body which is spinning freely is in equilibrium even though the first two conditions of (2.6) are met.

In order to appreciate the second part of (2.6), we derive the *moment transformation rule*. The result is similar to (2.5). Suppose we know the force resultant $\mathbf{F}$ and the moment resultant about some point A, $\mathbf{M}_A$. We wish to evaluate the moment resultant about another point B, $\mathbf{M}_B$. Consider Fig. 2.12. We denote the vector from A to P_i by $\mathbf{r}_{Ai}$ and the vector from B to P_i by $\mathbf{r}_{Bi}$. The vector from A to B is denoted by $\mathbf{r}_{AB}$.

Writing the moment resultant about A and then about B, we have

$$\mathbf{M}_A = \mathbf{r}_{A1} \times \mathbf{F}_1 + \mathbf{r}_{A2} \times \mathbf{F}_2 + \cdots + \mathbf{r}_{An} \times \mathbf{F}_n = \sum_{i=1}^{n} \mathbf{r}_{Ai} \times \mathbf{F}_i,$$

$$\mathbf{M}_B = \mathbf{r}_{B1} \times \mathbf{F}_1 + \mathbf{r}_{B2} \times \mathbf{F}_2 + \cdots + \mathbf{r}_{Bn} \times \mathbf{F}_n = \sum_{i=1}^{n} \mathbf{r}_{Bi} \times \mathbf{F}_i.$$

(2.7)

The vectors $\mathbf{r}_{Ai}$ and $\mathbf{r}_{Bi}$ are related, however. We have, for example,

$$\mathbf{r}_{A1} = \mathbf{r}_{AB} + \mathbf{r}_{B1},$$

since $\mathbf{r}_{A1}$ is the vector from A to P_1. But this is the same as the sum of the vector from A to B ($\mathbf{r}_{AB}$) plus the vector from B to P_1 ($\mathbf{r}_{B1}$). In general, we have

$$\mathbf{r}_{Ai} = \mathbf{r}_{AB} + \mathbf{r}_{Bi}.$$

Inserting this into the expression for $\mathbf{M}_A$, (2.7), gives

$$\mathbf{M}_A = \sum_{i=1}^{n} \mathbf{r}_{Ai} \times \mathbf{F}_i = \sum_{i=1}^{n} \mathbf{r}_{AB} \times \mathbf{F}_i + \sum_{i=1}^{n} \mathbf{r}_{Bi} \times \mathbf{F}_i.$$

The term

$$\sum_{i=1}^{n} \mathbf{r}_{AB} \times \mathbf{F}_i = \mathbf{r}_{AB} \times \left(\sum_{i=1}^{n} \mathbf{F}_i \right) = \mathbf{r}_{AB} \times \mathbf{F},$$

where $\mathbf{F} = \sum_{i=1}^{n} \mathbf{F}_i$ is the force resultant of the forces on the body.

The final term is just $\mathbf{M}_B$, as can be seen from the second equation of (2.7). Thus

$$\mathbf{M}_A = \mathbf{r}_{AB} \times \mathbf{F} + \mathbf{M}_B, \tag{2.8}$$

which is the rule for the moment resultant transformation.

Now suppose we consider a rigid body in equilibrium. Then by (2.6), we have

$$\mathbf{F} = 0 \quad \text{and} \quad \mathbf{M}_O = 0.$$

Note that (2.8) now tells us that in this circumstance,

$$\mathbf{M}_O = \mathbf{r}_{OB} \times \mathbf{F} + \mathbf{M}_B$$

or

$$0 = \mathbf{r}_{OB} \times 0 + \mathbf{M}_B,$$

and thus $\mathbf{M}_B = 0$.

There are two ways to view this result. First, if the force resultant $\mathbf{F}$ is zero and the moment resultant about any point $\mathbf{M}_O$ is zero, we are assured that the moment resultant about every point is zero.

The second way of viewing this result is from the point of view of independent equilibrium equations. Suppose we have the two vector equations

$$\mathbf{F} = 0 \quad \text{and} \quad \mathbf{M}_O = 0.$$

Can we find any other equilibrium equations which contain information not already in these equations? The answer is no, because as we have seen, these two conditions imply that the moment resultant about any point is zero, by (2.8).

Thus equations (2.6) give two independent vector equations of equilibrium, and no more. And in general these vector equations are equivalent to at most six independent scalar equations.

What we have said about independent equations is true, but can be misleading. There are better and worse equilibrium equations. If they have a simple form, they are easier to handle than if they contain all the unknowns of the problem. Thus the choice of which moment equation we write can be important.

Example 3 (The Lever)

Consider the weightless bar in Fig. 2.13(a). The forces on it are shown in Fig. 2.13(b). Suppose the force P_2 is given as 1000 lb. What force P_1 is required for equilibrium?

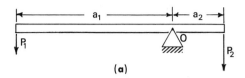

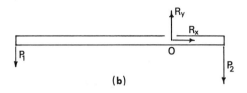

Figure 2.13

Solution

Note that there are three unknowns, P_1 and the pin reactions at the fulcrum, R_x and R_y. We have the equations of equilibrium:

$$F_x = 0: \qquad R_x = 0$$

$$F_y = 0: \qquad R_y - P_1 - P_2 = 0$$

$$M_{Oz} = 0: \qquad P_1 a_1 - P_2 a_2 = 0$$

Note that we wrote the equation for the moment about the fulcrum point, O. This meant that the moment relation did not involve the unknowns R_x and R_y. From the moment equation, we get

$$P_1 = \left(\frac{a_2}{a_1}\right) P_2.$$

With $P_2 = 1000$ lb, we can adjust the distances a_1 and a_2 so that P_1 is a reasonable value. For example, if

$$a_1 = 10 \text{ ft}, \qquad a_2 = 1 \text{ ft},$$

$$P_1 = \left(\frac{1 \text{ ft}}{10 \text{ ft}}\right) (1000) \text{ lb} = 100 \text{ lb}.$$

······································

Example 4 (Other Levers)

There are other physical systems which act like the lever in Example 3. We deal with three others here.

a) *Lever.* The lever system in Fig. 2.14(a) consists of a cylinder with strings wrapped around it at two radii, r_1 and r_2. Taking moments about O, the pin joint, we get

$$M_{Oz} = 0: \qquad P_1 r_1 - P_2 r_2 = 0 \qquad \text{or} \qquad P_1 = \left(\frac{r_2}{r_1}\right) P_2. \tag{a}$$

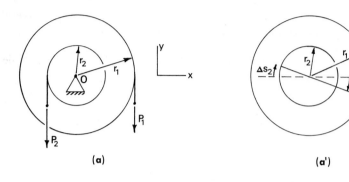

(a)   (a')

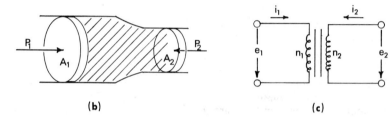

(b)   (c)

Figure 2.14

In the systems in Examples 3 and 4(a) (as well as 4b and 4c below), we seem to get something for nothing. That is, by applying a small force P_1 we can generate a large force P_2 (if $r_1 > r_2$). In fact, these are all examples of what are called *power-preserving elements*. We shall see more of this concept of power in dynamics. But here it is sufficient to say that the *distance* which P_1 moves is inversely proportional to the distance P_2 moves. Suppose the cylinder in Fig. 2.14(a) rotates an angle $\Delta\theta$ (Fig. 2.14a').

We see that

$$\Delta s_1 = r_1 \, \Delta\theta, \qquad \Delta s_2 = r_2 \, \Delta\theta.$$

From (a), we note that

$$P_1 \, \Delta s_1 = P_1 r_1 \, \Delta\theta = \left(\frac{r_2}{r_1}\right) P_2 r_1 \, \Delta\theta = P_2 \, \Delta s_2.$$

Thus,

$$P_1 \, \Delta s_1 = P_2 \, \Delta s_2,$$

and we see that we don't get something for nothing. The quantity $P \, \Delta s$ is the same at each side of the lever.

b) *Hydraulic transformer.** Figure 2.14(b) shows a hydraulic transformer, a type of system which has wide applications. Think, for example, of the hydraulic lift in automobile service stations.

The pistons have areas A_1 and A_2. Let p be the pressure in the fluid and P_1 and P_2 be the total forces exerted on the pistons. Then

$$\begin{matrix} pA_1 = P_1 \\ pA_2 = P_2 \end{matrix} \quad \text{or} \quad p = \frac{P_1}{A_1} = \frac{P_2}{A_2}$$

and

$$P_1 = \left(\frac{A_1}{A_2}\right) P_2.$$

c) *Electrical transformer.* Figure 2.14(c) shows an electrical transformer. Let n_1 be the number of windings on the input coil and n_2 be the number of windings on the output coil. Then

$$e_1 = \left(\frac{n_1}{n_2}\right) e_2 \qquad\qquad (b)$$

for the voltage drop across the input and output of the transformer. In doorbell transformers, for example, it is common to desire that

$$e_1 = 110 \text{ volts}, \qquad e_2 = 15 \text{ volts}.$$

Thus

$$\frac{e_1}{e_2} = 7.3 = \frac{n_1}{n_2}.$$

Since there is a net voltage drop going from the input to the output, it seems that this transformer gives "nothing for something." Again this is misleading, because the current relation for a transformer is

$$i_1 = \left(\frac{n_2}{n_1}\right) i_2. \qquad\qquad (c)$$

(Current can be taken to be analogous to velocities in Examples 3 and 4a.) Again from (a) and (c), we note that this is a power-preserving element:

$$e_1 i_1 = \left(\frac{n_1}{n_2}\right) e_2 \left(\frac{n_2}{n_1}\right) i_2 = e_2 i_2.$$

....

* We don't pretend to give the theory behind Examples 4(b) and 4(c), only to show the similarity to Examples 3 and 4(a).

Before we talk about any further examples of the equilibrium of rigid bodies, we must discuss four topics:

   (i) Couples

  (ii) Equivalent force systems

 (iii) Forces in statics

 (iv) Free-body diagrams

5. COUPLES

Suppose we consider two forces which are equal and opposite, but which act on parallel lines which are separated by a distance d (Fig. 2.15).

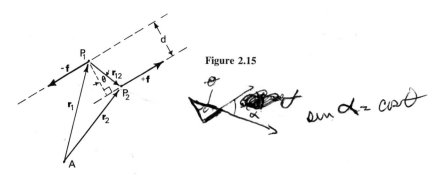

Figure 2.15

We compute the moment resultant of the two forces about some point A:

$$\mathbf{M}_A = \mathbf{r}_1 \times (-\mathbf{f}) + \mathbf{r}_2 \times (+\mathbf{f}) = (\mathbf{r}_2 - \mathbf{r}_1) \times \mathbf{f}.$$

Thus

$$\mathbf{M}_A = \mathbf{r}_{12} \times \mathbf{f}, \tag{a}$$

where $\mathbf{r}_{12}$ is the vector running from P_1 to P_2. Thus the moment resultant of these two forces does not depend on the point A. From (a), we see that the moment is perpendicular to the plane of the two forces and has magnitude

$$M = fd, \tag{2.9}$$

where d is the perpendicular distance between the line of $\mathbf{f}$ and the line of $-\mathbf{f}$. The sense is, of course, given by the right-hand rule.

We define a *couple* to be a force system which has a constant moment resultant field. The forces in Fig. 2.15 are a representation of a couple.

By (2.8), we see that the requirement that a couple have a constant moment resultant field implies that the force resultant of the couple be zero. We denote a couple by **C**. There is no subscript on **C**, since the position of a couple is irrelevant.

> A couple **C** is defined as a force system which results in a uniform moment resultant field. The force resultant of a couple is zero.

Example 5

Determine the moment and force resultants about the points $O = (0,0,0)$, $A_1 = (a,0,0)$, $A_2 = (0,b,0)$, and $A_3 = (0,0,c)$ of the forces and couple in Fig. 2.16.

Figure 2.16

Solution

The force resultant is

$$\mathbf{F} = F_1\mathbf{i} + F_2\mathbf{k}.$$

The moment resultant about the origin is

$$\mathbf{M}_O = (C_1\mathbf{i} - 2F_1\mathbf{k}).$$

We can compute the moment resultants about A_1, A_2, and A_3 from (2.8) if we use

$$\mathbf{r}_{A1} = -a\mathbf{i}, \quad \mathbf{r}_{A2} = -b\mathbf{j}, \quad \mathbf{r}_{A3} = -c\mathbf{k}.$$

Then

$$\mathbf{M}_{A1} = F_2a\mathbf{j} + \mathbf{M}_O,$$

$$\mathbf{M}_{A2} = F_1b\mathbf{k} - F_2b\mathbf{i} + \mathbf{M}_O,$$

$$\mathbf{M}_{A3} = -F_1c\mathbf{j} + \mathbf{M}_O.$$

Thus

$$\mathbf{M}_{A1} = C_1\mathbf{i} + F_2a\mathbf{j} - 2F_1\mathbf{k},$$

$$\mathbf{M}_{A2} = (C_1 - F_2b)\mathbf{i} + F_1(b-2)\mathbf{k},$$

$$\mathbf{M}_{A3} = C_1\mathbf{i} - F_1c\mathbf{j} - 2F_1\mathbf{k}.$$

Note, again, that the effect of the couple is the same at each point. We could have computed the moment at each point directly, noting, in particular, that the effect of the couple is the same for each point: $C_1\mathbf{i}$.

....

We end this section by writing the equations of equilibrium for a rigid body subject to applied forces and applied couples. Figure 2.17 shows a rigid body subject to forces $\mathbf{F}_1, \mathbf{F}_2, \ldots, \mathbf{F}_n$ and couples $\mathbf{C}_1, \mathbf{C}_2, \ldots, \mathbf{C}_m$. The conditions of equilibrium are again (2.6), but the expression for the moment resultant is somewhat different.

Since the force resultant of a couple is zero, the force resultant of the forces and couples acting on the rigid body is the sum of the applied forces alone:

$$\mathbf{F} = \mathbf{F}_1 + \mathbf{F}_2 + \cdots + \mathbf{F}_n = \sum_{i=1}^{n} \mathbf{F}_i. \tag{2.10}$$

For the moment resultant, we note that the moment field of a couple is uniform. That means the moment of, for example, $\mathbf{C}_1$ about any point is $\mathbf{C}_1$. Thus the moment resultant of all the forces and couples about the point O is

$$\mathbf{M}_O = \sum_{i=1}^{m} \mathbf{C}_i + \sum_{i=1}^{n} \mathbf{r}_i \times \mathbf{F}_i. \tag{2.11}$$

Thus the conditions of equilibrium for a rigid body can be written as follows.

Necessary and sufficient conditions for the equilibrium of a rigid body are:

Zero force resultant: $\displaystyle\sum_{i=1}^{n} \mathbf{F}_i = 0, \quad t > t_0$

Zero moment resultant: $\displaystyle\sum_{i=1}^{m} \mathbf{C}_i + \sum_{i=1}^{n} \mathbf{r}_i \times \mathbf{F}_i = 0, \quad t > t_0 \tag{2.12}$

Initial conditions: The rigid body is at rest in some inertial reference frame at time $t = t_0$.

And finally, we verify that the rule for moment transformation, (2.8), is still valid for a body subject to both forces and couples. Equations (2.7) are now:

$$\mathbf{M}_A = \sum_{i=1}^{m} \mathbf{C}_i + \sum_{i=1}^{n} \mathbf{r}_{Ai} \times \mathbf{F}_i, \qquad \mathbf{M}_B = \sum_{i=1}^{m} \mathbf{C}_i + \sum_{i=1}^{n} \mathbf{r}_{Bi} \times \mathbf{F}_i.$$

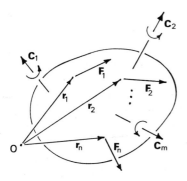

Figure 2.17

Subtracting, we get

$$\mathbf{M}_A - \mathbf{M}_B = \sum_{i=1}^{n} (\mathbf{r}_{Ai} - \mathbf{r}_{Bi}) \times \mathbf{F}_i = \mathbf{r}_{AB} \times \left(\sum_{i=1}^{n} \mathbf{F}_i \right)$$

and

$$\mathbf{M}_A = \mathbf{r}_{AB} \times \mathbf{F} + \mathbf{M}_B \tag{2.13}$$

as before, where $\mathbf{F}$ is the force resultant of the forces applied to the body.

6. TRANSFORMATION OF FORCE SYSTEMS; EQUIVALENT FORCE SYSTEMS

Suppose we have a rigid body subject to various forces and couples. We might ask whether another distribution of forces and couples exists which is equivalent to the first and, in particular, if there is a distribution which might, in some sense, be simpler than the original. Before answering these questions we must define what we mean by an equivalent force system.

We say that two sets of forces and couples are "statically equivalent" if the resultant force $\mathbf{F}$ and the resultant moment $\mathbf{M}_A$ are the same at every point A for the two systems.

How can we test to see whether two sets of forces and couples are equivalent? Must we march about to each point checking $\mathbf{F}$ and $\mathbf{M}$, or do we take the advice of the man who says "check a couple of points, and if $\mathbf{M}$ and $\mathbf{F}$ are the same, you can be pretty certain that the systems are equivalent." Fortunately we have neither to perform the (infinite) tedium of the first method nor to settle for the implied ambiguity of the second. Equivalence can be determined simply by means of Eq. (2.13).

i) Check whether $\mathbf{F}$ is the same for each system.

ii) Pick some convenient point A. Check to see that $\mathbf{M}_A$ is the same for both systems.

If the above conditions are met, the two systems of forces are equivalent. We can be certain of this by (2.13), for if the two systems agree on $\mathbf{F}$ and $\mathbf{M}_A$ at some point, (2.13) says that $\mathbf{M}_B$ will be the same for all other points B.

Suppose that the force resultant for a system is $\mathbf{F}$ and the moment resultant at A is $\mathbf{M}_A$. An equivalent system consists of a force $\mathbf{F}$ acting at A plus a couple $\mathbf{C} = \mathbf{M}_A$. Since the system has the same force resultant $\mathbf{F}$, and agrees with respect to the moment at one point, A (since $\mathbf{F}$ passes through A, the moment at A is entirely due to $\mathbf{C}$), the two systems are equivalent. Thus any system is equivalent to a single force plus a couple.

We can go further than this, however.* We can try to find a point A such that the resultant force and the couple defined above share the same direction. Suppose at some point O, we have $\mathbf{M}_O$ for a given force system with nonzero force resultant $\mathbf{F}$. Then we seek a point A such that

$$\mathbf{M}_A = \lambda \mathbf{F} = \mathbf{r}_{AO} \times \mathbf{F} + \mathbf{M}_O, \tag{a}$$

where λ is some scalar. If we dot (a) with $\mathbf{F}$, we get

$$\lambda \mathbf{F} \cdot \mathbf{F} = \mathbf{F} \cdot \mathbf{M}_O,$$

since $(\mathbf{r}_{AO} \times \mathbf{F})$ is perpendicular to $\mathbf{F}$. Since $\mathbf{F} \neq 0$, we have

$$\lambda = \frac{\mathbf{F} \cdot \mathbf{M}_O}{F^2},$$

and then

$$\mathbf{C} = \left(\frac{\mathbf{F} \cdot \mathbf{M}_O}{F^2} \right) \mathbf{F}. \tag{2.14}$$

If we form the cross product of (a) with $\mathbf{F}$, we get

$$0 = (\mathbf{r}_{AO} \times \mathbf{F}) \times \mathbf{F} + \mathbf{M}_O \times \mathbf{F}.$$

The vector triple product† of three vectors, $\mathbf{A}$, $\mathbf{B}$, and $\mathbf{C}$, has the expansion $(\mathbf{A} \times \mathbf{B}) \times \mathbf{C} = (\mathbf{A} \cdot \mathbf{C})\mathbf{B} - (\mathbf{B} \cdot \mathbf{C})\mathbf{A}$. Thus we have

$$(\mathbf{r}_{AO} \times \mathbf{F}) \times \mathbf{F} = (\mathbf{r}_{AO} \cdot \mathbf{F})\mathbf{F} - (\mathbf{F} \cdot \mathbf{F})\mathbf{r}_{AO}.$$

And

$$(\mathbf{r}_{AO} \cdot \mathbf{F})\mathbf{F} - F^2 \mathbf{r}_{AO} = -\mathbf{M}_O \times \mathbf{F}. \tag{b}$$

---

* Because one can go further, it does not follow that the trip is worthwhile. One can skip the remainder of this section without seriously impairing his career. The material on center of gravity which follows Eq. (2.16) is important, however.

† See Appendix 1, Eq. (A1.6).

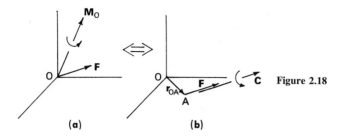

Figure 2.18

If we choose $\mathbf{r}_{AO}$ to be perpendicular to $\mathbf{F}$, the first term is zero and

$$\mathbf{r}_{AO} = \frac{\mathbf{M}_O \times \mathbf{F}}{F^2}.$$

This is the vector from A to O. Thus the position of A relative to O is $\mathbf{r}_{OA}$ or

$$\mathbf{r}_{OA} = \left(\frac{\mathbf{F} \times \mathbf{M}_O}{F^2}\right). \tag{c}$$

Thus we have Fig. 2.18.

The position given by (c) is not the only position at which the new force system can be placed. For example, consider the force in Fig. 2.19.

We consider two positions of the force, $\mathbf{F}$: $\mathbf{r}_1$ describes the original position and $\mathbf{r}_2$ gives a position in which $\mathbf{F}$ has been moved along its own line of action l. Suppose we compute $\mathbf{M}_A$ for the two positions.

1:    $\mathbf{M}_{A1} = \mathbf{r}_1 \times \mathbf{F}$,

2:    $\mathbf{M}_{A2} = \mathbf{r}_2 \times \mathbf{F}$.

The difference between these values is

$$\mathbf{M}_{A2} - \mathbf{M}_{A1} = (\mathbf{r}_2 - \mathbf{r}_1) \times \mathbf{F}.$$

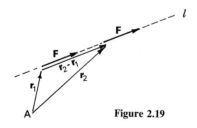

Figure 2.19

But $\mathbf{r}_2 - \mathbf{r}_1$ lies along $\mathbf{F}$ in Fig. 2.19. Thus the term on the right side is zero, and

$$\mathbf{M}_{A2} = \mathbf{M}_{A1}.$$

This means that so far as a force in statics is concerned, it can be moved arbitrarily along its line of action without any effect.

Returning to (c), we would expect from the above that we could take any position given by

$$\mathbf{r}_{OA} = \left(\frac{\mathbf{F} \times \mathbf{M}_O}{F^2}\right) + \alpha\mathbf{F} \tag{2.15}$$

for the position of the equivalent force and couple system, where α is an arbitrary scalar. In fact, insertion of (2.15) into (b) shows that $\mathbf{r}_{OA}$ is a solution, as anticipated.

An arbitrary force system equivalent to a force resultant $\mathbf{F}$ and moment resultant $\mathbf{M}_O$ at point O is, in turn, equivalent to the force $\mathbf{F}$ and couple $\mathbf{C}$,

$$\mathbf{C} = \left(\frac{\mathbf{F} \cdot \mathbf{M}_O}{F^2}\right)\mathbf{F},$$

located at position A, given by

$$\mathbf{r}_{OA} = \left(\frac{\mathbf{F} \times \mathbf{M}_O}{F^2}\right) + \alpha\mathbf{F},$$

where α is arbitrary. The combination of a force $\mathbf{F}$ and a couple $\mathbf{C}$ which have the same direction is called a *wrench*.

We have reduced an arbitrary force system to a wrench. The fact of the matter is that in practice one rarely feels compelled to make this reduction. It would be totally misleading to give the impression that a practicing engineer spends a good portion of his professional life finding wrenches. In fact, there are many successful engineers of great talent and reputation who have never even heard of a wrench!

One common use of force and moment resultants comes in describing distributed loads on a body, such as the gravity load. Figure 2.20 shows a rigid body in the gravity field. Let $\rho(x,y,z)$ be the mass per unit volume of the body. (In Chapter 4, Section 6, we shall discuss the nature of the gravity force in detail.) In particular, for objects close to the earth's surface, we write [see (4.21)]

$$\mathbf{F} = -mg\mathbf{k}, \tag{2.16}$$

where $\mathbf{k}$ is a unit vector along the local vertical, m is the mass of the body, and g is the "acceleration due to gravity." Suppose we examine the force of gravity on an element of volume of the body, dV. If the element is located at position (x,y,z), the mass will be

$$dm = \rho(x,y,z)\ dV.$$

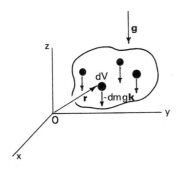

Figure 2.20

Thus the force on this element is

$$dF = -(dm)g\mathbf{k} = -\rho(x,y,z)g \; dV \; \mathbf{k}. \tag{a}$$

Thus the gravity load on a rigid body is a distribution of parallel loads. The force resultant of these loads is given by the summation of (a) over the body

$$\mathbf{F} = -\mathbf{k}g \int_V \rho(x,y,z) \; dV = -mg\mathbf{k}, \tag{b}$$

where m is the total mass of the body. (We have used a notation here for the volume integral which might be somewhat confusing. You might prefer

$$\iiint_V \rho \; dV.$$

We shall indicate volume integrals and area integrals by

$$\int_V \cdots \, dV = \iiint_V \cdots \, dV, \qquad \int_A \cdots \, dA = \iint_A \cdots \, dA.)$$

The moment of the elemental force (a) about O is given by

$$d\mathbf{M}_O = \mathbf{r} \times d\mathbf{F}, \tag{c}$$

where $\mathbf{r} = x\mathbf{i} + y\mathbf{j} + z\mathbf{k}$. Thus (c) becomes

$$d\mathbf{M}_O = \rho(x,y,z)g(-y\mathbf{i} + x\mathbf{j}) \; dV. \tag{d}$$

And the moment resultant about O is

$$\mathbf{M}_O = -g\mathbf{i} \int_V y\rho \; dV + g\mathbf{j} \int_V x\rho \; dV. \tag{e}$$

In this particular case, it makes sense to derive the wrench* equivalent to (b) and (e). The couple portion of the wrench is, by (2.14), equal to zero, since

___

* If you are one of the few who skipped the material on the wrench, we are seeking here the point A such that $\mathbf{M}_A = 0$, using (2.23). See Problems 2.26.

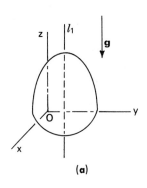

(a)

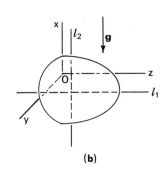

(b)

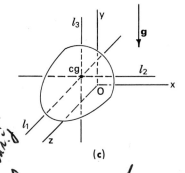

(c)

Figure 2.21

$\mathbf{F} \cdot \mathbf{M}_O = 0$. From (2.15), the point of location of the wrench A is

$$\mathbf{r}_{OA} = \frac{1}{m} \left(\mathbf{i} \int_V x\rho \, dV + \mathbf{j} \int_V y\rho \, dV \right) + \alpha \mathbf{k},$$

where α is arbitrary. Denote this set of points by the line l_1. Suppose we take the body and the axis system and rotate both until the x axis is vertical (Fig. 2.21b). We now obtain

$$\mathbf{r}_{OA} = \beta \mathbf{i} + \frac{1}{m} \left(\mathbf{j} \int_V y\rho \, dV + \mathbf{k} \int_V z\rho \, dV \right),$$

where β is arbitrary. Denote this set of points by the line l_2.

Finally rotate to the position in Fig. 2.21(c). The y axis is now vertical, and

$$\mathbf{r}_{OA} = \frac{\mathbf{i}}{m} \int_V x\rho \, dV + \gamma \mathbf{j} + \frac{\mathbf{k}}{m} \int_V z\rho \, dV,$$

where γ is arbitrary. This is the line l_3.

It is clear from the above equations for $\mathbf{r}_{OA}$ that there is always one point common to l_1, l_2, and l_3 through which the gravity load will pass no matter what the orientation of the body is. This point is called the *center of gravity:*

$$\mathbf{r}_{cg} = \frac{1}{m}\left(\mathbf{i}\int_V x\rho\; dV + \mathbf{j}\int_V y\rho\; dV + \mathbf{k}\int_V z\rho\; dV\right),$$

or, in components:

$$x_g = \frac{1}{m}\int_V x\rho\; dV,$$

$$y_g = \frac{1}{m}\int_V y\rho\; dV, \qquad\qquad\qquad (2.17)$$

$$z_g = \frac{1}{m}\int_V z\rho\; dV.$$

> For a rigid body, the resultant of the gravity load is a single force $\mathbf{F} = -mg\mathbf{k}$, where m is the total mass and $\mathbf{k}$ is a unit vector along the local vertical. The force, called the weight, passes through the center of gravity of the body.

In many bodies, the position of the center of gravity can be determined by inspection. This is true if the mass distribution is symmetrical.

Example 6

Find the center of gravity of a semicircular plate of uniform density (Fig. 2.22).

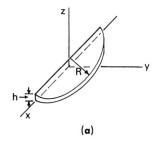

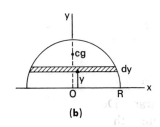

(a)   (b)   **Figure 2.22**

Solution

By symmetry, the center of gravity is located at

$$z_g = \frac{h}{2} \qquad \text{and} \qquad x_g = 0.$$

We need determine only the coordinate, y_g. There are two ways to approach this problem. We can write the triple integral,

$$y_g = \left(\frac{\rho}{m}\right) \int_{-R}^{+R} \int_0^{\sqrt{R^2-x^2}} \int_0^h y \, dz \, dy \, dx = \left(\frac{\rho}{m}\right) \tfrac{2}{3} h R^3.$$

But $m = \rho V = \rho \tfrac{1}{2} \pi R^2 h$. Thus

$$y_g = \frac{4R}{3\pi}.$$

A second solution can be obtained by considering the semicircle to be broken into strips as in Fig. 2.22(b). The strip at position y has the dimensions

$$h \quad \text{by} \quad dy \quad \text{by} \quad 2\sqrt{R^2 - y^2}.$$

Thus the mass of the strip is $2\rho h\sqrt{R^2 - y^2} \, dy = dm$. Then summing $(1/m)\,(y \, dm)$ over all such strips, we have

$$y_g = \frac{2\rho h}{m} \int_0^R y\sqrt{R^2 - y^2} \, dy = \frac{2\rho h}{m} (\tfrac{1}{3}R^2) = \frac{4R}{3\pi}.$$

. . . .

A third method of locating the center of gravity is to consult a table of centroids. The term *centroid* is somewhat less general than center of gravity (and related concepts). Suppose we have a body which has a constant density. Then Eqs. (2.17) become (writing x_C instead of x_g),

$$x_C = \frac{1}{V}\left(\int_V x \, dV\right), \qquad y_C = \frac{1}{V}\left(\int_V y \, dV\right), \qquad z_C = \frac{1}{V}\left(\int_V z \, dV\right) \qquad (2.18)*$$

Note that the computations involved in (2.18) are entirely geometrical. If we have a certain area, we can define its centroid by

$$x_C = \frac{1}{A}\left(\int_A x \, dA\right), \qquad y_C = \frac{1}{A}\left(\int_A y \, dA\right), \qquad z_C = \frac{1}{A}\left(\int_A z \, dA\right). \qquad (2.19)*$$

For a curve in space,

$$x_C = \frac{1}{L}\left(\int_L x \, ds\right), \qquad y_C = \frac{1}{L}\left(\int_L y \, ds\right), \qquad z_C = \frac{1}{L}\left(\int_L z \, ds\right). \qquad (2.20)$$

All these computations depend solely on the geometry of the object involved. Figure 2.23 shows the point C for three such geometrical objects. It should be

---

* Equations (2.18) and (2.19) involve volume and surface integrals. We have used an abbreviation for, say

$$x_C = \frac{1}{V} \iiint_V x \, dV \quad \text{or} \quad x_C = \frac{1}{A} \iint_A x \, dA, \quad \text{etc.}$$

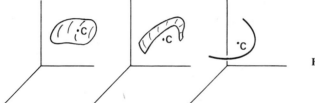

Figure 2.23

clear that the centroid need not lie on a point of the geometric object under consideration. For example, where is the centroid of a doughnut?

If the geometric object represents a distribution of mass in which the mass per volume (or mass per area or mass per length) is constant, the centroid can be interpreted as the center of gravity. In other physical problems, the centroid may play a different role.

We have seen that the distribution of the gravity forces of a rigid body is equivalent to a single force – the weight – which acts through the center of gravity. We might ask under what circumstances a given force system can be reduced to (i) a single couple or (ii) a single force. If the system is equivalent to a single couple, the force resultant $\mathbf{F}$ must be zero. On the other hand, if $\mathbf{F} \cdot \mathbf{M}_O = 0$, then by (2.14), the system is equivalent to a single force $\mathbf{F}$ passing through any of the points given by (2.15).

7. FORCES IN STATICS

In this section we shall list some of the common forces encountered in statics. All these forces will be encountered again in dynamics.

a) Gravity

We have seen that the gravity load on a rigid body is equivalent to a single force $\mathbf{W}$, the total weight, whose direction is opposite to the local vertical and which passes through the center of gravity of the body.

b) Restraining Forces

When constraints are placed on the motion of a rigid body, certain forces are generated to effect the restriction. We idealize several of the restraints, by means of the following.

i) *Pin joint*. Figure 2.24(a) shows a pin joint. This joint allows the body to rotate freely about the joint (in the absence of additional restraints). However, the point of the body at the pin can have no motion. The pin

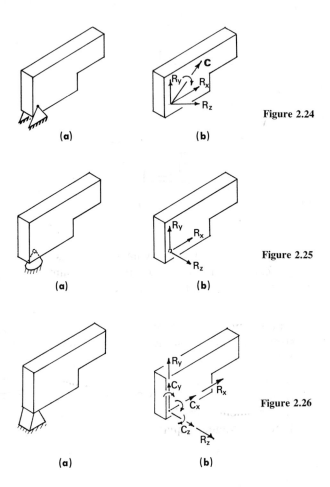

Figure 2.24

Figure 2.25

Figure 2.26

can exert forces in any direction, and a moment in the plane perpendicular to the axis of the pin. This force system is shown in Fig. 2.24(b).

ii) *Socket.* The pin joint restricts motion to two dimensions. The *socket*
restricts the motion of the point of application only. The forces corresponding to this restriction can be in any direction in space. However,
the socket is assumed to be unable to exert any moment on the rigid
body. The socket and its possible forces are shown in Figs. 2.25(a)
and (b).

iii) *Built-in constraint.* We say that a body is *built in* at some point *A* if that
point and the points in the immediate neighborhood of *A* can have no
motion (Fig. 2.26). Thus if a rigid body is built in at some point, the

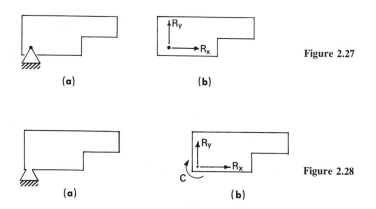

Figure 2.27

(a)   (b)

Figure 2.28

(a)   (b)

body can have no motion no matter what other loads are on the body. The resultant force and moment due to this restraint can be a force with components in all directions and a couple with moments in all directions.

Although the above is literally true, the force systems exerted by the pin joint and the built-in constraint may be simpler than indicated for many cases, particularly for plane problems.

iv) *Pin-joint–plane case.* The important forces for plane motion are the pin reactions R_x and R_y (Fig. 2.27b). This does not mean that the pin does not exert a moment, only that this moment won't ordinarily enter into either static or dynamic equations.

v) *Built-in constraint–plane case.* Although a built-in constraint can exert moments and forces in all directions, in the case of plane motion, the important forces and moment are the force components in the plane, R_x and R_y, and the moment perpendicular to the plane C (Fig. 2.28b).

c) Static Friction

Suppose a particle rests on a surface, as shown in Fig. 2.29(a). Suppose the total force applied to the body, including gravity, is $\mathbf{F}$. If the body is to remain stationary, the surface must exert an equal and opposite force $-\mathbf{F}$ on the body. We break this force into two components, a normal force $\mathbf{N}$ and a tangential force $\mathbf{f}$. For equilibrium, we have then

$$\mathbf{f} + \mathbf{N} = -\mathbf{F}.$$

It is clear that a solid surface can generate a normal force $\mathbf{N}$. However, it is by no means clear that an arbitrarily large force $\mathbf{f}$ can be produced. Suppose the surface has been lubricated. In this case the maximum force $\mathbf{f}$ is not very large.

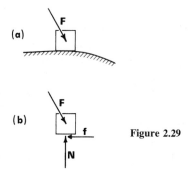

Figure 2.29

The force **f** which is tangential to the two contact surfaces is called *friction*. If the system is in equilibrium, it is called *static friction*. If the two surfaces have relative motion, the tangential force is called *sliding friction*.

Suppose the particle in Fig. 2.29 is in equilibrium. Then the magnitude of **f** is determined by the other forces on the block. The frictional force **f**, like the normal force **N**, is whatever it has to be to ensure equilibrium. Forces such as **N** are often referred to as *constraint forces,* in that their magnitudes are determined by the restrictions they put on the motion of a body. We shall deal extensively with constraint forces in the dynamics portion of this book.

There is a maximum value for the magnitude of **f**. Experiments have shown that

$$\max |\mathbf{f}| \propto |\mathbf{N}|.$$

The constant of proportionality depends on the materials of the adjacent surfaces. We usually write

$$f \le \mu N, \tag{2.21}$$

where $f = |\mathbf{f}|$, $N = |\mathbf{N}|$, and μ is called the *coefficient of static friction*. The maximum friction force does not depend on the area of surface contact.*

If the inequality (2.21) is violated, adjacent surfaces slide, and we then have sliding friction. The force of sliding friction has the magnitude

$$f = \mu N, \tag{2.22}$$

where μ is now called the *coefficient of sliding friction*. The values of μ in (2.21) and (2.22) are different. The force of sliding friction is generally less than the maximum force of static friction.

---

* This law of friction has to be taken with a grain of salt. If it were literally true, dragsters could use bicycle tires as well as wide tires. But this would be ludicrous. What it means is that this law of friction is inadequate to describe friction between soft rubber and other materials.

One of the common mistakes students make is to confuse (2.21) and (2.22). Static friction can have any value from 0 up to its maximum, μN. It is totally incorrect to begin a problem involving static friction by assuming that $f = \mu N$. Instead one must solve for f subject to equilibrium and then check to see that the inequality (2.21) is satisfied. If not, equilibrium is violated.

If we know that the surfaces slide on each other, we must then write equations in which the friction force is directed in such a way as to oppose the sliding of the surfaces, and in which the friction force has magnitude μN, where μ is the coefficient of sliding friction.

Often in problems of statics, we don't know the direction of the friction force between surfaces. Hence we simply assume a direction for the frictional force (in the free-body diagram; see the next section). If our assumption is incorrect, the subsequent analysis will indicate this fact by giving the frictional force a negative value. Thus the actual direction of the force is opposite to the direction assumed initially.

In cases involving sliding friction, this situation does not prevail. Sliding friction acts to oppose the slip between adjacent surfaces. If we assume the wrong direction for a force of sliding friction, we will not be corrected later by either equilibrium or dynamic equations, whichever apply. Our results will simply be incorrect.

d) Strings and Weightless Rods

In statics, strings are often considered to be weightless. Thus a string merely transmits a force from one point to another. A string cannot support a moment. Thus we have the forces $\mathbf{F}$ and $-\mathbf{F}$ at the two ends of the string, as shown in Fig. 2.30(b). A string cannot push an object, only pull or restrain it. If, in the solution of a problem, we find that a string is in compression and not tension, we can be sure that the system under consideration is *not* in equilibrium.

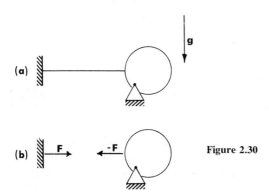

Figure 2.30

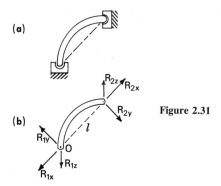

(a)

(b)

Figure 2.31

A generalization of the string is the weightless rod. Consider a rod with socket joints at its two ends (Fig. 2.31a). Let the x direction be determined by the line between the endpoints. We then decompose the end forces $\mathbf{R}_1$ and $\mathbf{R}_2$, as shown in Fig. 2.31(b). Suppose these are the only forces on the rod. Then by equilibrium we have

$$\mathbf{F} = 0: \qquad R_{1x} - R_{2x} = 0, \qquad R_{1y} - R_{2y} = 0, \qquad R_{1z} - R_{2z} = 0$$

and

$$\mathbf{M}_O = 0: \qquad R_{2y}l = 0, \quad R_{2z}l = 0.$$

Thus

$$R_{1x} = R_{2x} \qquad \text{and} \qquad R_{1y} = R_{2y} = R_{1z} = R_{2z} = 0.$$

Thus a weightless rod with socket ends can have forces only on its ends which lie along the line between its endpoints. Note that, unlike the case for the string, the forces can be either tensile or compressive.

If the rod has end conditions other than a socket, it can support moments and forces perpendicular to the line between the ends (suppose one end is built in, for example).

8. THE FREE-BODY DIAGRAM

We come now to a central idea of Newtonian mechanics, the free-body diagram. The idea is important in both statics and dynamics. Many engineering systems consist of a number of interconnected rigid bodies. We say that the system is in equilibrium if each individual body is itself in equilibrium. Thus, in order to study the equilibrium of the entire system, we need a method of investigating the equilibrium of subsystems. By means of a free-body diagram, we isolate the subsystem from its environment, replacing any constraints or connections that are broken (conceptually) by equivalent forces and moments.

The free-body diagram of a particle, rigid body, or system of rigid bodies consists of the system under consideration, with all external forces and couples on the system represented. The external forces and couples must include all possible reactions from connections the system has with its immediate environment.

Let us look at a number of examples which illustrate this concept.

Example 7

A body is suspended by two strings (Fig. 2.32a). Figure 2.32(b) shows the free-body diagram. The gravity load is equivalent to the weight W passing through the center of gravity C. The magnitudes of the string forces are not known, but their directions are given by the directions of the strings.

Find the equilibrium position of the body and the forces in the two strings.

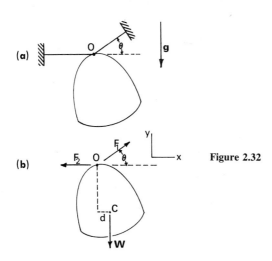

Figure 2.32

Solution

Equilibrium requires that

$$\mathbf{F} = 0: \quad (F_1 \cos \theta - F_2)\mathbf{i} + (F_1 \sin \theta - mg)\mathbf{j} = 0,$$

$$\mathbf{M}_O = 0: \quad -(mgd)\mathbf{k} = 0.$$

The moment equation says that $d = 0$. Thus the line OC must be vertical. The force equation is equivalent to

$$F_1 \cos \theta = F_2, \qquad F_1 \sin \theta = mg.$$

Thus

$$F_1 = mg \csc \theta, \qquad F_2 = mg \cot \theta.$$

which gives the forces in the strings.

Example 8

Romeo stands on a ladder of length l propped against Juliet's frictionless house at angle θ. The coefficient of friction at the base of the ladder is μ. Under what circumstances is Romeo in equilibrium?

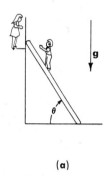

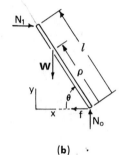

Figure 2.33

(a)   (b)

Solution

A free-body diagram of the ladder is shown in Fig. 2.33(b). We consider the ladder weightless. Since the upper end rests on a frictionless wall, we know that a normal force alone will act at that point. The weight of Romeo is shown at a position ρ, up the ladder. Equilibrium then requires:

$$F_x = 0: \qquad N_1 - f = 0$$

$$F_y = 0: \qquad N_0 - mg = 0,$$

$$M_{0z} = 0: \qquad -N_1 l \sin \theta + mg \rho \cos \theta = 0.$$

We have here unknowns N_1, N_0, and f in terms of the given parameters mg, θ, l, and ρ. Solving, we get

$$N_0 = mg, \qquad N_1 = mg(\rho/l) \cot \theta, \qquad f = N_1.$$

For what values of ρ will Romeo be assured that the ladder will not slip? We must check the maximum value the friction force f can have:

$$f \leq \mu N_0 = \mu mg.$$

Thus we have $mg (\rho/l) \cot \theta < \mu mg$ or

$$(\rho/l) \leq \mu \tan \theta.$$

If Romeo is to reach the window, we must have no slipping for the following values of ρ

$$(\rho/l) \leqslant 1.$$

Thus $\mu \tan \theta > 1$ or $\mu > \cot \theta$. Note that these inequalities do not involve the value of m. Thus, if the ladder is able to take Romeo all the way up to the window, it won't slip even if Romeo were to carry Juliet down the ladder.

....

Example 9

A cylinder of mass m_0 has a particle m attached to its rim (see Fig. 2.34a). The cylinder rests on a horizontal plane. Suppose that the coefficient of static friction is μ.

Find the minimum value of μ in order that the cylinder is stationary.

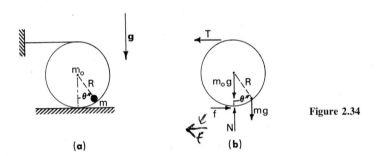

Figure 2.34

(a)   (b)

Solution

If m has an angle of inclination to the vertical of θ, the free-body diagram is as shown in Fig. 2.34(b). Here the string tension T, the normal force N, and the friction force f are constraint forces. They have whatever values are necessary for equilibrium. The essence of our problem consists of determining f and comparing it to its maximum allowable value, μN.

Equilibrium requires that

$$F_x = 0: \quad -T + f = 0,$$ (a)

$$F_y = 0: \quad N - m_0 g - mg = 0,$$ (b)

$$M_z = 0. \quad 2RT - mg\, R \sin \theta = 0.$$ (c)

because $T = f$

(Here we have taken moments about the point of contact of the cylinder.)

From (a) and (c), we get

$$f = \tfrac{1}{2}mg \sin \theta,$$

and from (b), we get

$$N = (m + m_0)g.$$

For static equilibrium, we must have $f \leqslant \mu N$. Thus

$$\mu \geqslant \frac{f}{N} = \frac{m \sin \theta}{2(m + m_0)}.$$

Example 10

Figure 2.35(a) shows a homogeneous cylinder with a string wound around the inner radius of the wheel. A force T is applied to the string at an angle θ. The coefficient of static friction is μ. Find the angle θ and range of forces for which the cylinder is in equilibrium.

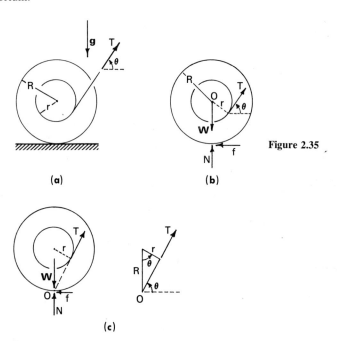

(a)

(b)

Figure 2.35

(c)

Solution

Figure 2.35(b) shows the free-body diagram for the cylinder. In particular, since equilibrium is assumed, the static friction force could be shown either as in the figure, or in the opposite direction.

Writing the equilibrium equations, we have

$$F_x = 0: \qquad -f + T \cos \theta = 0, \tag{a}$$

$$F_y = 0: \qquad N - W + T \sin \theta = 0, \tag{b}$$

$$M_{Oz} = 0: \qquad Tr - fR = 0. \tag{c}$$

Eliminating f between (a) and (c) gives us

$$\cos \theta = \frac{r}{R} \tag{d}$$

for the angle at which equilibrium is possible. If $\cos \theta = (r/R)$, we obtain $\sin \theta$:

$$\sin \theta = \frac{\sqrt{R^2 - r^2}}{R}. \tag{e}$$

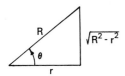

From (a) and (d), we get

$$f = T\frac{r}{R}, \tag{f}$$

and from (b) and (e), we get

$$N = W - T\left(\frac{\sqrt{R^2 - r^2}}{R}\right). \tag{g}$$

And finally, since μ is the coefficient of static friction,

$$f \leq \mu N.$$

From (f) and (g), we then obtain

$$T \leq \frac{\mu R W}{r + \mu \sqrt{R^2 - r^2}}. \tag{h}$$

Before leaving this example, we note that there is a simple explanation for result (d). Consider Fig. 2.35(c). We slide the weight through the point of contact. Note that all forces pass through the point of contact, with the possible exception of the string tension. But if T does not pass through O, there is no way in which $\mathbf{M}_O = 0$. On the other hand, if T does pass through O, the angle θ is determined as shown on the right side of Fig. 2.35(c), which confirms (d).

· ·

Example 11

A homogeneous cylinder which has a string wound around its inner radius (Fig. 2.36a) is supported by an incline. What is the minimum angle θ at which the cylinder will slip? Suppose that $r = \frac{1}{2}R$ and $\mu = 0.1$.

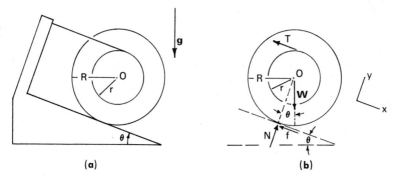

Figure 2.36

Solution

Figure 2.36(b) is a free-body diagram of the cylinder. We shall use equilibrium conditions to solve for the friction force f and then set f equal to its maximum static value to determine the angle θ.

Equilibrium requires the following:

$$F_x = 0: \qquad -f - T + W \sin\theta = 0, \tag{a}$$

$$F_y = 0: \qquad N - W \cos\theta = 0, \tag{b}$$

$$M_{Oz} = 0: \qquad Tr - fR = 0. \tag{c}$$

From (a) and (c), we get

$$f = \frac{r}{R+r} W \sin\theta \tag{d}$$

and from (b)

$$N = W \cos\theta. \tag{e}$$

For the maximum value of f, we have $f = \mu N$. From (d) and (e), we obtain

$$\frac{r}{R+r} W \sin\theta = \mu W \cos\theta$$

or

$$\tan\theta = \mu\left(\frac{R+r}{r}\right). \tag{f}$$

Finally, inserting the values $r = \frac{1}{2}R$ and $\mu = 0.1$, we obtain

$$\tan\theta = 0.1\left(\frac{\frac{3}{2}R}{\frac{1}{2}R}\right) = 0.3,$$

and thus

$\theta = 16.7°$.

Note that the result does not depend on the values of r and R, only on the ratio r/R.

....

Example 12

A simple plane truss is used to haul three banners that stream out behind an airplane (Fig. 2.37a).

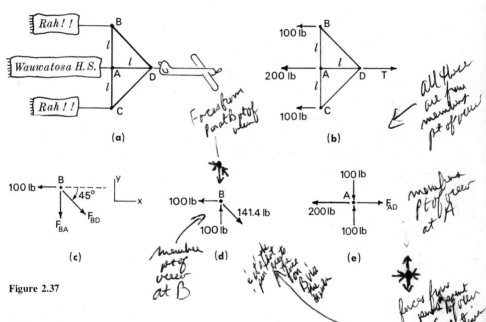

Figure 2.37

We suppose that the truss consists of weightless rods connected by pin joints (these are common assumptions in truss analysis). We are asked to determine the forces in each truss element, and also whether the forces are tensile or compressive. We are given that the drag forces caused by the banners are 100 lb, 200 lb, and 100 lb, respectively.

Solution

Equilibrium of the entire truss (Fig. 2.37b) shows that the force in the rope from the plane to the truss is

$T = 400$ lb.   (Why?)

We now wish to determine the forces in elements AB, BD, and AD. By symmetry of the loading and the truss, we know that the loads in AB and AC are equal; also that the forces in BD and CD are equal.

Consider the pin joint at B (Fig. 2.37c). Equilibrium of the pin B requires that

$$F_x = 0: \qquad -100 \text{ lb} + F_{BD} \cos 45° = 0, \tag{a}$$

$$F_y = 0: \qquad -F_{BA} - F_{BD} \sin 45° = 0. \tag{b}$$

Solving (a) and (b) for F_{BA} and F_{BD} gives

$$F_{BA} = -100.0 \text{ lb}, \qquad F_{BD} = +141.4 \text{ lb}.$$

The sign on F_{BA} means that we assumed the incorrect direction for F_{BA} in Fig. 2.37(c). It's not a bad idea to correct this mistake before proceeding (Fig. 2.37d).
Proceeding to point A, we now have Fig. 2.37(e). We get

$$F_{AD} = 200 \text{ lb}.$$

Thus our results are

$$F_{BA} = F_{CA} = 100.0 \text{ lb} \qquad \text{(compression)},$$

$$F_{BD} = F_{CD} = 141.4 \text{ lb} \qquad \text{(tension)},$$

$$F_{AD} = 200.0 \text{ lb} \qquad \text{(tension)}.$$

The forces at the pin are just opposite to the directions of the forces in the members. Thus a member is in *tension* if it *pulls* the pin; it is in *compression* if it *pushes* the pin.

Example 13

A thin, homogeneous rod rests on two frictionless surfaces (Fig. 2.38a). Find the tension in the string.

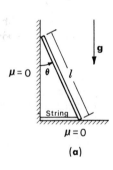

(a)

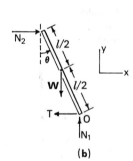

(b)

Figure 2.38

Solution

Figure 2.38(b) is the free-body diagram for the rod. Equilibrium requires that

$$F_x = 0: \qquad -T + N_2 = 0, \tag{a}$$

$$F_y = 0: \qquad N_1 - W = 0, \tag{b}$$

$$M_{Oz} = 0: \qquad -N_2 l \cos\theta + W\frac{l}{2}\sin\theta = 0. \tag{c}$$

From (a) and (c), we get

$$T = \tfrac{1}{2}W \tan \theta.$$

Note that T depends on the weight and inclination of the bar, but not on the length of the bar.

....

Example 14

Figure 2.39(a) shows a brake mechanism, consisting of three bars which we consider to be weightless. A mass m hangs from the end of bar 3. Find the forces in the three bars.

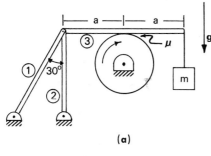

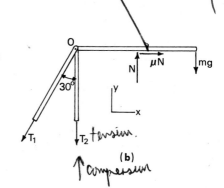

(a)   (b)

Figure 2.39

Solution

Figure 2.39(b) is the free-body diagram of the three bars.

Note that, since the cylinder is slipping, the friction force f must be written as μN, where μ is the coefficient of sliding friction. Forces T_1 and T_2 have the directions shown because bars 1 and 2 are weightless and transmit forces between two pin joints. Equilibrium requires that

$$F_x = 0: \qquad -T_1 \sin 30° + \mu N = 0, \tag{a}$$

$$F_y = 0: \qquad -T_1 \cos 30° - T_2 + N - mg = 0, \tag{b}$$

$$M_{Oz} = 0: \qquad aN - 2a\,mg = 0. \tag{c}$$

(The moment equation is written about the pin common to the three bars.) Solving (a), (b), and (c) gives us

$$T_1 = 4\mu\,mg, \qquad T_2 = mg(1 - 2\sqrt{3}\mu), \qquad N = 2\,mg.$$

This shows that bar 1 is in tension, bar 2 is in compression if $\mu > (\sqrt{3}/6)$ and in tension if $\mu < (\sqrt{3}/6)$.

What about bar 3?* A free-body diagram of bar 3 alone is shown in the first of Figs. 2.39(c), on page 58.

_____

* This part of the example will be meaningful to those who have studied strength of materials.

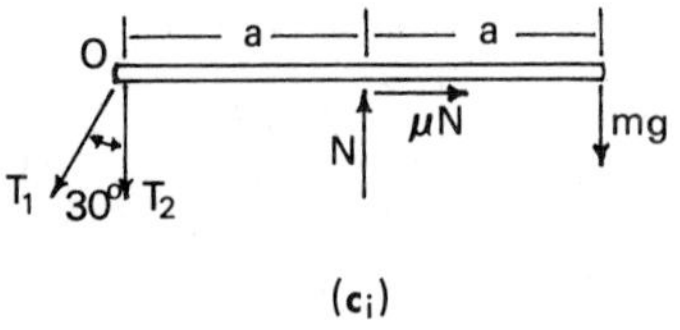

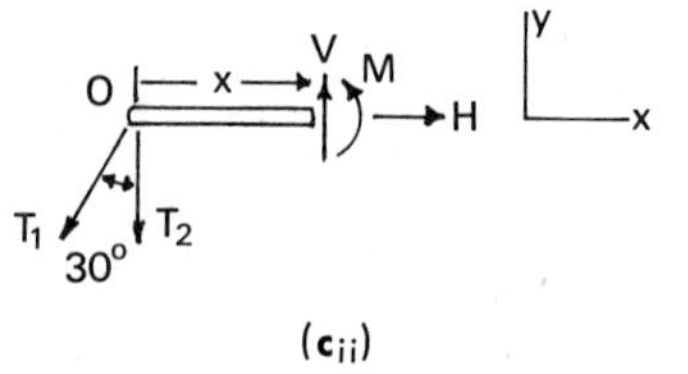

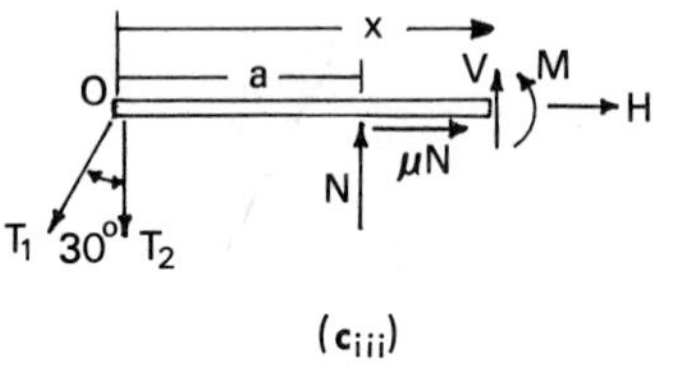

(c$_i$)

(c$_{ii}$)

Figure 2.39 (cont.)

(c$_{iii}$)

Bars 1 and 2 merely transmit forces. The situation in bar 3 is somewhat more complex. Suppose we make an imaginary cut in the bar at a position x from the left end of the bar [Figs. 2.39c, (ii) and (iii)]. We assume that an axial force H, a transverse force V (called a *shear force*), and a moment M act on this surface in order to maintain equilibrium.

The equilibrium equations depend on the region we are describing:

$$\left.\begin{array}{lll} F_x = 0: & -T_1 \sin 30° + H = 0 \\ F_y = 0: & -T_1 \cos 30° - T_2 + V = 0 \\ M_{Oz} = 0: & M + xV = 0 \end{array}\right\} \quad 0 \leqslant x < a$$

$$\left.\begin{array}{lll} F_x = 0: & -T_1 \sin 30° + \mu N + H = 0 \\ F_y = 0: & -T_1 \cos 30° - T_2 + N + V = 0 \\ M_{Oz} = 0: & M + aN + xV = 0 \end{array}\right\} \quad a \leqslant x \leqslant 2a$$

(In both cases the moment equation is written about the left end of the bar.)

We have already solved for T_1, T_2, and N. Thus we can now solve for H, V, and M:

$$\left.\begin{array}{l} H = 2\mu mg \\ V = mg \\ M = -xmg \end{array}\right\} \quad 0 \leqslant x < a$$

$$\left.\begin{array}{l} H = 0 \\ V = -mg \\ M = -(2a - x)mg \end{array}\right\} \quad a \leqslant x \leqslant 2a$$

The force resultants H and V and the moment resultant M are related to the stresses in the bar. This topic is studied in courses in strength of materials.

....

Example 15

When the car in Fig. 2.40(a) is parked on a level grade, 60% of the weight is on the front wheels. We are asked to determine (a) the normal force under the rear wheels when the

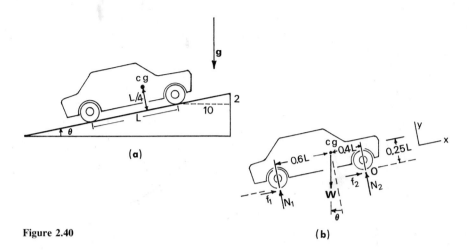

Figure 2.40   (b)

car is on a 20% grade, and (b) the minimum coefficient of friction when the car is parked using only the parking brake (on the rear wheels). We are given that the center of gravity is a distance $L/4$ above the pavement.

Solution

The center of gravity is $0.4L$ from the front wheels (can you show this?). Thus we have the free-body diagram in Fig. 2.40(b).

Our first problem is to determine the force N_1. Note that we have four unknowns in this problem: N_1, N_2, f_1, and f_2. In general, we would not expect to be able to solve this problem from statics alone, since we can write only three independent equilibrium equations (this is a plane problem). In this case, however, we can solve for N_1 and N_2. Taking moments about the front wheel, we have

$$M_{Oz} = 0: \quad -N_1 L + (mg \sin \theta)(0.25L) + (mg \cos \theta)(0.4L) = 0.$$

Noting that $\tan \theta = 0.200$, $\sin \theta = 0.196$, and $\cos \theta = 0.981$, we get $N_1 = mg(0.441)$. Setting $f_2 = 0$, we can solve for f_1:

$$F_x = 0: \quad f_1 - mg \sin \theta = 0.$$

Thus $f_1 = mg(0.196)$. And since $f_1 \leqslant \mu N_1$, we get

$$\mu \geqslant 0.45.$$

As a point of interest, what are the chances for parking a car safely on a 20% grade?

....

Example 16

In Contrast to Example 15, we are asked to determine the position of the center of gravity of some large body, such as an automobile.

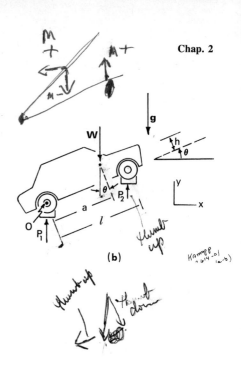

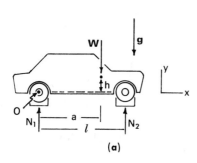

Figure 2.41

Solution

The scheme we shall use is to measure the normal forces under the front and back wheels in two orientations (Figs. 2.41a and 2.41b).

Suppose we measure the weight of the car W and the force under the front tires in each orientation, N_2 and P_2.

Equilibrium in Fig. 2.41(a) gives

$$F_y = 0: \qquad N_1 + N_2 - W = 0, \tag{a}$$

$$M_{Oz} = 0: \qquad N_2 l - Wa = 0, \tag{b}$$

and equilibrium in Fig. 2.41(b) gives

$$F_y = 0: \qquad P_1 + P_2 - W = 0 \tag{c}$$

$$M_{Oz} = 0: \qquad P_2 l \cos \theta + (W \sin \theta) h - (W \cos \theta) a = 0. \tag{d}$$

From (b) we have

$$a = \left(\frac{N_2}{W}\right) l. \tag{e}$$

And from (d) and (e), we have

$$h = \left(\frac{N_2}{W} - \frac{P_2}{W}\right) l \cot \theta. \tag{f}$$

To take a specific example, suppose that $l = 8.2$ ft, $\theta = 10°$, and suppose that when the car is level 58.0% of the weight is on the front wheels, while in the inclined position 53.5% of the weight is on the front wheels. Thus

$$\frac{N_2}{W} = 0.580, \qquad \frac{P_2}{W} = 0.535.$$

From (e), we obtain

$a = (0.580)(8.2 \text{ ft}) = 4.76 \text{ ft},$

$h = (0.580 - 0.535)(8.2 \text{ ft})(5.67) = 2.09 \text{ ft}.$

. . . .

9. CONCLUSION

At this point it is worth reviewing the methods and procedures in statics. Suppose we have a system subject to certain forces, couples, and constraints. We might be asked to find out whether the system is in equilibrium, or if it *is* in equilibrium, what the reaction forces and couples are.

In any event, we begin by identifying the nature of the forces and couples acting on the body. We then translate this information into a free-body diagram which indicates the known forces and couples and all possible forces and couples at any constraint (pin joint, socket, etc.). After that, we write equilibrium equations for the body. This means that $\mathbf{F} = 0$, the force resultant vanishes, and $\mathbf{M} = 0$, the moment resultant about some point vanishes. We need evaluate the moment about only one point. Thus it is worthwhile to pick a convenient point.

In the process of writing equilibrium equations, we may well wish to reduce some force distributions to simpler, statically equivalent force systems. Considering the weight of the body as a single force through the center of gravity is an example of this.

There are many concepts from statics which we have not covered here; for example, we haven't studied trusses or beams, or deformable bodies in detail. You encounter these in courses in strength of materials and elasticity. However, we have presented here the statics of these problems.

10. SUMMARY

A. A Single Particle

A single particle is in equilibrium if and only if the force resultant on it is zero.

In equilibrium, the moment resultant about any point is zero. The moment condition is necessary, but not sufficient for equilibrium of a particle.

B. Rigid Body

Necessary and sufficient conditions for the equilibrium of a rigid body are

i) $\mathbf{F} = \sum_{i=1}^{n} \mathbf{F}_i = 0$ Zero force resultant, all times $t \geq t_0$

ii) $\mathbf{M}_O = \sum\limits_{i=1}^{m} \mathbf{C}_i + \sum\limits_{i=1}^{n} \mathbf{r}_i \times \mathbf{F}_i = 0$ Zero moment resultant,
all times $t \geq t_0$.

iii)   The rigid body is at rest with respect to some inertial reference frame
at $t = t_0$.

C. Equivalent Force Systems

Moment transformation rule:

$$\mathbf{M}_A = \mathbf{r}_{AB} \times \mathbf{F} + \mathbf{M}_B.$$

Two force systems are equivalent if they have the same force resultant and
agree on the moment resultant at any one point. If a given force system has

Force resultant, $\mathbf{F}$ and   Moment resultant at A, $\mathbf{M}_A$,

then an equivalent force system consists of

i) a force $\mathbf{F}$ through A

ii) a couple $\mathbf{C} = \mathbf{M}_A$.

The center of gravity of a body is the point through which the weight
vector always passes.

For a body with constant density, the center of gravity is located at the
centroid of the body.

D. Forces in Statics

i) *Gravity*
Particle: $-mg\mathbf{k}$ . . . $\mathbf{k}$ vertical
Rigid body: $-mg\mathbf{k}$ . . . through the center of gravity.

ii) *Restraint Forces.* Forces and moments are produced to restrict the
movement of the point of application. These forces and moments must
be determined by equilibrium conditions in statically determinant
problems.

iii) *Static Friction.* $f < \mu N$; μ is the coefficient of static friction, N is the
magnitude of the force normal to the adjacent surfaces.

iv) *Sliding Friction.* $f = \mu N$; μ is the coefficient of sliding friction, N is the
magnitude of the force normal to the adjacent surfaces.

v) *String.* A string can transmit forces only, and only in the direction of
its length. A string can support only tensile forces.

vi) *Weightless Bar.* A weightless bar with pins at its ends (2-dimensional problem) or sockets at its ends (3-dimensional problem) can transmit forces only along the line between its ends. These forces can be either tensile or compressive.

E. Free-Body Diagram

The free-body diagram shows a system isolated from its constraints. All given forces and moments and all possible constraint forces and moments are represented.

PROBLEMS

Section 2.3

Problem 2.1

Given that a force $\mathbf{F} = 200\mathbf{j}$ (newtons) is applied at the position

$$\mathbf{r} = 10\mathbf{i} - 20\mathbf{j} + 5\mathbf{k} \text{ in.,}$$

find $\mathbf{M}_O$ in ft-lb.

Problem 2.2

Suppose that, for a given force system,

$$\mathbf{M}_O = 100\mathbf{k} \text{ (ft-lb)}, \qquad \mathbf{F} = 10\mathbf{i} - 20\mathbf{j} \text{ (lb)}.$$

a) Find $\mathbf{M}_A$ where $\mathbf{r}_{OA} = x\mathbf{i} + y\mathbf{j}$ (ft).
b) Find $\mathbf{M}_A$ where $\mathbf{r}_{OA} = 10\mathbf{k}$ (ft).

For Problems 2.3 through 2.6, suppose that a particle is located at the position $\mathbf{r}_P = 3\mathbf{i} + 4\mathbf{j} + 5\mathbf{k}$ (ft). The following forces act on the particle:

a) $\mathbf{F}_1 = \mathbf{i} + 2\mathbf{j}$ (lb)   c) $\mathbf{F}_3 = -\mathbf{i} - \mathbf{j} - \mathbf{k}$ (lb)
b) $\mathbf{F}_2 = -\mathbf{j} + \mathbf{k}$ (lb)   d) $\mathbf{F}_4 = 2\mathbf{i} + \frac{3}{2}\mathbf{j} + \mathbf{k}$ (lb)

Problem 2.3

Find the moment of each force about the point $\mathbf{r}_A = -\mathbf{i} + \mathbf{j} + 3\mathbf{k}$ (ft).

Problem 2.4

Find the total moment of the forces about the point $\mathbf{r}_A$.

Problem 2.5

Determine whether the particle is in equilibrium

Problem 2.6

Determine the total moment of the forces about the point $r_B = i + j + k$ (ft).

Problem 2.7

A particle which weighs 10 lb is to be in equilibrium on a sphere. The maximum possible friction force between the sphere and the particle is $0.1N$, where N is the normal force between the particle and the sphere. Find the maximum angle θ for which equilibrium is possible.

Figure 2.42

Problem 2.8

The particle shown in Fig. 2.43 weighs 100 lb and is suspended by a string and rests on a smooth sphere. Find the tension in the string.

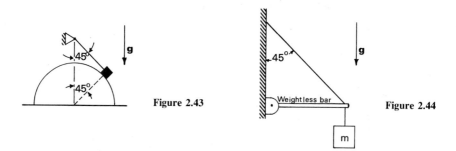

Figure 2.43   Figure 2.44

Problem 2.9

A mass m which weighs 320 lb is suspended by means of a weightless bar and a string as shown in Fig. 2.44.

   a) Find the tension in the string above the bar.

   b) Find the force in the bar. Is the bar in tension or compression?

Problem 2.10

A mass m is suspended by a weightless bar and a string as shown.

a) Find the tension in the string above the bar as a function of the weight of m, W, the length L, and the height H.

b) The weight of the mass is 100 lb, $L = 10$ ft, and the string above the bar will break at the tension of 250 lb. Find the minimum height H.

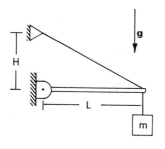

Figure 2.45

Problem 2.11

The mass m in the figure weighs 200 lb. It is suspended by two strings of equal length. The movable support on the right is restrained by a horizontal force P. Assume that the system is in equilibrium.

a) Find the force P as a function of the angle θ.

b) Find the tension in each string as a function of θ.

c) Suppose that the strings will break at a tension of 400 lb. Find the range of angles for which the strings will support the mass.

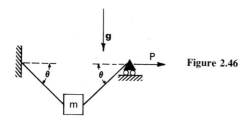

Figure 2.46

Section 2.4

For Problems 2.12 and 2.13, suppose that the force $F_1 = 5i + 3j + k$ (lb) acts at $r = 0$, and that the force $F_2 = -5i - 3j - k$ (lb) acts at $r = i + 2j + 3k$ (ft).

Problem 2.12

Find (a) the force resultant of F_1 and F_2, (b) the moment resultant abour $r_A = 0$, and (c) the moment resultant about $r_B = i + j + k$ (ft).

Problem 2.13

Find the perpendicular distance between the line of F_1 and the line of F_2. Verify that the magnitude of the moments determined in Problem 2.12 can be written $|F_1|d$, where d is the distance determined above.

Problem 2.14

In the system shown in the figure, consider the pulleys to be weightless. Find the force F such that the system is in equilibrium.

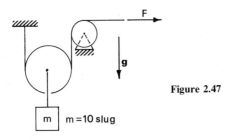

Figure 2.47

Problem 2.15

For the system shown in the figure, find the tension of the string in regions A, B, C, and D. The masses of the various parts of the system, including that of the uniform beam at the bottom, are indicated in the figure.

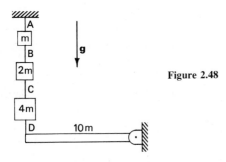

Figure 2.48

Problem 2.16

The lever shown in Fig. 2.49 supports a mass m, which weighs 1000 lb.

    a) If $a_2 = 1.0$ ft, how long will a_1 have to be in order that a force $P = 50$ lb will support the mass?

    b) For the system in part (a), if the mass is raised 1.0 in., through what distance will the force P move?

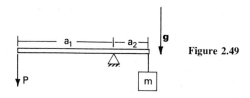

Figure 2.49

Problem 2.17

For the system shown in Fig. 2.50:

  a) Determine the force P required to maintain equilibrium.
  b) Determine the reaction forces at O.

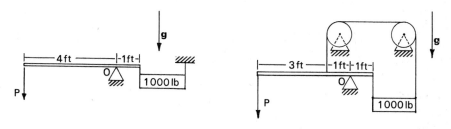

Figure 2.50   Figure 2.51

Problem 2.18

For the system shown in Fig. 2.51:

  a) Determine the force, P, required to maintain equilibrium.
  b) Determine the reactions at O.

Problem 2.19

For the system shown in the figure:

  a) Find the force P required to maintain equilibrium.
  b) Find the reactions at O.

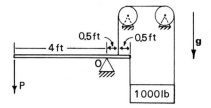

Figure 2.52

Problem 2.20

The fluid system shown in the figure is used to support a 1000-lb weight. Neglect the weight of the fluid.

    a) Find the force P required for equilibrium.

    b) If the piston on the left side of the system moves 1.0 in., how far does the weight move? (Assume that the fluid is incompressible.)

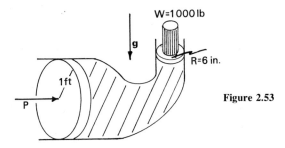

Figure 2.53

Section 2.5

Problem 2.21

Consider the forces applied to the disk as shown in the figure:

    a) What is the force resultant?

    b) What is the moment resultant about the origin (0,0)?

    c) What is the moment resultant about the point (1,0) ft?

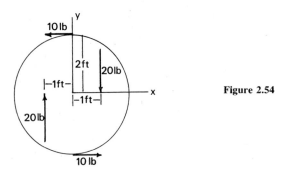

Figure 2.54

Problem 2.22

Consider the force system shown in Fig. 2.55. What force and couple must be added to the system at the point (0,1,0) in order that the new force resultant and the new moment resultant about (0,0,0) are zero?

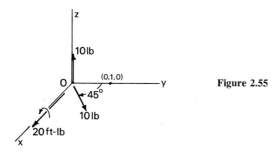

Figure 2.55

Problem 2.23

For the system shown in the figure, show that the pin reaction forces form a couple. Determine the pin forces. Neglect the weight of the bar and assume there is no axial force in the bar.

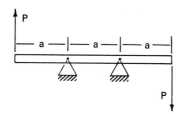

Figure 2.56

Problem 2.24

A bent bar is built into the wall on the left of the figure. If the forces applied to the bar are as shown, find the force and couple which the wall exerts on the bar at O so that the force resultant $\mathbf{F}$ and the moment resultant $\mathbf{M}_O$ are zero.

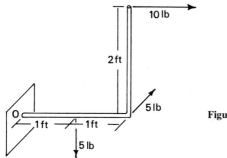

Figure 2.57

Problem 2.25

A beam is built in at its left end as shown in the figure. The forces indicated are applied to the beam. Neglecting the weight of the beam, determine the force system exerted on the beam by the wall at the left end.

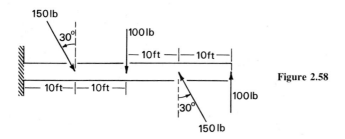

Figure 2.58

Section 2.6

Problem 2.26

A body is subject to gravity forces. Suppose the force resultant is

$$\mathbf{F} = -mg\mathbf{k},$$

and the moment resultant about the point O is

$$\mathbf{M}_O = -g\mathbf{i} \int_V y\rho \, dV + g\mathbf{j} \int_V x\rho \, dV.$$

If

$$\mathbf{r}_{OA} = \frac{1}{m} \left(\mathbf{i} \int_V x\rho \, dV + \mathbf{j} \int_V y\rho \, dV \right) + \alpha\mathbf{k},$$

where α is an arbitrary constant, show that $\mathbf{M}_A = 0$. [*Hint:* Use (2.13).]

Problem 2.27

Find the centroid of the shaded area in the figure.

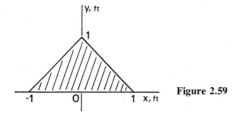

Figure 2.59

Problem 2.28

Find the centroid of the shaded area in the figure. [*Hint:* Recall the results of Problem 2.27.]

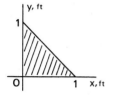

Figure 2.60

Problem 2.29

Show that the position of the centroid of the thin, circular arc shown in the figure is given by

$$x_C = \frac{R \sin \alpha}{\alpha}, \qquad y_C = 0.$$

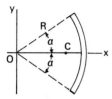

Figure 2.61

Problem 2.30

Show that the position of the centroid of the circular sector shown in the figure is given by

$$x_C = \frac{2R \sin \alpha}{3\alpha}.$$

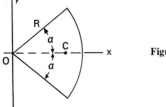

Figure 2.62

Problem 2.31

The object shown in the figure has a uniform thickness and density. Find the position of its center of gravity.

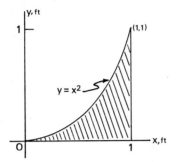

Figure 2.63

Problem 2.32

The various parts of the object shown in the figure are thin, uniform, and have the weights indicated. Find the position of the center of gravity of the entire system. [*Hint:* Use the result of Problem 2.31.]

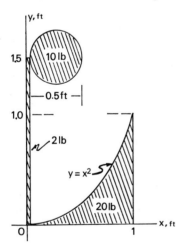

Figure 2.64

Problem 2.33

Show that the position of the center of gravity of a right circular cone of height h and base radius R (Fig. 2.65) is $x_{cg} = y_{cg} = 0$, $z_{cg} = \frac{1}{4}h$. Assume that the density is constant.

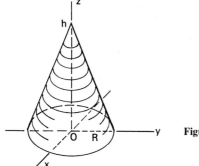

Figure 2.65

Problem 2.34

A thin bar is made in two sections, each of length a. Suppose that the density is given by

$$\rho = \rho_1, \quad 0 \leqslant x \leqslant a,$$

$$\rho = \rho_2, \quad a < x \leqslant 2a.$$

Determine x_{cg} for the combined bar.

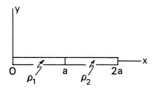

Figure 2.66

Problem 2.35

The density of a thin bar of length L varies along its length according to

$$\rho = \rho_1 + \left(\frac{\rho_2 - \rho_1}{L}\right) x, \quad \rho = \text{mass/length}$$

a) Determine x_{cg} of the bar.

b) If $\rho_1 = \rho_2$, determine x_{cg}.

Problem 2.36

Consider a plate with a constant thickness and constant density, and suppose that it has a hole cut in it. Let A be the area of the plate (with the hole), and let A_h be the area of the hole. Show that

$$Ax_{cg} = (A + A_h)x_1 - A_h x_h, \qquad Ay_{cg} = (A + A_h)y_1 - A_h y_h,$$

where (x_{cg}, y_{cg}) is the center of gravity of the plate with the hole, (x_1, y_1) is the center of gravity of a plate with the hole filled in, and (x_h, y_h) is the center of gravity of the hole.

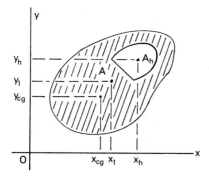

Figure 2.67

Problem 2.37

The plate shown has a constant thickness and constant density. Find the position of its center of gravity. [*Hint:* Refer to Problem 2.36.]

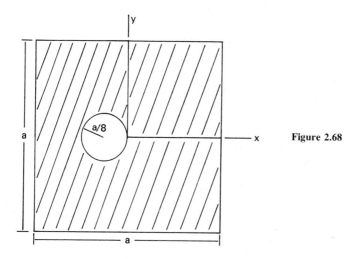

Figure 2.68

Problem 2.38

Consider a rectangular plate of constant thickness and constant density. At what position (x_1, y_1) should we drill a hole of radius $\frac{1}{4}a$ in order to shift the center of gravity from $x = y = 0$ to $x_{cg} = d_1$, $y_{cg} = -d_2$. [*Hint:* Refer to Problem 2.36.]

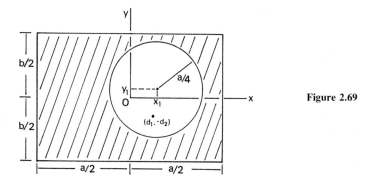

Figure 2.69

Problem 2.39

A rectangular parallelepiped of height h and constant density ρ has a right circular conical hole, also of height h, drilled out of it. Determine the position of the center of gravity. [*Hint:* Refer to Problems 2.33 and 2.36.]

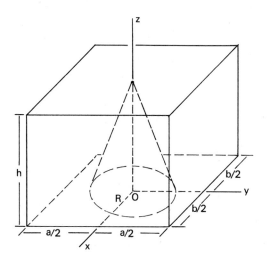

Figure 2.70

Section 2.7

Problem 2.40

A block of mass m is placed on an incline as shown in the figure. To determine the coefficient of static friction μ, the angle of the incline is increased until the block slides. Suppose this occurs at angle $\theta = \theta_0$. Given θ_0, determine μ.

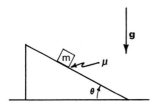

Figure 2.71

Problem 2.41

Determine the range of values for the force F such that the system in the figure will be in equilibrium.

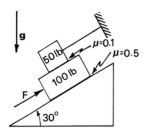

Figure 2.72

Problem 2.42

Consider the system of Problem 2.41. Suppose the coefficient of static friction *between* the blocks is $\mu = 0.7$, but the other parameters of the problem are unchanged. Determine the maximum value of the force F for which the system will be in equilibrium.

Problem 2.43

Determine the range of values for the force F such that the system in Fig. 2.73 will be in equilibrium.

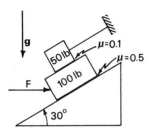

Figure 2.73

Problem 2.44

Consider the system of Problem 2.43. Suppose that the coefficient of static friction *between* the blocks is $\mu = 0.7$, but the other parameters of the problem are unchanged. Determine the maximum value of the force F which the system will be in equilibrium.

Section 2.8

For Problems 2.45 through 2.49, sketch the free-body diagram for each shaded body.

Problem 2.45

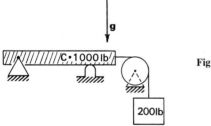

Figure 2.74

Problem 2.46

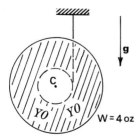

Figure 2.75

Problem 2.47   *Problem 2.48*

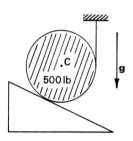

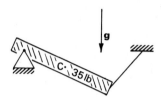

Figure 2.76   Figure 2.77

Problem 2.49

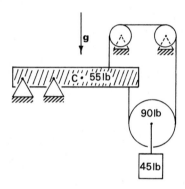

Figure 2.78

Problem 2.50

In the system shown in the figure, the 20-lb block can slide on a weightless beam. For what position x will the system be in equilibrium?

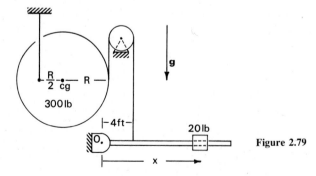

Figure 2.79

Problem 2.51

The doughnut in Fig. 2.80 has a radius of 5 ft and weighs 100 lb. The forces shown are applied to it.

    a) Calculate the wrench which is equivalent to the force system on the doughnut.

    b) What force system (force and couple) would have to be applied at the point A in order to have the doughnut in equilibrium?

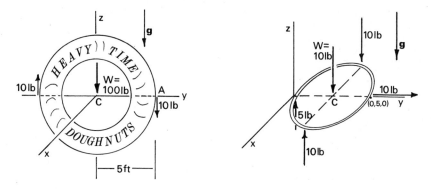

Figure 2.80   **Figure 2.81**

Problem 2.52

A thin ring has a diameter of 5 ft and weighs 10 lb. The ring is subjected to the weight W and the other forces shown in Fig. 2.81.

    a) Find the wrench equivalent to the forces acting on the ring.

    b) What force system would have to be applied to the ring at the point $\mathbf{r} = 5\mathbf{j}$ (ft) in order that the ring be in equilibrium?

Problem 2.53

A bar is placed against a vertical wall as shown in the figure. The horizontal plane is smooth. Show that equilibrium is impossible no matter what the coefficient of static friction might be at the vertical wall.

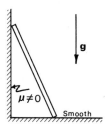

Figure 2.82

Problem 2.54

The mass m shown in the figure weighs 100 lb and is supported by means of a bar which weighs 50 lb.

a) Find the pin reactions at point O.

b) Find the normal force acting at the lower end of the bar.

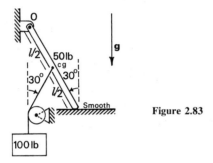

Figure 2.83

Problem 2.55

For the system shown in the figure, assume all surfaces are smooth. The bar weighs 100 lb and the disk weighs 300 lb.

a) Find the force P for equilibrium.

b) Find the force exerted by the bar on the disk.

c) Find the force exerted by the horizontal plane on the disk.

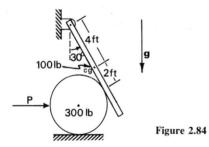

Figure 2.84

Problem 2.56

The system shown in Fig. 2.85 is to be in equilibrium.

a) Find the minimum coefficient of static friction between the bar and the disk μ_1.

b) Find the minimum coefficient of static friction between the disk and the horizontal plane μ_2.

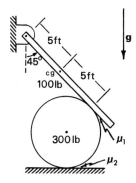

Figure 2.85

Problem 2.57

The bar shown in the figure has weight W. It is supported by three smooth walls. Find the forces exerted by the walls on the bar.

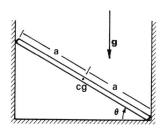

Figure 2.86

Problem 2.58

The homogeneous cylinder of weight W lb in the figure is in equilibrium under the action of a force P. Determine the wall reactions.

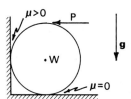

Figure 2.87

Problem 2.59

The homogeneous cylinder of weight W lb in the figure is in equilibrium under the action of a force Q. Determine the wall reactions.

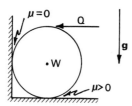

Figure 2.88

Problem 2.60

The homogeneous cylinder of weight W lb shown in the figure just begins to slip under the influence of the force R. Determine the magnitude of the force R.

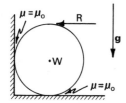

Figure 2.89

Problem 2.61

For the cylinder shown in the figure, find the minimum force F required to initiate slip. The coefficient of sliding friction between the cylinder and both walls is given as μ. The weight of the cylinder is W lb.

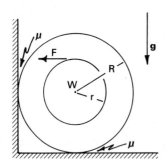

Figure 2.90

Problem 2.62

For the cylinder shown, find the minimum force F required to initiate slip. The string shown is wrapped around the inner radius of the cylinder of weight W lb.

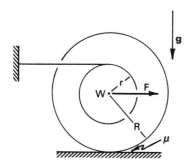

Figure 2.91

Problem 2.63

For the cylinder shown in Fig. 2.92 find the minimum force F required to initiate slip. The string shown is wrapped around the inner radius of the cylinder of weight W lb.

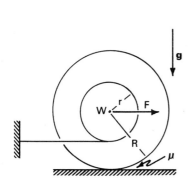

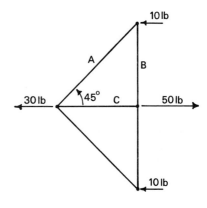

Figure 2.92   Figure 2.93

Problem 2.64

Thus truss shown in Fig. 2.93 consists of rigid, weightless bars connected by pin joints. Determine the forces in members A, B, and C. In each case, state whether the member is in tension or compression.

Problem 2.65

Consider the members of the structure shown in the figure to be weightless. Find the forces in members AB, BC, CD, and EB. Use a positive sign to denote a member in tension, a negative sign for a member in compression.

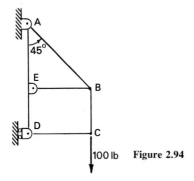

100 lb   **Figure 2.94**

Problem 2.66

The structure shown in the figure consists of weightless members. The structure is subjected to the force P. Determine:

   a) the force in member AB,

   b) the pin reactions at C.

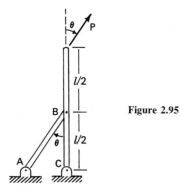

Figure 2.95

Problem 2.67

The system shown in Fig. 2.96 consists of weightless bars connected by pin joints. The system supports a 50-lb weight.

   a) Find the force in member BC.

   b) Find the pin reactions at points A and C.

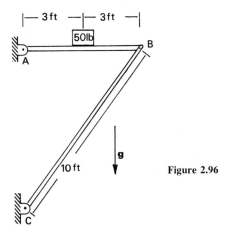

Figure 2.96

Problem 2.68

The bar of weight W shown in Fig. 2.97 rests on two smooth walls supported by the string. Determine:

a) the tension in the string,

b) the normal forces exerted by the horizontal and vertical walls on the bar.

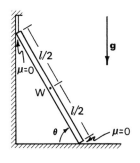

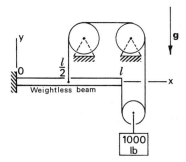

Figure 2.97   Figure 2.98

Problem 2.69

Consider the beam and the pulleys in the system shown in Fig. 2.98 to be weightless. The system supports a 1000-lb weight.

a) Find the bending moment $M(x)$ for the regions (i) $0 \leqslant x < \frac{1}{2}l$, (ii) $\frac{1}{2}l \leqslant x \leqslant l$.

b) Find the shear force $V(x)$, for the regions (i) $0 \leqslant x < \frac{1}{2}l$, (ii) $\frac{1}{2}l \leqslant x \leqslant l$.

Problem 2.70

The *nonhomogeneous* cylinder of mass m and radius R shown in the figure has its center of gravity located a distance r from its geometric center. The cylinder is restrained by a string. For the position shown, determine:

  a) the string tension,

  b) the force(s) acting at the contact surface of the cylinder.

  c) Is there a minimum value of the coefficient of static friction necessary to ensure equilibrium?

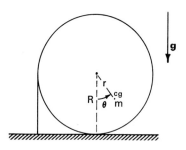

Figure 2.99

Problem 2.71

A *nonhomogeneous* cylinder has its center of gravity at the position shown in the figure. A string restrains the top of the cylinder. The coefficient of static friction between the cylinder and the horizontal plane is μ. Find the maximum angle θ for equilibrium.

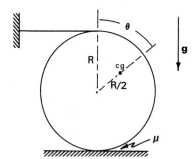

Figure 2.100

Problem 2.72

Consider the system shown in Fig. 2.101. The *nonhomogeneous* cylinder has its center of gravity at the position indicated. Suppose that $\mu_2 = 0$ and $\mu_1 = 0.25$. Calculate the range of angles θ for which the cylinder will be in equilibrium.

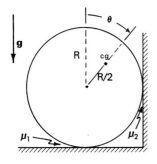

Figure 2.101

Problem 2.73

For the system in Problem 2.72, suppose that $\mu_1 = 0$ and $\mu_2 > 0$. Show that equilibrium is possible only if $\theta = 0$ or $\theta = 180°$.

Problem 2.74

A semicircular and a circular disk have the positions shown in the figure. The weights of the disks are $\frac{1}{2}W$ and W, respectively. Both walls are smooth. Find the force P in order that the two disks be in equilibrium.

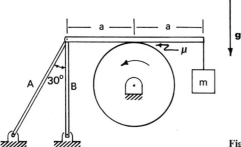

Figure 2.102

Problem 2.75

For the system shown in the figure, consider the three bars to be weightless. Determine the forces in bars A and B and note whether the bars are in tension or compression.

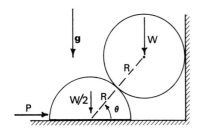

Figure 2.103

Problem 2.76

The beam in the figure weighs 67.8 lb, is pinned at its left end, and supports a mass of 1.0 slug on the right end. Below the middle of the beam is a spinning disk. The coefficient of sliding friction between the disk and the beam is $\mu = 0.1$. Determine the pin reactions.

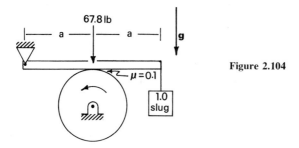

Figure 2.104

Problem 2.77

Consider the beam in Problem 2.76. Could the pin joint at the left end of the beam be replaced by a string anchored to some point? If so, describe the orientation of the string and determine the tension in the string.

Problem 2.78

For the system shown in the figure, determine the minimum mass m_0 such that the cylinder will slip. Consider the bar to be weightless.

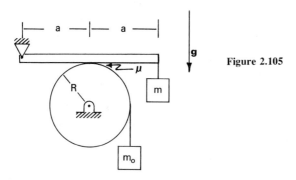

Figure 2.105

Problem 2.79

The brake in Fig. 2.106 is used to slow the spinning disk. Find the pin reactions at point A. Neglect the weight of the brake mechanism.

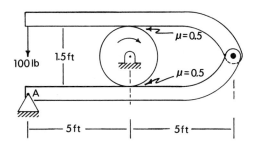

Figure 2.106

Problem 2.80

For the brake mechanism in Problem 2.79, determine the pin reactions at the center of the spinning disk.

Problem 2.81

Show that the shear force $V(x)$ and the bending moment $M(x)$ in the cantilever beam shown in the figure can be written

$$V(x) = \int_x^L f(u) \, du, \qquad M(x) = \int_x^L (u - x)f(u) \, du.$$

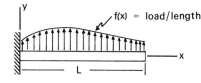

Figure 2.107

Problem 2.82

Determine the shear force $V(x)$ and the bending moment $M(x)$ for the cantilever beam shown in the figure. (Neglect the weight of the beam.)

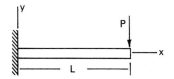

Figure 2.108

Problem 2.83

For the beam shown in the figure, determine:

    a) the pin reactions at A and B,

    b) the shear force $V(x)$,

    c) the bending moment $M(x)$.

(Neglect the weight of the beam.)

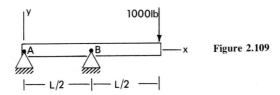

Figure 2.109

Problem 2.84

The weightless beam shown in the figure supports a 1000-lb weight. Determine the moment, shear, and axial resultants, $M(x)$, $V(x)$, and $H(x)$ for the regions: (i) $0 \leqslant x < 10$ ft, (ii) 10 ft $\leqslant x \leqslant 20$ ft.

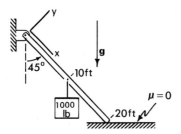

Figure 2.110

Problem 2.85

When the car shown in the figure is parked on a level street, 40% of the weight W is supported by the rear wheels. The center of gravity is located a distance $0.5L$ above the ground.

    Suppose the car is parked heading up a 30° hill. Find the normal force beneath the two rear wheels (the sum of the normal forces under the two rear wheels).

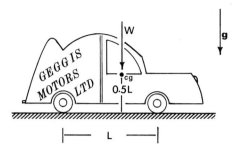

Figure 2.111

Problem 2.86

Again suppose that the car in Problem 2.85 is parked heading up a 30° hill. Suppose the rear wheels are locked and the front wheels are free. Find the minimum coefficient of friction between the pavement and the rear tires in order that the car will not slide.

Problem 2.87

Suppose the car in Problem 2.85 is parked heading downhill.

a) Find the total normal force under the rear wheels.

b) Assume that the parking brake operates on the rear wheels and that the front wheels are free. Find the minimum coefficient of friction in order that the car will not slip.

Problem 2.88

Again suppose the car in Problem 2.85 is parked heading up a 30° hill. Suppose that the parking brake operates on the front wheels (as in a Saab 99) and that the rear wheels are free. Find the minimum coefficient of friction between the pavement and the front wheels in order that the car will not slide.

Problem 2.89

Suppose that the car in Problem 2.85 is parked heading downhill.

a) Find the total normal force under the front wheels.

b) Assume that the parking brake operates on the front wheels and that the rear wheels are free. Find the minimum coefficient of friction in order that the car will not slip.

3

KINEMATICS OF
A PARTICLE

This reality is mobility. Not *things* made,
but things in the making, not self-main-
taining *states,* but only changing states,
exist.

An Introduction to Metaphysics
Henri Bergson

1. INTRODUCTION

Thus far we have seen that the basis of our study of dynamics is contained in
the equation

$$\mathbf{F} = m\mathbf{a}, \tag{3.1}$$

where $\mathbf{F}$ is the vector sum of the forces acting on a particle m, and $\mathbf{a}$ is the
acceleration of m measured relative to an inertial reference frame. The
question of whether a given reference frame is inertial or not is by no means
trivial.

To describe an inertial reference as one which is fixed in space begs the
question. However, to say that a reference frame is inertial if Eq. (3.1) can be
written in the frame has a very circular ring to it. In fact, the existence of an
inertial reference frame is one of the postulates of Newton's laws. The question of whether a given reference frame is inertial or not is then left to the un-
settling task of experimental verification.

If we can convince ourselves that one reference frame is inertial, we can
then determine an arbitrary number of other inertial reference frames. Whenever a reference frame is in uniform motion with respect to the inertial frame
(that is, whenever it has a uniform translation with respect to the inertial
frame with no change in its orientation), it is also an inertial reference frame.
It is easy to show that the acceleration computed in the two frames is the
same. Thus, if (3.1) is valid in one frame, it is valid in both.

To begin to write the equations of dynamics, we first need a reference
frame R, an inertial reference frame.* Suppose a particle moves along a path

---

* We may choose to write our equations in a noninertial reference frame. To do
this properly, however, we must still know the motion of the noninertial frame relative
to an inertial frame.

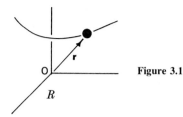

Figure 3.1

in space (Fig. 3.1). If O is the origin of R, we let $\mathbf{r}$ be the vector from O to the particle. The vector $\mathbf{r}$ is then the *position vector* of the particle relative to the inertial reference frame R.

We can now compute the velocity and acceleration vectors relative to R:

$$\mathbf{v} = \frac{d\mathbf{r}}{dt}, \tag{3.2}$$

$$\mathbf{a} = \frac{d\mathbf{v}}{dt} = \frac{d^2\mathbf{r}}{dt^2}. \tag{3.3}$$

To be perfectly precise, we should always say that $\mathbf{v}$ and $\mathbf{a}$ are the velocity and acceleration *relative to the frame* R. In most problems, there is no confusion if we drop the final qualifying phrase. When we speak of velocity and acceleration without a qualifying statement, we mean the velocity and acceleration relative to an inertial reference frame.

2. NATURAL COORDINATES

Any coordinate system brings something of its own personality to the description of a problem. In fact, because of this, the theory of dynamics is best developed with vectors alone — not the components of vectors. On the other hand, an actual problem is rarely formulated without the aid of some coordinate system.

To understand the notions of velocity and acceleration, we begin by discussing natural coordinates. These are coordinates defined in terms of the particular motion under consideration. Natural coordinates are occasionally used in the formulation of problems, but more important, the study of velocity and acceleration in natural coordinates gives us insight into the nature of these quantities. This will be particularly valuable when we come to investigate the other coordinate systems presented in this chapter.

There are many problems in dynamics in which the path of a particle is a known curve. The particle might be constrained to remain on a given curve — a straight line or a circle, for example, as in Figs. 3.2(a) and (b).

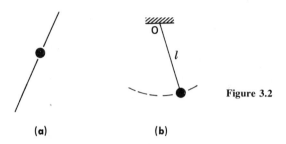

Figure 3.2

(a)   (b)

In (a), the particle is restricted to move on a line. This could be, for example, a bead on a stiff, straight wire. In (b), the particle is attached to the point O by an inextensible string of length l. If the particle is required to remain in a plane, the path of the particle is the circle of radius l with center at O.

In using natural coordinates, we assign three unit vectors to each point of the path. These are (1) the unit tangent vector e_t, (2) the unit normal vector e_n, and (3) the unit binormal vector e_b. (Actually, if we are concerned only with the position, velocity, and acceleration vectors, we do not need the binormal vector. See Problem 3.10.) These vectors are defined, as their names suggest, in terms of the properties of the actual path the particle travels. It is in this sense that the coordinates are "natural."

a) Velocity

Suppose a particle is in motion on a path C. Let O be the origin of an inertial reference frame and take $\mathbf{r}$ to be the vector from O to the particle (Fig. 3.3). The velocity is then

$$\mathbf{v} = \frac{d\mathbf{r}}{dt} = \lim_{\Delta t \to 0} \frac{\mathbf{r}(t + \Delta t) - \mathbf{r}(t)}{\Delta t}.$$

If we write $\mathbf{r}(t + \Delta t) = \mathbf{r}(t) + \Delta \mathbf{r}$, then

$$\mathbf{v} = \lim_{\Delta t \to 0} \frac{\Delta \mathbf{r}}{\Delta t}, \tag{3.4}$$

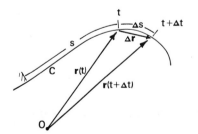

Figure 3.3

where $\Delta\mathbf{r}$ is as shown in Fig. 3.3. From the figure, it is clear that as $\Delta t \to 0$, $\Delta\mathbf{r}$ becomes tangent to the path C.

Suppose that, from some convenient point on C, we begin measuring arc length along the path s. Let the arc length traveled from time t until time $(t + \Delta t)$ be denoted by Δs. Then we can write (3.4) as

$$\mathbf{v} = \lim_{\Delta t \to 0} \left(\frac{\Delta\mathbf{r}}{\Delta s}\right) \left(\frac{\Delta s}{\Delta t}\right) = \left(\frac{d\mathbf{r}}{ds}\right) \frac{ds}{dt}.$$

The scalar quantity (ds/dt) is just the speed along the curve (the reading of your speedometer if you were driving along the curve). Also, as $\Delta t \to 0$, we see (Fig. 3.3) that

$$\left|\frac{\Delta\mathbf{r}}{\Delta s}\right| \to 1.$$

Thus $d\mathbf{r}/ds$ is a unit tangent vector, and we denote it by $\mathbf{e}_t$:

$$\frac{d\mathbf{r}}{ds} = \mathbf{e}_t. \tag{3.5}$$

Thus, writing* $ds/dt = \dot{s}$, we have the velocity

$$\mathbf{v} = \dot{s}\mathbf{e}_t. \tag{3.6}$$

> The velocity of a particle is tangent to its path and has the magnitude of the speed $\dot{s}$ along the path.

b) Acceleration

To find the acceleration vector $\mathbf{a}$, we differentiate $\mathbf{v}$ in (3.6) with respect to time:

$$\mathbf{a} = \ddot{s}\mathbf{e}_t + \dot{s}\dot{\mathbf{e}}_t. \tag{3.7}$$

Many students argue that since $\mathbf{e}_t$ has a constant length, its derivative must be zero. This is not, in general, the case. Consider the identity

$$\mathbf{e}_t \cdot \mathbf{e}_t \equiv 1.$$

Differentiating, we get

$$\dot{\mathbf{e}}_t \cdot \mathbf{e}_t + \mathbf{e}_t \cdot \dot{\mathbf{e}}_t = 0,$$

and since for any two vectors $\mathbf{a} \cdot \mathbf{b} = \mathbf{b} \cdot \mathbf{a}$, we have

$$2\dot{\mathbf{e}}_t \cdot \mathbf{e}_t = 0.$$

---

* When time derivatives are not ambiguous, we often use the dot notation for differentiation with respect to time.

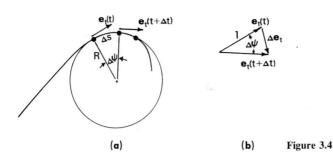

(a)   (b)   Figure 3.4

There are two possibilities:

i) $\dot{\mathbf{e}}_t = 0$.

ii) $\dot{\mathbf{e}}_t$ is perpendicular to $\mathbf{e}_t$.

We shall encounter both in this chapter.

Let us return to the path of the particle and consider three neighboring positions of the particle (Fig. 3.4).

Through these three points we fit a circle. The radius of this circle (in the limit as the points come together) is the radius of curvature R of the path at that point. We consider the unit tangent vector $\mathbf{e}_t$ at two positions separated by arc length Δs. From Fig. 3.4(b), we see that $\Delta \mathbf{e}_t$ is perpendicular to $\mathbf{e}_t$, as we have shown. In particular, the length of $\Delta \mathbf{e}_t$ is approximately

$$|\Delta \mathbf{e}_t| = 1 \cdot \Delta \psi;$$

on the other hand

$$\Delta s = R\, \Delta \psi,$$

where $\Delta \psi$ is the angle subtended at the center of the circle by the two positions under consideration. Thus

$$\frac{|\Delta \mathbf{e}_t|}{\Delta s} = \frac{1}{R}.$$

Thus the vector $d\mathbf{e}_t/ds$ is normal to the path (directed toward the center of the circle) and has magnitude $(1/R)$. We can now write

$$\frac{d\mathbf{e}_t}{ds} = \frac{1}{R}\, \mathbf{e}_n, \tag{3.8}$$

where $\mathbf{e}_n$ is the unit normal vector pointing toward the center of the circle fitted to the path.

Returning to Eq. (3.7), we can now evaluate $\dot{\mathbf{e}}_t$ by the chain rule and (3.8),

$$\frac{d\mathbf{e}_t}{dt} = \frac{d\mathbf{e}_t}{ds}\frac{ds}{dt} = \frac{1}{R}\frac{ds}{dt}\, \mathbf{e}_n.$$

Thus (3.7) becomes

$$\mathbf{a} = \ddot{s}\mathbf{e}_t + \frac{(\dot{s})^2}{R}\mathbf{e}_n.$$

(3.9)

> The acceleration of a particle can be decomposed into two components: the acceleration along the path ($\ddot{s}$), and the acceleration due to the change of direction of the velocity ($\dot{s}^2/R$).

In this chapter we have begun what is called mathematically *differential geometry*, and in particular the Frenet–Serret formulas.*

Example 1 (Circular Motion)

The particle in Fig. 3.5 moves along a circle of radius R. The angle which the line from O to the particle makes with the horizontal axis is denoted by θ. Find the velocity and acceleration of the particle.

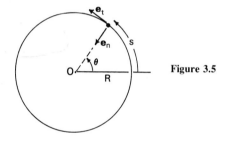

Figure 3.5

Solution

Here $s = R\theta$, $\dot{s} = R\dot{\theta}$, $\ddot{s} = R\ddot{\theta}$. In this problem, R is the radius of curvature. Thus from (3.6) and (3.9) we get

$$\mathbf{v} = R\dot{\theta}\,\mathbf{e}_t,$$

$$\mathbf{a} = R\ddot{\theta}\,\mathbf{e}_t + R\dot{\theta}^2\,\mathbf{e}_n.$$

· ·

Example 2 (Motion on a Parabola)

Suppose the particle moves along the parabola $y = x^2$ in Fig. 3.6 with constant speed $\dot{s} = v_0$. Find the acceleration of the particle.

---

* See D. J. Struik, *Differential Geometry*, Reading, Mass.: Addison-Wesley, 1961, pages 1–20.

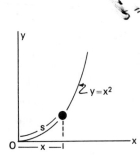

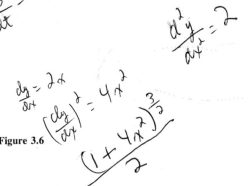

Figure 3.6

Solution

The radius of curvature is given by*

$$R = \frac{[1 + (dy/dx)^2]^{3/2}}{|d^2y/dx^2|}.$$

In the present case we obtain

$$R(x) = \tfrac{1}{2}(1 + 4x^2)^{3/2}$$

for the radius of curvature at position x. Thus we obtain the acceleration at x from (3.9), noting that $\ddot{s} = 0$:

$$\mathbf{a} = \frac{2v_0^2}{(1 + 4x^2)^{3/2}}\,\mathbf{e}_n.$$

You are encouraged to attempt to find **a** directly by using Cartesian coordinates. See Problems 3.5 and 3.6.

. . . .

Example 3

Figure 3.7(a) shows a particle restrained by an inextensible string which wraps around a fixed cylinder of radius a. Suppose the speed of the particle is constant, $\dot{s} = v_0$, ... constant.

Find the acceleration of the particle at the point P if the free length of the string at point P_0 is l.

Solution

This is one problem for which natural coordinates clearly provide the best means for analyzing the kinematics.

The problem becomes simple once one notes that the radius of curvature equals the length of free string at any given position. Thus the radius of curvature at P_0 is

$$R = l$$

---

* See Appendix 1, Eq. (A1.29).

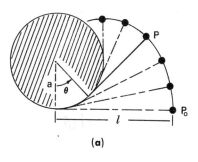

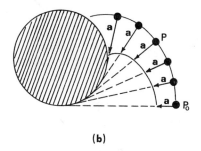

(a)   (b)

Figure 3.7

and at P is

$R = l - a\theta.$

From (3.9), noting that $\dot{s} = v_0$ and $\ddot{s} = 0$, we obtain the acceleration at P:

$$\mathbf{a} = \left(\frac{v_0^2}{l - a\theta}\right)\mathbf{e}_n.$$

Thus, as the string wraps around the cylinder, the magnitude of the acceleration grows, becoming infinite when the particle collides with the cylinder $[(l - a\theta) \to 0]$. Figure 3.7(b) indicates the acceleration at various points.

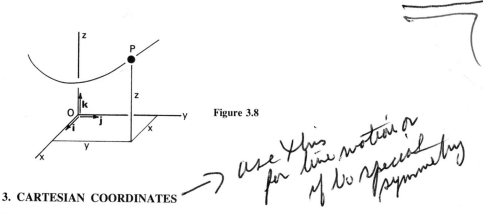

Figure 3.8

3. CARTESIAN COORDINATES

use this for line motion or of no special symmetry

Let the reference frame in Fig. 3.8 be inertial. We label the axes x, y, and z. Let $\mathbf{i}$, $\mathbf{j}$, $\mathbf{k}$ be unit vectors along the x, y, and z axes, respectively. Since the axis system is inertial, it has a constant orientation. The position vector $\mathbf{r}$ can be written

$$\mathbf{r} = x\mathbf{i} + y\mathbf{j} + z\mathbf{k}, \qquad\qquad (3.10)$$

$$d\mathbf{r} = \frac{dx}{dt}\mathbf{i} + \frac{dy}{dt}\,dt\,\mathbf{j} + \frac{dz}{dt}\,dt\,\mathbf{k}$$

and thus

$$\mathbf{v} = \frac{d\mathbf{r}}{dt} = \dot{x}\mathbf{i} + \dot{y}\mathbf{j} + \dot{z}\mathbf{k} + (x\dot{\mathbf{i}} + y\dot{\mathbf{j}} + z\dot{\mathbf{k}}).$$

The final set of terms is zero, since **i**, **j**, **k** have both constant length and constant direction. [Hence, for example, $\mathbf{i}(t + \Delta t) = \mathbf{i}(t)$.] Thus

$$\mathbf{v} = \dot{x}\mathbf{i} + \dot{y}\mathbf{j} + \dot{z}\mathbf{k}, \tag{3.11}$$

and the acceleration has the following form:

$$\mathbf{a} = \ddot{x}\mathbf{i} + \ddot{y}\mathbf{j} + \ddot{z}\mathbf{k}. \tag{3.12}$$

Example 4

The position of a particle has the following cartesian components:

$$x(t) = (10t)\ \text{ft}, \qquad y(t) = (48t - 16t^2)\ \text{ft, where } t \text{ is in seconds.}$$

i) Find the times and positions of the particle when $y = 0$.

ii) Find the velocity and acceleration of the particle as a function of time.

iii) Find the position of the particle when $\dot{y} = 0$.

iv) Find the trajectory of the particle, i.e., the path $y = y(x)$.

Solution

i) From $y = 0$, we have

$$48t - 16t^2 = 0 \qquad \text{or} \qquad t(48 - 16t) = 0.$$

The times are $t = 0$ and $t = 3$ sec. At $t = 0$, $\mathbf{r} = 0$; at $t = 3$ sec, $\mathbf{r} = 30\mathbf{i}$ ft.

ii) $\dot{x} = 10$ ft/sec, $\qquad \dot{y} = (48 - 32t)$ ft/sec,

$$\ddot{x} = 0, \qquad \ddot{y} = -32\ \text{ft/sec}^2.$$

Thus $\mathbf{v} = 10\mathbf{i} + (48 - 32t)\mathbf{j}$ ft/sec, $\mathbf{a} = -32\mathbf{j}$ ft/sec².

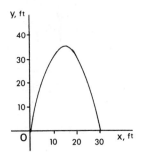

Figure 3.9

iii) From part (ii), $\dot{y} = (48 - 32t)$ ft/sec. Thus $\dot{y} = 0$ gives $t = 1.5$ sec; and at $t = 1.5$ sec, $\mathbf{r} = 15\mathbf{i} + 36\mathbf{j}$ ft.

iv) From $x = 10t$ ft and $y = 48t - 16t^2$ ft, we eliminate the time parameter t. Writing $t = x/10$ sec, we obtain

$$y = (4.8x - 0.16x^2) \text{ ft.}$$

This is the equation of a parabola. The trajectory is shown in Fig. 3.9.

Example 5 (Rectilinear Motion with Constant Acceleration)

Suppose that

$$\ddot{x} = a_0, \ldots \text{ constant} \tag{a}$$

and $y = z = 0$. Find expressions for $\dot{x}(t)$, $x(t)$, and $v(x)$, where $v = \dot{x}$. Let $x = x_0$ and $\dot{x} = v_0$ at $t = 0$.

Solution

Integrating (a), we obtain

$$\dot{x} = a_0 t + c_1, \tag{b}$$

and again

$$x = \tfrac{1}{2}a_0 t^2 + c_1 t + c_2. \tag{c}$$

We have to evaluate the integration constants. From (b) and (c), we see that, at $t = 0$,

$$\dot{x} = c_1, \qquad x = c_2.$$

Thus c_1 and c_2 give the speed and position of the point at $t = 0$. By the initial conditions, we have

$$\dot{x}(0) = v_0, \qquad x(0) = x_0.$$

Then (b) and (c) can be written

$$\dot{x} = a_0 t + v_0, \tag{d}$$

$$x = \tfrac{1}{2}a_0 t^2 + v_0 t + x_0. \tag{e}$$

To obtain the expression for $v = v(x)$, let $v = \dot{x}$. Then

$$\frac{dv}{dt} = \ddot{x},$$

and we can write dv/dt by the chain rule,

$$\frac{dv}{dt} = \frac{dv}{dx}\frac{dx}{dt} = \frac{dv}{dx}v. \; \approx \; a_0$$

In our particular case $\ddot{x} = a_0$, . . . constant. Thus

$$v\frac{dv}{dx} = a_0 \quad \text{or} \quad v\,dv = a_0\,dx.$$

Integrating, we have

$$\int_{v_0}^{v} v\,dv = \int_{x_0}^{x} a_0\,dx$$

or

$$\tfrac{1}{2}(v^2 - v_0^2) = a_0(x - x_0)$$

or

$v(x) =$ $$v = \sqrt{v_0^2 + 2a_0(x - x_0)}\,,$$ (f)

which gives the speed $v = \dot{x}$ at a given position x.

We note again that the results (d), (e), and (f) apply *only* to the case of constant acceleration.

Example 6

Suppose that $x(t) = (4t^3 - 48t^2 + 180t)$ ft.

   i) Find x at $t = 0$ and $t = 10$ sec.

   ii) Find the times and values of x for which $\dot{x} = 0$.

   iii) Find the range of times for which $\dot{x} > 0$ and for which $\dot{x} < 0$.

   iv) Find the total distance traveled between times $t = 0$ and $t = 10$ sec.

Solution

   i) $x(0) = 0$ ft, $x(10) = 1000$ ft.

   ii) $\dot{x} = (12t^2 - 96t + 180)$ ft/sec; thus $\dot{x} = 0$ gives

$$12t^2 - 96t + 180 = 0,$$

   or, factoring, we obtain

$$(3t - 9)(4t - 20) = 0.$$

   Thus

$\dot{x} = 0$ at $t = 3$ sec, $x = 144$ ft,

$\dot{x} = 0$ at $t = 5$ sec, $x = -420$ ft.

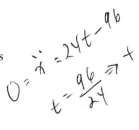

$$\ddot{x} = 24t - 96$$
$$0 = \ddot{x} = 24t - 96$$
$$t = \frac{96}{24}$$

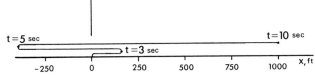

Figure 3.10

iii) At $t = 0$, $\dot{x} = 180$ ft/sec > 0 and $\dot{x} = 0$ at 3 sec and 5 sec. Thus

$$\dot{x} > 0, \qquad 0 \leqslant t < 3 \text{ sec}, 5 < t \text{ sec},$$

$$\dot{x} < 0, \qquad 3 < t < 5 \text{ sec}.$$

In Fig. 3.10 we see what happens in the motion.

iv) The total distance traveled from $t = 0$ until $t = 10$ sec can be determined from Fig. 3.10.

0 to 3 sec	· · ·	144 ft
3 to 5 sec	· · ·	564 ft
5 to 10 sec	· · ·	1420 ft
Total distance		2128 ft

Example 7

A particle has motion described by

$$x = a \sin \omega t, \qquad y = b \cos \omega t, \qquad a, b, \omega, \ldots \text{ constants}.$$

i) Determine the trajectory of the particle.

ii) Show that the acceleration is always directed toward the origin, $x = y = 0$.

Solution

i) We want to eliminate the parameter t between x and y. Note that

$$\sin \omega t = (x/a), \qquad \cos \omega t = (y/b).$$

Thus from the trigonometric identity

$$\sin^2 \omega t + \cos^2 \omega t = 1,$$

we get

$$\frac{x^2}{a^2} + \frac{y^2}{b^2} = 1.$$

The trajectory is an ellipse with semi-major axis a and semi-minor axis b. Figure 3.11 shows the trajectory.

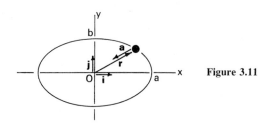

Figure 3.11

ii) We have the position of the particle as

$\mathbf{r} = x\mathbf{i} + y\mathbf{j}$ and $\mathbf{a} = \ddot{x}\mathbf{i} + \ddot{y}\mathbf{j}$.

But with

$x = a \sin \omega t$ and $y = b \cos \omega t$,

we get

$\ddot{x} = -\omega^2 a \sin \omega t$ and $\ddot{y} = -\omega^2 b \cos \omega t$.

Thus

$\ddot{x} = -\omega^2 x$ and $\ddot{y} = -\omega^2 y$.

Finally, comparing

$\mathbf{r} = x\mathbf{i} + y\mathbf{j}$ and $\mathbf{a} = -\omega^2(x\mathbf{i} + y\mathbf{j})$,

we see that

$\mathbf{a} = -\omega^2 \mathbf{r}$.

Thus the acceleration vector passes through the origin, as shown in Fig. 3.11.

Example 8

The piston in Fig. 3.12 is connected to a flywheel of radius R by means of a connecting rod of length l, where $l > 2R$. Find $x(t)$, given that $\theta = \omega_0 t$, $\omega_0 = $ constant.

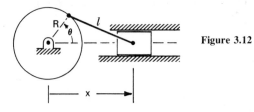

Figure 3.12

Solution

From Fig. 3.12 and the law of cosines, we have

$$l^2 = x^2 + R^2 - 2Rx \cos \theta$$

or

$$x^2 - (2R \cos \theta)x - (l^2 - R^2) = 0.$$

Solving for x, we get

$$x = R \cos \theta \pm \sqrt{l^2 - R^2 \sin^2 \theta}.$$

At $\theta = 0$, $x = R + l$. Thus the minus sign is extraneous. We have

$$x(t) = R \cos \omega_0 t + \sqrt{l^2 - R^2 \sin^2 \omega_0 t}.$$

From the equation for $x(t)$, we need only differentiate to find the velocity and acceleration.

Example 9

A disk of radius R rolls without slipping on a line. Find the velocity and acceleration of a point Q at a radius r from the center of the disk. See Fig. 3.13.

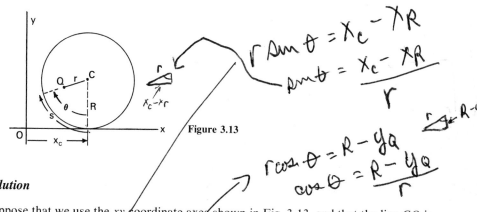

Figure 3.13

Solution

Suppose that we use the xy coordinate axes shown in Fig. 3.13, and that the line CQ is vertical when $x_C = 0$. We can write the x and y coordinates of Q after the wheel has rolled through the angle θ:

$$x_Q = x_C - r \sin \theta, \qquad y_Q = R - r \cos \theta.$$

If the disk rolls without slip, the arc length s equals the distance x_C. Thus $x_C = R\theta$. Thus

$$x_Q = R\theta - r \sin \theta, \qquad y_Q = R - r \cos \theta. \tag{a}$$

Differentiating, we get

$$\dot{x}_Q = \dot{\theta}(R - r \cos \theta), \qquad \dot{y}_Q = \dot{\theta} r \sin \theta. \qquad\qquad (b)$$

The formulas (b) give the velocity components of a point Q which has the position described by the parameters r and θ. Consider the point which has zero velocity. From (b), we have

$$R = r \cos \theta \qquad \text{and} \qquad r \sin \theta = 0.$$

This gives $\theta = 0$ and $r = R$. Thus the point of the disk which is instantaneously in contact with the plane has zero velocity. (We shall encounter this point, which is called the *instant center*, again in our discussion of the plane motion of rigid bodies. See Chapter 8.)

To determine the acceleration of Q, we differentiate (b):

$$\ddot{x}_Q = \ddot{\theta}(R - r \cos \theta) + r\dot{\theta}^2 \sin \theta,$$

$$\ddot{y}_Q = \ddot{\theta} r \sin \theta + r\dot{\theta}^2 \cos \theta. \qquad\qquad (c)$$

It is natural to ask what acceleration the point of contact experiences. Again we take $r = R$ and $\theta = 0$. Then, calling this point P, we have

$$\ddot{x}_P = 0, \qquad \ddot{y}_P = R\dot{\theta}^2. \qquad\qquad (d)$$

Equations (d) express the fact that the acceleration of the point of contact is not zero. In fact, (d) states that P has "nowhere to go but up."

4. CYLINDRICAL COORDINATES

In cylindrical coordinates, we describe the position of an arbitrary point by means of the parameters r, θ, z, as shown in Fig. 3.14. Here r is the distance from the z axis to the point, θ is the angle which the plane through P and the z axis makes with the x axis, and z is the distance of P from the xy plane. Again we assume that the base reference frame (labeled here by x, y, z) is inertial.

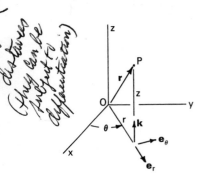

Figure 3.14

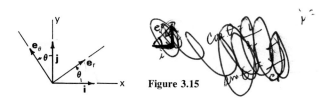

Figure 3.15

We introduce unit vectors $\mathbf{e}_r$, $\mathbf{e}_\theta$, and $\mathbf{k}$, as shown. The position of P is given by

$$\mathbf{r} = r\mathbf{e}_r + z\mathbf{k}. \tag{3.13}$$

It is crucial to distinguish between the scalar distance r and the position vector $\mathbf{r}$. Anyone who does not distinguish between scalars and vectors in the equations he writes will get into hot water when he writes (3.13).*

To find the velocity and acceleration of P, we have to compute the derivatives of $\mathbf{e}_r$ and $\mathbf{e}_\theta$. In order to do this, we write $\mathbf{e}_r$ and $\mathbf{e}_\theta$ in terms of $\mathbf{i}$ and $\mathbf{j}$ (see Fig. 3.15):

$$\mathbf{e}_r = \mathbf{i}\cos\theta + \mathbf{j}\sin\theta, \qquad \mathbf{e}_\theta = -\mathbf{i}\sin\theta + \mathbf{j}\cos\theta. \tag{3.14}$$

Differentiating, we get (noting that $\mathbf{i}$ and $\mathbf{j}$ are zero):

$$\frac{d\mathbf{e}_r}{dt} = \dot{\theta}(-\mathbf{i}\sin\theta + \mathbf{j}\cos\theta), \qquad \frac{d\mathbf{e}_\theta}{dt} = \dot{\theta}(-\mathbf{i}\cos\theta - \mathbf{j}\sin\theta). \tag{3.15}$$

Noting (3.14), we can write (3.15) as

$$\dot{\mathbf{e}}_r = \dot{\theta}\mathbf{e}_\theta, \qquad \dot{\mathbf{e}}_\theta = -\dot{\theta}\mathbf{e}_r. \tag{3.16}$$

We see that the derivative with respect to time of a vector of unit length is a vector of length $\dot{\theta}$ at an angle increased by $(\pi/2)$ radians from the original vector.

Returning to (3.13), we now compute

$$\mathbf{v} = \dot{r}\,\mathbf{e}_r + r\,\dot{\mathbf{e}}_r + \dot{z}\,\mathbf{k}$$

or by (3.16)

$$\mathbf{v} = \dot{r}\,\mathbf{e}_r + r\,\dot{\theta}\mathbf{e}_\theta + \dot{z}\,\mathbf{k}; \tag{3.17}$$

and for the acceleration vector:

$$\mathbf{a} = \ddot{r}\,\mathbf{e}_r + \dot{r}\,\dot{\mathbf{e}}_r + \dot{r}\,\dot{\theta}\mathbf{e}_\theta + r\,\ddot{\theta}\,\mathbf{e}_\theta + r\,\dot{\theta}\,\dot{\mathbf{e}}_\theta + \ddot{z}\,\mathbf{k}$$

---

* The notation here is certainly subject to severe criticism. Unfortunately this is the standard notation.

or

$$\mathbf{a} = (\ddot{r} - r\dot{\theta}^2)\mathbf{e}_r + (r\ddot{\theta} + 2\dot{r}\dot{\theta})\mathbf{e}_\theta + \ddot{z}\mathbf{k}. \tag{3.18}$$

Example 10 (Circular Motion)

The point P at a radius R from O moves in a circle (Fig. 3.16). What are the velocity and acceleration of P?

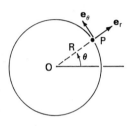

Figure 3.16

Solution

We have $r = R$, so that $\dot{r} = \ddot{r} = 0$. Also $z = 0$; thus $\dot{z} = \ddot{z} = 0$. And therefore, from (3.17) and (3.18), we have

$$\mathbf{v} = R\dot{\theta}\mathbf{e}_\theta, \qquad \mathbf{a} = -R\dot{\theta}^2\mathbf{e}_r + R\ddot{\theta}\mathbf{e}_\theta. \tag{a}$$

[Compare (a) to the results in Example 1 of this chapter.]

Example 11

A disk rotates about O with $\dot{\theta} = \omega_0, \; \ldots$ constant (Fig. 3.17). In a slot through O, a particle moves with speed $v_0 = $ constant relative to the track. Find the velocity and acceleration of the particle.

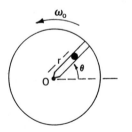

Figure 3.17

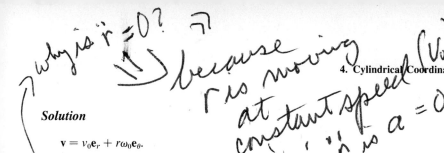

Solution

$\mathbf{v} = v_0\mathbf{e}_r + r\omega_0\mathbf{e}_\theta.$

$\mathbf{a} = -r\omega_0^2\mathbf{e}_r + 2v_0\omega_0\,\mathbf{e}_\theta.$

Example 12

A disk which rotates in a horizontal plane with $\dot\theta = \omega_0$, . . . constant (Fig. 3.18) has a frictionless track in which a particle can slide. It can be shown that the radial component of acceleration is zero. At $t = 0$, $r = 0$ and the speed of the particle is v_0. Show that

$$r(t) = \left(\frac{v_0}{\omega_0}\right) \sinh \omega_0 t \tag{a}$$

is the motion of the particle.

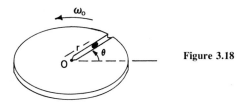

Figure 3.18

Solution

The acceleration of the particle, by (3.18), is

$$\mathbf{a} = (\ddot r - r\dot\theta^2)\mathbf{e}_r + (r\ddot\theta + 2\dot r\dot\theta)\mathbf{e}_\theta.$$

Since the radial component is zero and we note that $\dot\theta = \omega_0$, we have

$$\ddot r - \omega_0^2 r = 0. \tag{b}$$

To demonstrate that (a) is the motion of the particle, we must show two things:

i) (a) must satisfy (b),

ii) (a) must satisfy the initial conditions:

$$\left.\begin{array}{l} r = 0 \\ \dot r = v_0 \end{array}\right\} \quad \text{at } t = 0.$$

Differentiation of (a) gives

$$\ddot{r} = \omega_0 v_0 \sinh \omega_0 t.$$

Substituting in (b) gives

$$\omega_0 v_0 \sinh \omega_0 t - \omega_0^2 \left(\frac{v_0}{\omega_0} \right) \sinh \omega_0 t = 0,$$

as required. For the initial conditions, we check that

$$r(0) = \left(\frac{v_0}{\omega_0} \right) \sinh (0) = 0, \qquad \dot{r}(0) = v_0 \cosh (0) = v_0,$$

as required.

Example 13

Figure 3.19 shows a horizontal, frictionless table, with a small hole in the center, through which a string passes. Attached to the end of the string is a particle.

We use cylindrical coordinates to describe the motion of the particle. It is given that

$$r^2 \dot{\theta} = C, \; \ldots \text{ constant} \quad \text{and} \quad \text{at } t = 0, \, r = l. \tag{a}$$

Suppose that the string is pulled in at a constant speed v_0. Evaluate the velocity and acceleration of the particle as functions of time.

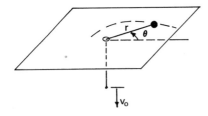

Figure 3.19

Solution

We have

$$\mathbf{v} = \dot{r} \, \mathbf{e}_r + r \dot{\theta} \, \mathbf{e}_\theta,$$

$$r = (l - v_0 t), \qquad \dot{r} = -v_0, \qquad \text{and} \qquad \dot{\theta} = \frac{C}{r^2} = \frac{C}{(l - v_0 t)^2}.$$

Thus

$$\mathbf{v} = -v_0 \, \mathbf{e}_r + \frac{C}{(l - v_0 t)} \, \mathbf{e}_\theta.$$

For the acceleration, we have

$$\mathbf{a} = (\ddot{r} - r\dot{\theta}^2)\mathbf{e}_r + (r\ddot{\theta} + 2\dot{r}\dot{\theta})\mathbf{e}_\theta. \tag{b}$$

Since $\dot{r} = -v_0$, . . . constant, $\ddot{r} = 0$. We have determined

$$r = (l - v_0 t), \qquad \dot{\theta} = \frac{C}{(l - v_0 t)^2}, \qquad \dot{r} = -v_0.$$

Differentiating (a), we can find $\ddot{\theta}$:

$$2r\dot{r}\dot{\theta} + r^2\ddot{\theta} = 0 \qquad \text{or} \qquad r\ddot{\theta} + 2\dot{r}\dot{\theta} = 0.$$

Note that this is the second term of equation (b). Thus we have

$$\mathbf{a} = -\frac{C^2}{(l - v_0 t)^3}\mathbf{e}_r.$$

[handwritten annotation: for symmetry about a pt in 3 dimensions]

5. SPHERICAL COORDINATES*

In spherical coordinates, the position of each point P is defined in terms of the parameters ρ, θ, and ϕ (see Fig. 3.20). Here ρ is the distance of P from O, θ is the angle between the z–OP plane and the xz plane (the same angle used in cylindrical coordinates), and ϕ is the angle between the z axis and the line OP.

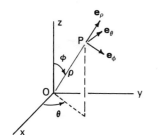

Figure 3.20

The unit vectors $\mathbf{e}_\rho$, $\mathbf{e}_\phi$, and $\mathbf{e}_\theta$ are defined as follows.

$\mathbf{e}_\rho$ = unit vector along OP.

$\mathbf{e}_\phi$ = unit vector perpendicular to $\mathbf{e}_\rho$ and in the $(OP$–$z)$ plane in the direction of increasing ϕ.

$\mathbf{e}_\theta$ = unit vector perpendicular to both $\mathbf{e}_\rho$ and $\mathbf{e}_\phi$ and having the direction of increasing θ.

---

* This section is included primarily for your future reference, but it can be omitted without loss.

The directions of these vectors are defined by considering increments in the parameter associated with the vector's name while holding the remaining parameters constant. Thus $\mathbf{e}_\rho$ is defined by changing ρ while keeping ϕ and θ fixed, etc.

Again we assume that the base reference frame x, y, z is inertial.

The position of P is given by

$$\mathbf{r} = \rho \mathbf{e}_\rho. \tag{3.19}$$

To find the velocity and acceleration, we need to compute

$$\dot{\mathbf{e}}_\rho, \dot{\mathbf{e}}_\phi, \dot{\mathbf{e}}_\theta.$$

To begin, we express $\mathbf{e}_\rho$, $\mathbf{e}_\phi$, and $\mathbf{e}_\theta$ in terms of $\mathbf{i}$, $\mathbf{j}$, and $\mathbf{k}$:

$$\mathbf{e}_\rho = \mathbf{i} \sin \phi \cos \theta + \mathbf{j} \sin \phi \sin \theta + \mathbf{k} \cos \phi,$$

$$\mathbf{e}_\phi = \mathbf{i} \cos \phi \cos \theta + \mathbf{j} \cos \phi \sin \theta - \mathbf{k} \sin \phi, \tag{3.20}$$

$$\mathbf{e}_\theta = -\mathbf{i} \sin \theta + \mathbf{j} \cos \theta.$$

Thus

$$\dot{\mathbf{e}}_\rho = \mathbf{i}(\dot{\phi} \cos \phi \cos \theta - \dot{\theta} \sin \phi \sin \theta)$$
$$+ \mathbf{j}(\dot{\phi} \cos \phi \sin \theta + \dot{\theta} \sin \phi \cos \theta) + \mathbf{k}(-\dot{\phi} \sin \phi),$$

$$\dot{\mathbf{e}}_\phi = \mathbf{i}(-\dot{\phi} \sin \phi \cos \theta - \dot{\theta} \cos \phi \sin \theta) \tag{3.21}$$
$$+ \mathbf{j}(-\dot{\phi} \sin \phi \sin \theta + \dot{\theta} \cos \phi \sin \theta - \mathbf{k}\dot{\phi} \cos \phi),$$

$$\dot{\mathbf{e}}_\theta = -\mathbf{i}\dot{\theta} \cos \theta - \mathbf{j}\dot{\theta} \sin \theta.$$

Comparing* (3.21) and (3.20), we get

$$\dot{\mathbf{e}}_\rho = \dot{\phi}\mathbf{e}_\phi + \dot{\theta} \sin \phi \mathbf{e}_\theta,$$

$$\dot{\mathbf{e}}_\phi = -\dot{\phi}\mathbf{e}_\rho + \dot{\theta} \cos \phi \mathbf{e}_\theta, \tag{3.22}$$

$$\dot{\mathbf{e}}_\theta = -\dot{\theta} \sin \phi \mathbf{e}_\rho - \dot{\theta} \cos \phi \mathbf{e}_\phi.$$

Another way of obtaining (3.22) is given in Problem 3.34.

Once we have obtained (3.22), determining the velocity and acceleration components in spherical coordinates is child's play (somewhat tedious child's play!). Differentiating (3.19), we obtain

$$\mathbf{v} = \dot{\rho}\mathbf{e}_\rho + \rho \dot{\mathbf{e}}_\rho.$$

---

* This step is not trivial. It is easier to insert (3.20) into (3.22) and verify that (3.21) results.

Thus, by (3.22), we have

$$\mathbf{v} = \dot{\rho}\mathbf{e}_\rho + \rho\dot{\phi}\mathbf{e}_\phi + \rho\dot{\theta}\sin\phi\,\mathbf{e}_\theta, \tag{3.23}$$

and similarly from (3.23)

$$\mathbf{a} = \ddot{\rho}\mathbf{e}_\rho + \dot{\rho}\dot{\mathbf{e}}_\rho + (\dot{\rho}\dot{\phi} + \rho\ddot{\phi})\mathbf{e}_\phi + \rho\dot{\phi}\dot{\mathbf{e}}_\phi$$
$$+ (\dot{\rho}\dot{\theta}\sin\phi + \rho\ddot{\theta}\sin\phi + \rho\dot{\theta}\dot{\phi}\cos\phi)\mathbf{e}_\theta + \rho\dot{\theta}\sin\phi\,\dot{\mathbf{e}}_\theta.$$

Thus by (3.22)

$$\mathbf{a} = (\ddot{\rho} - \rho\dot{\phi}^2 - \rho\dot{\theta}^2\sin^2\phi)\mathbf{e}_\rho$$
$$+ (2\dot{\rho}\dot{\phi} + \rho\ddot{\phi} - \rho\dot{\theta}^2\sin\phi\cos\phi)\mathbf{e}_\phi \tag{3.24}$$
$$+ (2\dot{\rho}\dot{\theta}\sin\phi + 2\rho\dot{\phi}\dot{\theta}\cos\phi + \rho\ddot{\theta}\sin\phi)\mathbf{e}_\theta.$$

Example 14

Show that if $\phi = \pi/2$ in Eqs. (3.23) and (3.24) the equations for $\mathbf{v}$ and $\mathbf{a}$ reduce to those in cylindrical coordinates $(z = 0)$, Eqs. (3.17) and (3.18).

Solution

If $\phi = \pi/2$, then $\sin\phi = 1$, $\cos\phi = 0$ and

$$\dot{\phi} = \ddot{\phi} = 0.$$

Thus (3.23) and (3.24) become

$$\mathbf{v} = \dot{\rho}\mathbf{e}_\rho + \rho\dot{\theta}\mathbf{e}_\theta, \qquad \mathbf{a} = (\ddot{\rho} - \rho\dot{\theta}^2)\mathbf{e}_\rho + (2\dot{\rho}\dot{\theta} + \rho\ddot{\theta})\mathbf{e}_\theta. \tag{a}$$

In the above it is clear that ρ is now the parameter r of cylindrical coordinates and $\mathbf{e}_\rho$ is $\mathbf{e}_r$. The parameter θ is the same in both systems.

....

We have yet to specify when to use one coordinate system in preference to another. In theory any system can be used. In practice, the simplicity of formulation of a problem and the likelihood that a solution to it can be found readily depend largely on the coordinate system used to describe it.

There are two considerations in the choice of a coordinate system: (i) description of the forces, (ii) description of the kinematics. If a rule had to be given, one could say that description of kinematics probably has priority over description of forces. If a problem has symmetry about a line, it would be best to use cylindrical coordinates. If the problem is three-dimensional with symmetry about a point, spherical coordinates would probably be in order. If the motion occurs along a line or if it has no special symmetry, cartesian coordinates would probably be as good as anything else. If the trajectory of the problem is known, natural coordinates may be useful.

We have not completed our discussion of the kinematics of a particle. Suppose the base reference frame in our analyses was *not* inertial. We could still compute velocity and acceleration with respect to this noninertial frame – using the equations we have developed in this chapter – but the resulting quantities would be relative velocity and relative acceleration. These quantities alone are insufficient to write correct dynamic equations. We shall deal with this problem in Chapter 8. For a large number of interesting and important problems, we have enough kinematic equipment.

6. SUMMARY

Position vector $\mathbf{r}$ = vector from origin O of an inertial reference frame to the particle

Velocity vector $\mathbf{v} = \dfrac{d\mathbf{r}}{dt}$

Acceleration vector $\mathbf{a} = \dfrac{d\mathbf{v}}{dt} = \dfrac{d^2\mathbf{r}}{dt^2}$

Velocity and Acceleration Components

Natural coordinates:

$$\mathbf{v} = \dot{s}\,\mathbf{e}_t$$

$$\mathbf{a} = \ddot{s}\,\mathbf{e}_t + [(\dot{s})^2/R]\mathbf{e}_n$$

Cartesian coordinates:

$$\mathbf{v} = \dot{x}\,\mathbf{i} + \dot{y}\,\mathbf{j} + \dot{z}\,\mathbf{k}$$

$$\mathbf{a} = \ddot{x}\,\mathbf{i} + \ddot{y}\,\mathbf{j} + \ddot{z}\,\mathbf{k}$$

Cylindrical coordinates:

$$\mathbf{v} = \dot{r}\,\mathbf{e}_r + r\dot{\theta}\,\mathbf{e}_\theta + \dot{z}\,\mathbf{k}$$

$$\mathbf{a} = (\ddot{r} - r\dot{\theta}^2)\mathbf{e}_r + (r\ddot{\theta} + 2\dot{r}\dot{\theta})\mathbf{e}_\theta + \ddot{z}\,\mathbf{k}$$

Spherical coordinates:

$$\mathbf{v} = \dot{\rho}\,\mathbf{e}_\rho + \rho\dot{\phi}\,\mathbf{e}_\phi + \rho\dot{\theta}\sin\phi\,\mathbf{e}_\theta$$

$$\mathbf{a} = (\ddot{\rho} - \rho\dot{\phi}^2 - \rho\dot{\theta}^2\sin^2\phi)\mathbf{e}_\rho$$
$$+ (2\dot{\rho}\dot{\phi} + \rho\ddot{\phi} - \rho\dot{\theta}^2\sin\phi\cos\phi)\mathbf{e}_\phi$$
$$+ (2\dot{\rho}\dot{\theta}\sin\phi + 2\rho\dot{\phi}\dot{\theta}\cos\phi + \rho\ddot{\theta}\sin\phi)\mathbf{e}_\theta$$

PROBLEMS

Section 3.2

Problem 3.1

A point P moves on a circular path as shown in the figure.

    a) Write $\dot{x}$, $\dot{y}$ and $\ddot{x}$, $\ddot{y}$ as functions of θ, $\dot{\theta}$, $\ddot{\theta}$ to determine

$$\mathbf{v} = v_x\mathbf{i} + v_y\mathbf{j}$$

and

$$\mathbf{a} = a_x\mathbf{i} + a_y\mathbf{j}.$$

    b) Transform the results of (a) into natural coordinates.

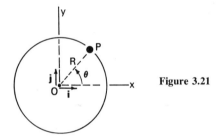

Figure 3.21

Problem 3.2

The particle in the figure moves along a circular path of radius R.

    a) Write the unit vectors of the natural coordinates $\mathbf{e}_t$ and $\mathbf{e}_n$ in terms of $\mathbf{i}$ and $\mathbf{j}$ and the arc length s.

    b) Compute $d\mathbf{e}_t/ds$, using the results of part (a), and show that

$$\frac{d\mathbf{e}_t}{ds} = \frac{1}{R}\,\mathbf{e}_n.$$

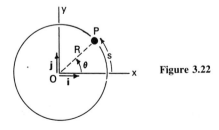

Figure 3.22

Problem 3.3

The racetrack shown in the figure consists of three circular sections of radius 100 ft, 200 ft, and 300 ft. Suppose that a car travels around the track at the constant speed of 60 mph. Find |a| on each portion of the track.

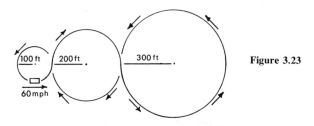

Figure 3.23

Problem 3.4

A car travels along the path indicated in the figure. The distance s along the path is given to be

$$s = 10t^2 \text{ ft.}$$

a) Find |v| as a function of s.

b) Find |a| as a function of s.

c) Find the point of the path at which the magnitude of the acceleration is a maximum. What is this maximum value?

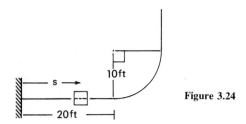

Figure 3.24

Problem 3.5

Transform the expression for **a** in Example 3.2 to give the acceleration components in the x and y directions.

Problem 3.6

A particle moves along the path $y = x^2$ with constant speed $\dot{s} = v_0$. Find $\mathbf{a} = \ddot{x}\mathbf{i} + \ddot{y}\mathbf{j}$ by differentiating $y = x^2$. That is, find the results of Problem 3.5. (The point of this problem is to contrast the above procedure with the procedure in Example 3.2.)

Problem 3.7

A particle moves along the trajectory

$$y = -\ln (\cos x).$$

a) Find the radius of curvature of the trajectory at a given value of x.

b) What does $R = \infty$ mean?

Problem 3.8

Consider the ellipse shown in the figure:

$$\left(\frac{x}{a}\right)^2 + \left(\frac{y}{b}\right)^2 = 1.$$

a) Find the radius of curvature at the point B: $x = 0$, $y = b$.

b) Use the result (a) to determine the radius of curvature at the point A: $x = a$, $y = 0$.

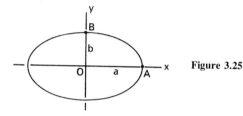

Figure 3.25

Problem 3.9

The particle P shown in the figure is attached to one end of an inextensible string of length 16 in. The other end of the string is attached to point A of a rectangular block. The particle P travels at a constant speed of 10 in./sec as the string wraps around the block. Find the magnitude of the acceleration of P as the string pivots about the point (a) A, (b) B, (c) C, (d) D.

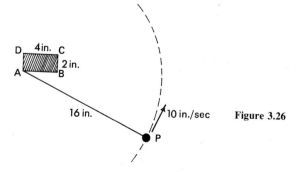

Figure 3.26

Problem 3.10

Equations (3.6) and (3.9) show that the velocity and acceleration of a particle lie in the plane determined by the unit vectors e_t and e_n. The unit binormal vector is defined by

$$e_b \equiv e_t \times e_n.$$

Find the component of the jerk da/dt in the binormal direction. [*Hint:* It can be shown that

$$\frac{de_t}{ds} = \frac{1}{R} e_n \quad \text{and} \quad \frac{de_n}{ds} = \frac{1}{\tau} e_b - \frac{1}{R} e_t,$$

where τ is called the torsion of the curve.]

Section 3.3

Problem 3.11

A particle has the position vector

$$r(t) = \left(10t - \frac{120}{\pi} \cos \frac{\pi}{6} t\right) i \text{ ft}, \qquad t \text{ in seconds.}$$

a) Determine v and a.

b) For the times $0 \leq t \leq 12$ sec, find the times at which $v = 0$. Evaluate a at these times.

c) Find the total distance traveled from $t = 0$ until $t = 11$ sec.

Problem 3.12

A particle has velocity

$$v = e^{10t} i + 6t j \text{ ft/sec.}$$

At $t = 0$, the position of the particle is

$$r(0) = 1i + 3j + 5k \text{ ft.}$$

a) Find $a(t)$.

b) Find $r(t)$.

Problem 3.13

The motion of a projectile is given as

$$x = v_0 t, \qquad y = v_0 t - \tfrac{1}{2} g t^2, \qquad v_0, g, \ldots \text{ positive constants.}$$

a) Find the times t at which $y = 0$.

b) Find the trajectory of the projectile $y = y(x)$.

c) Find the position x at which the trajectory $y = y(x)$ is flat (that is, $dy/dx = 0$). At what time t does this occur?

Problem 3.14

Suppose the acceleration of a particle in the x direction is given to be

$\ddot{x} = 2x^2$ (ft/sec²).

a) Write $v = dx/dt$. Then show that $\ddot{x} = v(dv/dx)$.

b) Suppose that at $x = x_0$, $\dot{x} = v_0$. Use the result of part (a) to find v as a function of x.

Problem 3.15

The particle in the figure is released from rest at a height h above the earth's surface. The acceleration of the particle is

$\ddot{y} = -32$ ft/sec².

What is the speed of the particle when it hits the earth? [*Hint:* At $y = h$, $\dot{y} = 0$. Use the result of Problem 3.14, part (a), and find $\dot{y}$ at $y = 0$.]

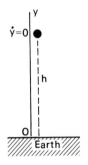

Figure 3.27

Problem 3.16

A particle has its motion defined by

$x = a \sin \omega t, \qquad y = b \cos \omega t,$

where a, b, ω are positive constants. (Assume $a < b$.)

a) Find the points on the trajectory (i) nearest, and (ii) farthest from $x = y = 0$. [*Hint:* Minimize or maximize $D^2 = x^2 + y^2$.]

b) Find the points on the trajectory at which the speed of the particle is (i) a minimum, (ii) a maximum.

Problem 3.17

The cylinder of radius R shown in Fig. 3.28 rolls without slip on the x axis. Denote the angle through which the cylinder rotates by θ. Suppose that $\theta = 0$ when the center of the cylinder is above $x = 0$.

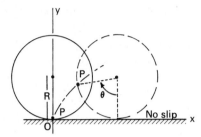

Figure 3.28

a) Find the equations for the path of the point P on the edge of the cylinder (that is, find $x(\theta)$, $y(\theta)$ for the point P).

b) Find the angles $0 \leqslant \theta < 2\pi$ at which the path of P has (i) vertical and (ii) horizontal tangents.

c) If $\dot{\theta} = \omega_0$, . . . constant, find the acceleration of P when $\theta = 0$.

Problem 3.18

The motion of a particle is specified to be

$$x(t), \ y(t), \ z(t).$$

Show that the total distance traveled by the particle between times t_0 and t_1 is given by

$$s = \int_{t_0}^{t_1} (\dot{x}^2 + \dot{y}^2 + \dot{z}^2)^{1/2} \, dt.$$

Problem 3.19

A particle has the motion

$$x = 10t \text{ ft}, \qquad y = 10t - 16t^2 \text{ ft}, \qquad z = 0.$$

Find the total distance traveled by the particle from $t = 0$ until $t = \frac{5}{8}$ sec. [*Hint:* See Problem 3.18.]

Problem 3.20

A particle moves along the trajectory $y = f(x)$ with its velocity in the x direction given as dx/dt. Show that the speed of the particle is

$$\frac{ds}{dt} = \left[1 + \left(\frac{dy}{dx}\right)^2\right]^{1/2} \frac{dx}{dt}.$$

Problem 3.21

A particle moves along the trajectory

$$y = x^2, \qquad z = 0.$$

Suppose that the velocity in the x direction is constant, $\dot{x} = 60$ mph, and at $t = 0$, $x = 0$.

    a) Find the velocity and acceleration as functions of time.

    b) Find the velocity and acceleration as functions of the position, x.

Problem 3.22

The projectile shown in the figure is propelled through the barrel of a light gas gun. The horizontal displacement is given as

$$x = \tfrac{1}{2} a_0 t^2, \qquad a_0, \ldots \text{ constant.}$$

Suppose the gun barrel has a shape given by $y = f(x)$.

    a) Find $\ddot{x}$ and $\ddot{y}$ as functions of time.

    b) Find $\ddot{x}$ and $\ddot{y}$ as functions of x.

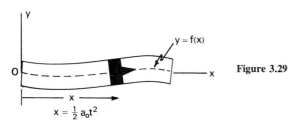

Figure 3.29

Section 3.4

Problem 3.23

The particle P shown in the figure moves in a track cut in a cylinder. The cylinder rotates in the plane about O so that $\dot{\theta} = \omega_0, \ldots$ constant. The distance of the particle from O is given by

$$r(t) = \cosh(\omega_0 t).$$

Find the velocity and acceleration of the particle. In particular, show that $a_r = 0$.

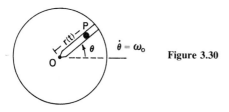

Figure 3.30

Problem 3.24

The particle P shown in the figure moves in the slot of a rotating cylinder so that

$$r = 10 + 5 \sin 4t \text{ ft}, \qquad t \text{ in seconds.}$$

The motion of the cylinder is given to be $\theta = 2t$ rad.
- a) Find the acceleration of P as a function of time.
- b) Find the acceleration of P as a function of θ.

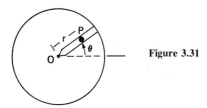

Figure 3.31

Problem 3.25

The particle P shown in the figure moves at a constant speed v_0 along the line L. The line L passes through O and is specified by the constant, direction cosines $l = \cos \alpha$, $m = \cos \beta$.
- a) Express the position of P in cylindrical coordinates when the distance $OP = d$. What is the value of the polar angle θ?
- b) Write the velocity of P in cylindrical coordinates.

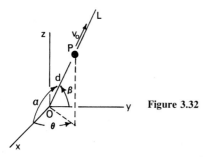

Figure 3.32

Problem 3.26

A particle moves such that

$$r = 10t \text{ (in.)}, \qquad \dot{\theta} = 5 \text{ rad/sec}, \qquad t \text{ in seconds}$$

- a) Find the velocity $\mathbf{v}$. Plot the components of $\mathbf{v}$ versus time.
- b) Find the acceleration $\mathbf{a}$. Plot the components of $\mathbf{a}$ versus time.

Problem 3.27

The machine element shown in the figure rotates about the fixed cylinder of radius R at the increasing angular rate $\dot{\theta} = \alpha_0 t$. A particle in the slot of the element is attached to a string which winds about the fixed inner cylinder. At $t = 0$, $\theta = 0$ and $r = 10R$.

a) Find the trajectory of the particle $r = r(\theta)$.

b) For an arbitrary position θ, find the acceleration of P as a function of θ.

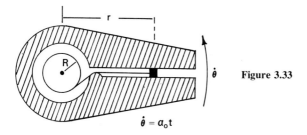

Figure 3.33

$\dot{\theta} = a_0 t$

Problem 3.28

A particle moves on a helix $z = 5\theta$ ft, $r = 10$ ft.

a) Write the speed of the particle v in terms of $\dot{\theta}$.

b) It is given that $v^2 = 64(100 - z)$ (ft/sec)². Use the result of part (a) to determine $\dot{\theta}$ as a function of θ.

Problem 3.29

In the automobile industry, passenger comfort is sometimes related to the "jerk," da/dt, which is the derivative of the acceleration. Write da/dt in cylindrical coordinates.

Problem 3.30

The motion of a particle is specified by cylindrical coordinates: $r = r(t)$, $\theta = \theta(t)$, $z = z(t)$. Show that the total distance traveled by the particle from $t = t_0$ until $t = t_1$ is

$$s = \int_{t_0}^{t_1} (\dot{r}^2 + r^2\dot{\theta}^2 + \dot{z}^2)^{1/2} \, dt.$$

Problem 3.31

The trajectory of a particle is specified by $r = r(\theta)$, $z = z(\theta)$, where θ is the polar angle. Show that the total distance traveled between the points given by $\theta = \theta_0$ and $\theta = \theta_1$ is

$$s = \int_{\theta_0}^{\theta_1} \left[\left(\frac{dr}{d\theta}\right)^2 + r^2 + \left(\frac{dz}{d\theta}\right)^2 \right]^{1/2} d\theta.$$

Problem 3.32

Suppose that the trajectory of a particle is given to be

$$r(\theta) = R(10 - \theta), \qquad \text{(Problem 3.27).}$$

Find the arc length of the trajectory between $\theta = 0$ and $\theta = 5$ rad. [*Hint:* See Problem 3.31.]

Problem 3.33

The trajectory of a particle is given as

$$r = 50\theta \text{ ft}, \qquad z = 0.$$

Find the total distance traveled between $\theta = 0$ and $\theta = \pi$ rad. [*Hint:* See Problem 3.31.]

Section 3.5

Problem 3.34

Consider the unit vectors of spherical coordinates, $\mathbf{e}_\rho$, $\mathbf{e}_\phi$, $\mathbf{e}_\theta$. Let the distance ρ and the angles ϕ and θ undergo the following changes:

$$\rho \to \rho + \Delta\rho, \qquad \phi \to \phi + \Delta\phi, \qquad \theta \to \theta + \Delta\theta.$$

Find the corresponding changes in the unit vectors $\mathbf{e}_\rho$, $\mathbf{e}_\phi$, $\mathbf{e}_\theta$. For example, show that

$$\Delta\mathbf{e}_\rho = \Delta\phi\,\mathbf{e}_\phi + \Delta\theta \sin\phi\,\mathbf{e}_\theta,$$

and thus verify Eqs. (3.22).

4

DYNAMICS OF
A PARTICLE

Canio: "La commedia è finita!"
I Pagliacci
R. Leoncavallo

1. INTRODUCTION

With this chapter we begin our study of the dynamics of mechanical systems. Here we treat the motion of the simplest mechanical system: a single particle. The basis of our analysis is Newton's second law,

a) $\mathbf{F} = \dot{\mathbf{p}} = m\mathbf{a}$

relating the force resultant $\mathbf{F}$ acting on the particle to the rate of change of the linear momentum $\mathbf{p}$. In principle, the analysis would seem to be straight-forward.

i) Analyze the forces on the particle. (Construct the free-body diagram.)

ii) Analyze the kinematics of the particle. (Write the acceleration $\mathbf{a}$ in some coordinate system.)

iii) The equation of motion is given by relating (i) and (ii) in

$\mathbf{F} = \dot{\mathbf{p}} = m\mathbf{a}$.

For many problems, this procedure is an entirely satisfactory means of analyzing the dynamics. However, there are problems which we can analyze more easily using another form of the fundamental equations. There are two other forms which we might contemplate:

b) $\mathbf{M}_O = \dot{\mathbf{H}}_O$

relating the moment of the forces acting on the particle to the rate of change of the moment of the linear momentum;

c) The work–energy relation.

There are problems which we can understand most easily using one of the last two forms of the fundamental equations. This, in fact, amounts to the heart of the difficulty of dynamics: For a given problem, which of equations (a), (b), or (c) is the most convenient for the analysis? There is no good answer to this question, so we shall proceed by simply giving a host of example problems to illustrate some of the considerations which one makes in using one of (a), (b), or (c) rather than another.

One final point should be made. We are concerned here with the dynamics of a particle. You have probably realized by now that the concept of a particle is a bit ambiguous. You may think that your experience with point masses is limited. You are more familiar with such objects as automobiles, which are hardly dimensionless.

The word *particle* can be properly applied to two classes of problems:

i) The motion of bodies for which rotational effects are negligible

ii) The motion of the center of mass of a body.

We shall prove this in Chapter 8. For the time being, let us say only that if the "particle" in an example does not appear to be dimensionless, we are still talking, quite properly, about the motion of the center of mass of the body.

2. $\mathbf{F} = \dot{\mathbf{p}}$

Suppose that the particle under consideration has velocity $\mathbf{v}$.* Then we define:

Linear momentum vector: $\mathbf{p} = m\mathbf{v}$.

Given that $\mathbf{F}$ is the vector sum of all forces on the particle, Newton's second law becomes

$$\mathbf{F} = \dot{\mathbf{p}} = m\mathbf{a}.* \tag{4.1}$$

Many students are confused by the form of (4.1). They reason that

$$\dot{\mathbf{p}} = \dot{m}\mathbf{v} + m\mathbf{a}$$

and that $\dot{m}$ could, for example, be associated with the rate of mass flow from a rocket. In fact, $\mathbf{F} = \dot{\mathbf{p}}$ in Newtonian mechanics is written for a single particle, whose mass is assumed to be constant. Thus $\dot{m} = 0$, and there is no confusion in (4.1). The only case in which the mass of a particle changes is in the theory of relativity. Here we still have $\mathbf{F} = \dot{\mathbf{p}}$, but

$$\mathbf{p} = \frac{m\mathbf{v}}{\sqrt{1 - (v/c)^2}},$$

---

* When we speak of velocity or acceleration in an unqualified context, we mean a velocity or acceleration with respect to an inertial reference frame.

where m is called the rest mass. Here the mass can be considered a function of velocity.

Rocket and other problems misleadingly called "variable-mass problems" will be considered in Chapter 7.

Example 1 (Motion on an Inclined Plane)

Figure 4.1(a) shows a particle sliding down an inclined plane. The coefficient of sliding friction is μ.* Find the acceleration of m down the incline.

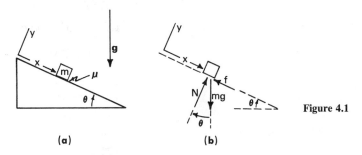

(a)   (b)   Figure 4.1

Solution

Figure 4.1(b) shows a free-body diagram for the particle.

We align the x axis along the incline. Then the force resultant on the particle is

$$\mathbf{F} = (-f + mg \sin \theta)\mathbf{i} + (N - mg \cos \theta)\mathbf{j}.$$

Commonly we shall write $\mathbf{F}$ directly in its components

$$\left.\begin{array}{l} F_x = -f + mg \sin \theta \\ F_y = N - mg \cos \theta \\ F_z = 0. \end{array}\right\} \tag{a}$$

For the acceleration of m we have

$$\mathbf{a} = \ddot{x}\,\mathbf{i} + \ddot{y}\,\mathbf{j} + \ddot{z}\,\mathbf{k},$$

but $y = 0$ and $z = 0$. Thus

$$\left.\begin{array}{l} a_x = \ddot{x} \\ a_y = 0 \\ a_z = 0 \end{array}\right\} \tag{b}$$

Now writing $\mathbf{F} = \dot{p}$, we have

$$F_x = ma_x: \quad -f + mg \sin \theta = m\ddot{x}$$

$$F_y = ma_y: \quad N - mg \cos \theta = 0. \tag{c}$$

---

* The coefficient of sliding friction has been introduced in Section 7 of Chapter 2 and will be reviewed in Section 6 of this chapter.

In (c) we have the following *known* quantities: m, g, θ. The other quantities are *unknown: f, x, N*.

The difficulty here is that we have three unknowns and only two equations. We clearly need one more equation. Since the particle is assumed to slide, we know from (2.22)

$$f = \mu N.$$

Now from (c) and (c'), we have

$$N = mg \cos \theta$$

and thus $f = \mu mg \cos \theta$, and finally

$$-mg(\mu \cos \theta + \sin \theta) = m\ddot{x}$$

or

$$\ddot{x} = -g(\mu \cos \theta + \sin \theta). \tag{d}$$

The term on the right side of (d) is a (complicated) constant. Thus this is a case of motion with constant acceleration. (Recall Example 5 of Chapter 3.)

(handwritten annotation: getting rid of unknown N as an unknown replacing f (c') ∴ 2 equations 2 unknowns)

Example 2 (Simple Pendulum)

Figure 4.2(a) shows a simple pendulum, consisting of a particle at the end of an inextensible string of length l. The particle is assumed to have motion in a single, vertical plane. Find the equation of motion for the pendulum.

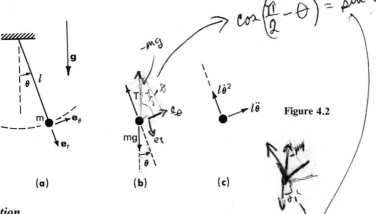

Figure 4.2

(handwritten: $\cos\left(\frac{\pi}{2} - \theta\right) = \sin \theta$)

(a)   **(b)**   **(c)**

Solution

In this case, we know that the particle will move in a circle of radius l. We use polar coordinates to describe the problem.

Figure 4.2(b) shows a free-body diagram. The forces on m are T, the string tension, and mg, the weight. Thus

$$\mathbf{F} = (-T + mg \cos \theta)\mathbf{e}_r - mg \sin \theta \, \mathbf{e}_\theta. \tag{a}$$

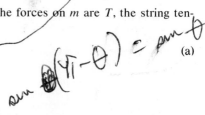

(handwritten: $\sin\left(\frac{\pi}{2} - \theta\right) = \sin \theta$)

From (3.18), the acceleration in polar coordinates is

$$\mathbf{a} = (\ddot{r} - r\dot{\theta}^2)\mathbf{e}_r + (r\ddot{\theta} + 2\dot{r}\dot{\theta})\mathbf{e}_\theta.$$

Here $r = l$ and $\dot{r} = \ddot{r} = 0$. Thus

$$\mathbf{a} = -l\dot{\theta}^2\,\mathbf{e}_r + l\ddot{\theta}\,\mathbf{e}_\theta. \tag{b}$$

From (a) and (b), we can write $\mathbf{F} = \dot{\mathbf{p}}$:

$$F_r = ma_r: \qquad -T + mg\cos\theta = -ml\dot{\theta}^2 \tag{c}$$

$$F_\theta = ma_\theta: \qquad -mg\sin\theta = ml\ddot{\theta}$$

Unless we are interested in the string tension T, the second equation of (c) is sufficient to determine the motion of the pendulum:

$$\ddot{\theta} + \left(\frac{g}{l}\right)\sin\theta = 0.$$

We shall investigate this sort of problem in detail in Chapter 6.

....

We shall give more examples of problems involving $\mathbf{F} = \dot{\mathbf{p}}$ (and the other methods of this chapter) at the end of the chapter.

3. $\mathbf{M} = \dot{\mathbf{H}}$

Now let us consider the generalization of the equation of moment equilibrium from statics, (2.6ii).

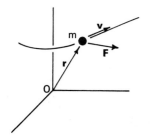

Figure 4.3

Let O in Fig. 4.3 be a fixed point of an inertial reference system. Let $\mathbf{F}$ be the total force on a particle m, and $\mathbf{v}$ the velocity of the particle. Let $\mathbf{r}$ be the position vector of m.

The moment of the force on m about O is

$$\mathbf{M}_O = \mathbf{r} \times \mathbf{F}. \tag{4.2}$$

We define the moment of the momentum vector about O by:

> Moment of momentum: $\mathbf{H}_O = \mathbf{r} \times \mathbf{p} = \mathbf{r} \times m\mathbf{v}$

The vector $\mathbf{H}_O$ is also called the *angular momentum vector*.

It is worth noting here that both the moment vector and the moment of momentum vector depend on the point about which they are written.

We can now show that, if O is a fixed point in an inertial reference frame,

$$\mathbf{M}_O = \dot{\mathbf{H}}_O. \tag{4.3}$$

The proof is quite straightforward. Writing $\mathbf{F} = \dot{\mathbf{p}}$ for the particle and forming the cross product of each side with $\mathbf{r}$, we obtain

$$\mathbf{r} \times \mathbf{F} = \mathbf{r} \times \dot{\mathbf{p}}.$$

The left side of the equation is $\mathbf{M}_O$. To show that the right side is $\dot{\mathbf{H}}_O$, we compute

$$\dot{\mathbf{H}}_O = \frac{d}{dt}(\mathbf{r} \times \mathbf{p}) = \frac{d\mathbf{r}}{dt} \times \mathbf{p} + \mathbf{r} \times \dot{\mathbf{p}}.$$

But $d\mathbf{r}/dt \times \mathbf{p} = \mathbf{v} \times m\mathbf{v} = 0$, since $\mathbf{v}$ and $m\mathbf{v}$ have the same direction. Thus (4.3) is established.

The form of (4.3) is analogous to the form of (4.1).

Example 3 (Simple Pendulum)

Using $\mathbf{M} = \dot{\mathbf{H}}$, find the equation of motion of the pendulum in Fig. 4.4(a).

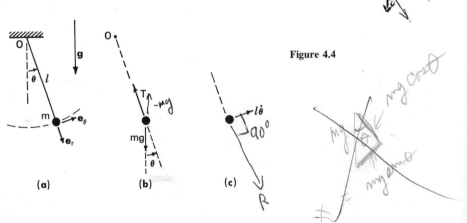

Figure 4.4

(a)   (b)   (c)

Solution

Figure 4.4(b) is a free-body diagram showing the forces on the particle. Figure 4.4(c) shows the velocity of the particle.

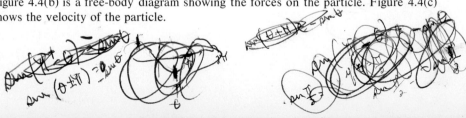

We will write $\mathbf{M} = \dot{\mathbf{H}}$ about the point O. To compute $\mathbf{M}_O$, we consider Fig. 4.4(b).

$$\mathbf{M}_O = -l(mg \sin \theta)\mathbf{k}. \tag{a}$$

Note that the string tension has no moment about the point O.

To compute $\mathbf{H}_O$, we recall that $\mathbf{H}$ is the moment of the linear momentum. That is, we take the moment of $m\mathbf{v}$ as if it were a force. From Fig. 4.4(c), we get

$$\mathbf{H}_O = l(ml\dot{\theta})\mathbf{k}.$$

Since $\dot{\mathbf{k}} = 0$, we compute

$$\dot{\mathbf{H}}_O = ml^2\ddot{\theta}\,\mathbf{k}. \tag{b}$$

Finally, from (a) and (b), we write

$$\mathbf{M}_O = \dot{\mathbf{H}}_O: \quad -mgl \sin \theta\, \mathbf{k} = ml^2\ddot{\theta}\,\mathbf{k},$$

or

$$\ddot{\theta} + \left(\frac{g}{l}\right) \sin \theta = 0.$$

Note that the above equation agrees with the result of Example 2. However, if we wanted to know the string tension T, the formulation of Example 2 would provide the information, whereas the above analysis would not.

···

Example 4

Figure 4.5(a) is a diagram of a frictionless, horizontal table with a hole in the center, through which a string passes. A particle m is attached to the end of the string. The string is pulled in by some means. Show that the quantity $(r^2\dot{\theta})$ is constant. Suppose that at $t = 0$, $r = 10$ ft, $v_r = -100$ ft/sec, and $v_\theta = 50$ ft/sec. Find $\dot{\theta}$ when $r = 2.5$ ft.

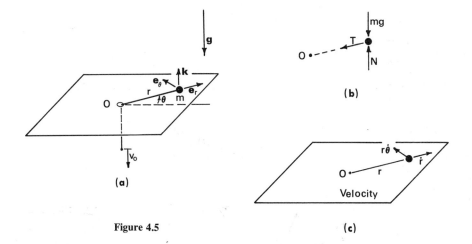

(a)   (b)   (c)

Velocity

Figure 4.5

Solution

Our first problem is to show that $(r^2\dot\theta)$ is constant. There are two approaches, one direct and another, somewhat indirect.

   i) Write $\mathbf{M}_O = \dot{\mathbf{H}}_O$.

We employ cylindrical coordinates. The forces on m add up to

$$\mathbf{F} = -T\mathbf{e}_r + (N - mg)\mathbf{k}.$$

Now if the particle remains on the plane, we have $z = 0$, and thus $\ddot{z} = 0$, so we get

$$F_z = ma_z: \quad N - mg = 0.$$

Thus the total force on m is

$$\mathbf{F} = -T\mathbf{e}_r, \tag{a}$$

and the moment about O is

$$\mathbf{M}_O = 0. \tag{b}$$

   The velocity of m is $\mathbf{v} = \dot{r}\,\mathbf{e}_r + r\dot\theta\,\mathbf{e}_\theta$, as shown in Fig. 4.5(c). Thus, with $\mathbf{r} = r\,\mathbf{e}_r$, we get

$$\mathbf{H}_O = \mathbf{r} \times m\mathbf{v} = rmr\dot\theta\,\mathbf{k} = mr^2\dot\theta\,\mathbf{k}. \tag{c}$$

From (b) and (c), we get

$$\mathbf{M}_O = \dot{\mathbf{H}}_O: \quad 0 = \frac{d}{dt}\,[(mr^2\dot\theta)\,\mathbf{k}]. \tag{d}$$

But (d) implies that

$$mr^2\dot\theta, \ \ldots \ \text{constant.}$$

And since m is a constant, we see that

$$r^2\dot\theta, \ \ldots \ \text{constant.} \tag{e}$$

   ii) An alternative way to obtain (e) is to write $\mathbf{F} = \dot{\mathbf{p}}$ for m.
The acceleration of m is

$$\mathbf{a} = (\ddot{r} - r\dot\theta^2)\mathbf{e}_r + (r\ddot\theta + 2\dot{r}\dot\theta)\mathbf{e}_\theta. \tag{f}$$

We get the force $\mathbf{F}$ from (a). Thus

$$F_r = ma_r: \quad -T = m(\ddot{r} - r\dot\theta^2), \tag{g}$$

$$F_\theta = ma_\theta: \quad 0 = m(r\ddot\theta + 2\dot{r}\dot\theta). \tag{h}$$

Notice (h). We will show that

$$m(r\ddot\theta + 2\dot{r}\dot\theta) = \frac{m}{r}\left[\frac{d}{dt}\,(r^2\dot\theta)\right]. \tag{i}$$

Performing the differentiation on the right side of (i) gives

$$\frac{m}{r}(2r\dot{r}\dot{\theta} + r^2\ddot{\theta}) = m(2\dot{r}\dot{\theta} + r\ddot{\theta}),$$

which proves (i). But now, from (h) and (i), we see that

$$\frac{m}{r}\left[\frac{d}{dt}(r^2\dot{\theta})\right] = 0$$

or

$r^2\dot{\theta}, \ldots$ constant.

Now, at $t = 0$,

$\mathbf{v} = -100\ \mathbf{e}_r + 50\ \mathbf{e}_\theta$ ft/sec $= \dot{r}\ \mathbf{e}_r + r\dot{\theta}\ \mathbf{e}_\theta$ ft/sec.

Thus $r\dot{\theta} = 50$ ft/sec at $t = 0$, and we know that $r = 10$ ft at $t = 0$. Thus, at $t = 0$,

$\dot{\theta} = 5$ rad/sec.

Finally we see that

$r^2\dot{\theta} = (100\ \text{ft}^2)(5\ \text{rad/sec}).$

Thus

$r^2\dot{\theta} = 500\ \text{ft}^2\text{-rad/sec}.$

Since this is a constant, with $r = 2.5$ ft, we have

$(2.5^2\ \text{ft}^2)\dot{\theta} = 500\ \text{ft}^2\text{-rad/sec}$

or

$\dot{\theta} = 80$ rad/sec at $r = 2.5$ ft.

4. THE WORK–ENERGY RELATION

We can best introduce the work–energy relation by speaking first of the power of a force. Let $\mathbf{F}$ be the sum of the forces on a particle m. We define the *power of* $\mathbf{F}$ to be the scalar quantity

$$P = \mathbf{F} \cdot \mathbf{v}, \tag{4.4}$$

where $\mathbf{v}$ is the velocity of m measured with respect to an inertial reference frame. Power is a scalar quantity because it is defined as the dot product of the two vectors, $\mathbf{F}$ and $\mathbf{v}$.

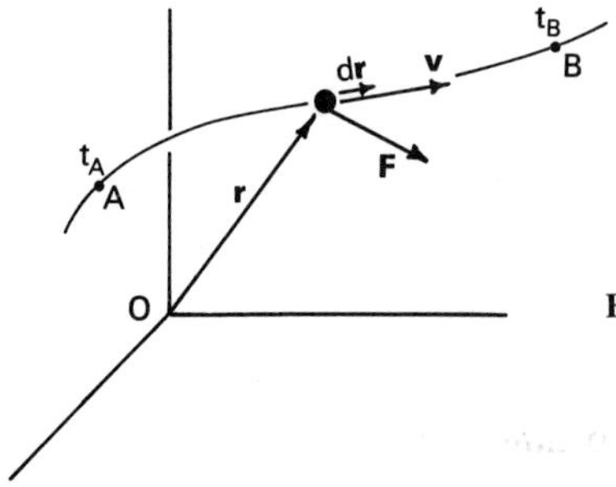

Figure 4.6

The *work done by* F on *m* in going from position A to position B (Fig. 4.6) is defined by

$$W_{AB} = \int_{t_A}^{t_B} P\, dt = \int_{t_A}^{t_B} \mathbf{F} \cdot \mathbf{v}\, dt,$$

where t_A and t_B are the times at which the particle is at positions A and B. Since $\mathbf{v} = d\mathbf{r}/dt$, $\mathbf{v}\, dt = d\mathbf{r}$, the work can be written as the line integral:*

$$W_{AB} = \int_A^B \mathbf{F} \cdot d\mathbf{r}. \tag{4.5}$$

Note that work (another scalar) is a quantity describing the action of F on *m* over an interval of space. Power, on the other hand, is an instantaneous quantity.

Traditionally, the kinetic energy of a particle *m* is defined by the quantity

$$\tfrac{1}{2}mv^2.$$

A definition of kinetic energy which is consistent with the definition of potential energy (Section 5 of this chapter) and which can be applied to non-mechanical systems (e.g. electrical circuits) is†

$$T = \int_0^P \mathbf{v} \cdot d\mathbf{p}. \tag{4.6}$$

This integral is assumed to be computed for a particle whose momentum is brought from zero to the level **p**.

We shall discuss the more explicit form which (4.6) takes in mechanics in a moment. For the time being, let us derive the work–energy relation only on the basis of the form (4.6).

---

* See Appendix 1, Eq. (A1.18).

† See Crandall, Karnopp, Kurtz, and Pridmore-Brown, *Dynamics of Mechanical and Electromechanical Systems,* New York: McGraw-Hill, 1968, pages 18–22.

From Newton's second law, we have

$$\mathbf{F} = \dot{\mathbf{p}}.$$

Forming the dot product of each side of this equation with the velocity $\mathbf{v}$, we obtain

$$\mathbf{F} \cdot \mathbf{v} = \dot{\mathbf{p}} \cdot \mathbf{v}. \qquad (a)$$

By (4.4), the left side of (a) is the power of $\mathbf{F}$.

If we differentiate (4.6) with respect to time,* we get

$$\frac{dT}{dt} = \mathbf{v} \cdot \frac{d\mathbf{p}}{dt} = \mathbf{v} \cdot \dot{\mathbf{p}}. \qquad (b)$$

Thus, from (a) and (b), we have

$$P = \frac{dT}{dt}. \qquad (4.7)$$

> The power of the forces acting on a particle equals the rate of change of the kinetic energy of the particle with respect to time.

To obtain the work–energy relation, we integrate (4.7) with respect to time, noting that by definition

$$\int_{t_A}^{t_B} P \, dt = W_{AB}. \qquad (c)$$

The integral of the power over time equals the work done on the particle. Here the times t_A and t_B are the times at which the particle has positions A and B. (Refer again to Fig. 4.6.) Integrating the right side of (4.7) gives

$$\int_{t_A}^{t_B} \frac{dT}{dt} \, dt = T(B) - T(A). \qquad (d)$$

Finally, from (c) and (d), the integral of (4.7) is

$$W_{AB} = T(B) - T(A). \qquad (4.8)$$

---

* Write

$$T = \int_0^{t_p} \left(\mathbf{v} \cdot \frac{d\mathbf{p}}{dt} \right) dt.$$

Then

$$\frac{dT}{dt} = \mathbf{v} \cdot \frac{d\mathbf{p}}{dt}.$$

> The work of the forces in moving a particle from position A to position B equals the change of the kinetic energy T between those positions.

Finally, let us write the kinetic energy in explicit terms. From (4.6), we had

$$T = \int_0^P \mathbf{v} \cdot d\mathbf{p}.$$

For Newtonian dynamics,

$$\mathbf{p} = m\mathbf{v}, \qquad \text{so} \qquad \mathbf{v} = \frac{\mathbf{p}}{m}, \tag{a}$$

and thus

$$T = \int_0^P \frac{1}{m} \mathbf{p} \cdot d\mathbf{p} = \frac{p^2}{2m} \Big|_0^P = \frac{p^2}{2m} - 0. \tag{b}$$

Thus

$$T = \frac{p^2}{2m}.$$

If we wish to express T in terms of the velocity variables, we note (a) and get

$$T = \tfrac{1}{2}mv^2. \tag{4.9}$$

It is this expression for kinetic energy which we shall use most often in our study of dynamics.

Equation (4.8) can now be written in its most useful form,

$$W_{AB} = \tfrac{1}{2}mv_B^2 - \tfrac{1}{2}mv_A^2. \tag{4.10}$$

Example 5

A slider which weighs 10 lb is pulled along a frictionless rod by a constant force acting at 60° to the rod. Find the force P required to take the slider from a velocity of $2\mathbf{i}$ ft/sec at A to a velocity of $10\mathbf{i}$ ft/sec at B, 10 ft from A.

Solution

Since the slider remains on the rod,

$$d\mathbf{r} = dx\,\mathbf{i}.$$

Figure 4.7(b) is a free-body diagram of the forces acting on the slider.

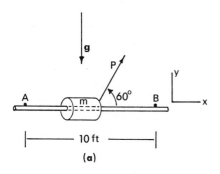

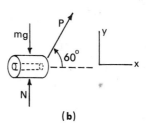

(a)   (b)   Figure 4.7

The resultant force on m is

$$\mathbf{F} = (P \cos \theta)\mathbf{i} + (P \sin \theta + N - mg)\mathbf{j}, \quad \text{where } \theta = 60°.$$

Thus

$$\mathbf{F} \cdot d\mathbf{r} = P \cos \theta \, dx.$$

And

$$W_{AB} = \int_A^B P \cos \theta \, dx = P \cos \theta \int_A^B dx$$

or

$$W_{AB} = P \cos \theta (x_B - x_A).$$

And thus, from (4.10), we have

$$P \cos \theta (x_B - x_A) = \tfrac{1}{2}mv_B^2 - \tfrac{1}{2}mv_A^2.$$

Inserting the numerical data, we get

$$P(\tfrac{1}{2})(10 \text{ ft}) = \tfrac{1}{2}(\tfrac{10}{32} \text{ slug})(100 \text{ ft}^2/\text{sec}^2) - \tfrac{1}{2}(\tfrac{10}{32} \text{ slug})(4 \text{ ft}^2/\text{sec}^2).$$

Thus

$$P = 3 \text{ slug-ft/sec}^2 = 3 \text{ lb}.$$

· ·

Example 6

Figure 4.8(a) shows a slider of mass m which is pulled upward by means of a string. The force in the string is $P = $ constant. At A, the velocity of the slider is zero. Find the velocity at point B, given that there is no friction between the slider and the rod.

Solution

There are two ways to compute the work done on the slider.

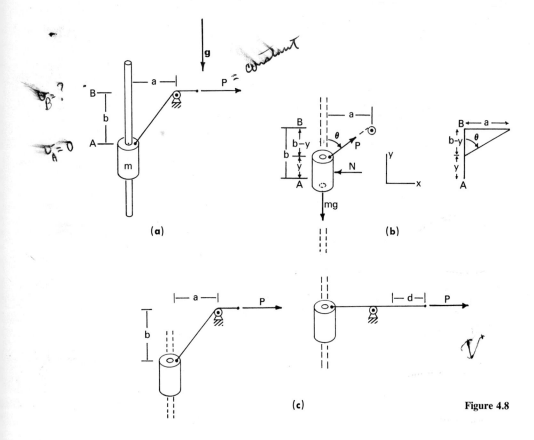

(a)

(b)

(c)

Figure 4.8

i)   Figure 4.8(b) shows the free-body diagram of the slider. The force resultant acting
on the slider is

$$\mathbf{F} = (P \sin \theta - N)\mathbf{i} + (P \cos \theta - mg)\mathbf{j} \quad \text{and} \quad d\mathbf{r} = dy\,\mathbf{j}. \quad (p\,160)$$

Thus

$$\mathbf{F} \cdot d\mathbf{r} = (P \cos \theta - mg)\,dy. \quad (\cos 0° = 1)$$

Before we can use the above, we must write $\cos \theta$ as a function of y. From the geometry of the problem, we see that

$$\cos \theta = \frac{b - y}{\sqrt{a^2 + (b - y)^2}}.$$

Thus

$$W_{AB} = \int_0^b \left[\frac{P(b - y)}{\sqrt{a^2 + (b - y)^2}} - mg \right] dy.$$

It is an exercise in freshman calculus* to evaluate the integral

$$\int_0^b \frac{b - y}{\sqrt{a^2 + (b - y)^2}}\, dy = \sqrt{a^2 + b^2} - a.$$

Thus

$$W_{AB} = P(\sqrt{a^2 + b^2} - a) - mgb.$$

Then, by (4.10), noting that $v_A = 0$, we get

$$P(\sqrt{a^2 + b^2} - a) - mgb = \tfrac{1}{2}mv_B^2,$$

from which we can determine v_B.

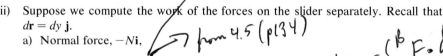

ii) Suppose we compute the work of the forces on the slider separately. Recall that $d\mathbf{r} = dy\,\mathbf{j}$.

a) Normal force, $-N\mathbf{i}$,

$$dW = -N\mathbf{i} \cdot dy\,\mathbf{j} = 0, \qquad \text{since} \qquad \mathbf{i} \cdot \mathbf{j} = 0.$$

b) Gravity force, $-mg\mathbf{j}$,

$$dW = -mg\mathbf{j} \cdot dy\,\mathbf{j} = -mg\,dy,$$

$$W_{AB} = \int_0^b - mg\,dy = -mgb.$$

c) String tension, P: Since the force P is constant, we need only find the distance which P moves to compute the work. (See Fig. 4.8c.)

The distance d is seen to be

$$\sqrt{a^2 + b^2} - a.$$

Thus the work of P is

$$W_{AB} = P(\sqrt{a^2 + b^2} - a).$$

Adding the work terms, we get (4.10) as

$$-mgb + P(\sqrt{a^2 + b^2} - a) = \tfrac{1}{2}mv_B^2.$$

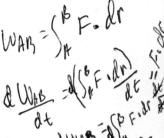

....

The most common use of the work–energy relation is concerned with a special class of forces called *conservative forces*. The next example will serve as an introduction to such forces.

Example 7

A particle m is acted on by a force field which is specified as a function of position, as

---

* Write $u = b - y$, $du = -dy$. Then write $v = a^2 + u^2$, etc.

follows:

$$\mathbf{F} = \frac{F_0}{a^3} (xy^2\mathbf{i} + x^2 y\mathbf{j}).$$

Suppose that the particle moves from $A = (0,0,0)$ to $B = (a,a,0)$ (i) along the line $x = y$, and (ii) along the parabola $y = (x^2/a)$. Suppose the speed of the particle at A is v_0. Find the speed at B in each case (see Fig. 4.9).

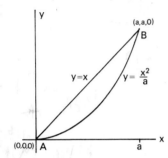

Figure 4.9

Solution

We can solve each problem by means of the work–energy relation if we can compute the work of $\mathbf{F}$.

i) $\mathbf{r} = x\mathbf{i} + y\mathbf{j}, \qquad d\mathbf{r} = dx\,\mathbf{i} + dy\,\mathbf{j}.$

Thus

$$W_{AB} = \int_A^B \mathbf{F} \cdot d\mathbf{r} = \frac{F_0}{a^3} \int_A^B (xy^2\,dx + x^2 y\,dy). \qquad \text{(a)}$$

In the present case $x = y$. Thus $dx = dy$ along this path, and (a) becomes

$$W_{AB} = \frac{F_0}{a^3} \int_0^a 2x^3\,dx = \tfrac{1}{2} F_0\, a.$$

Thus writing

$$W_{AB} = T(B) - T(A), \qquad \text{(b)}$$

we have

$$\tfrac{1}{2} F_0 a = \tfrac{1}{2} m v_B^2 - \tfrac{1}{2} m v_0^2$$

or

$$v_B = \sqrt{\left(\frac{F_0 a}{m} + v_0^2\right)}. \qquad \text{(c)}$$

ii)   With $y = (x^2/a)$, we have

$$dy = \frac{2x}{a}\, dx$$

along the second path. Thus (a) becomes

$$W_{AB} = \frac{F_0}{a^3} \int_0^a \left(\frac{x^5}{a^2}\, dx + \frac{2x^5}{a^2}\, dx \right) = \tfrac{1}{2} F_0 a.$$

Since the work term here is the same as in (i), (b) will yield the same value (c) for the final speed v_B.

   We should not in general expect that the computation W_{AB} would turn out to be the same for every path between A and B. Forces for which this is true are called *conservative*. The force $\mathbf{F}$ of this example is a conservative force, as we shall see from the next section.

....

5. CONSERVATIVE FORCES

In the last section, we established the work–energy relation. The result said that the work of the force on a particle equals the change in its kinetic energy. For certain forces it is possible to define energy functions such that the work done by the force can be described in terms of changes in the functions. These forces are called conservative. We shall see the rationale for this name shortly.

   Consider a particle which can occupy various positions in space, and suppose that $\mathbf{F}$ is a force on the particle whose magnitude and direction are given functions of position in space. Finally suppose that $\mathbf{F}$ can be written

$$\mathbf{F} = \nabla \psi,$$ is usually V (4.11)

where ψ is a scalar function of position. [In cartesian coordinates, (4.11) reads

$$\mathbf{F} = \mathbf{i}\, \frac{\partial \psi}{\partial x} + \mathbf{j}\, \frac{\partial \psi}{\partial y} + \mathbf{k}\, \frac{\partial \psi}{\partial z}$$ ← gradient

or in components

$$F_x = \frac{\partial \psi}{\partial x}, \qquad F_y = \frac{\partial \psi}{\partial y}, \qquad F_z = \frac{\partial \psi}{\partial z}.]$$

If such a function ψ exists, we say that the force $\mathbf{F}$ is conservative, and define the *potential energy* of $\mathbf{F}$ to be*

$$V = -\psi.$$ (4.12)

---

* The minus sign in V enables us to add potential and kinetic energies.

Let us investigate briefly some of the important properties of conservative forces. Consider a particle acted on by a conservative force **F**. Suppose
that the particle moves from point A to point B, as in Fig. 4.6. The work of **F**
is given by

$$W_{AB} = \int_A^B \mathbf{F} \cdot d\mathbf{r}.$$

But by (4.11) we can write

$$\mathbf{F} \cdot d\mathbf{r} = \nabla\psi \cdot d\mathbf{r}.$$

If we introduce the arc length s and write

$$\nabla\psi \cdot \frac{d\mathbf{r}}{ds}\, ds = \nabla\psi \cdot \mathbf{e}_t\, ds$$

[by (3.5), $d\mathbf{r}/ds$ is $\mathbf{e}_t$ is the unit tangent vector], we recognize

$$\nabla\psi \cdot \mathbf{e}_t = \frac{d\psi}{ds},$$

the directional derivative* of ψ; that is, the derivative of ψ in the direction of
the tangent to the path. Thus we have

$$W_{AB} = \int_A^B \frac{d\psi}{ds}\, ds = \int_A^B d\psi = \psi(B) - \psi(A).$$

And by (4.12), we obtain

$$W_{AB} = V(A) - V(B). \tag{4.13}$$

(If you find directional derivatives somewhat mysterious, you might do
well to consider the above development in cartesian coordinates. Then

$$\mathbf{r} = x\mathbf{i} + y\mathbf{j} + z\mathbf{k} \qquad \text{and} \qquad d\mathbf{r} = \mathbf{i}\, dx + \mathbf{j}\, dy + \mathbf{k}\, dz.$$

And

$$\mathbf{F} \cdot d\mathbf{r} = \frac{\partial\psi}{\partial x}\, dx + \frac{\partial\psi}{\partial y}\, dy + \frac{\partial\psi}{\partial z}\, dz = d\psi.\Big)$$

We note that (4.13) says that the work of **F** depends only on the end
points A and B. Thus if we compute the work of **F** along any of the paths
shown in Fig. 4.10, the result will always be the same, $V(A) - V(B)$. In particular, if the path begins and ends at the same point A, we have the work

$$W = V(A) - V(A) = 0.$$

Thus the work of a conservative force about a closed path is zero.

---

* See Appendix 1, Eq. (A1.16).

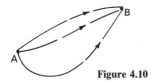

Figure 4.10

Let us now consider the motion of a particle under the influence of several conservative forces. Each force has a potential energy. We can denote these by $V_1, V_2, \ldots$, and then write the work–energy relation as

$$T(B) - T(A) = [V_1(A) - V_1(B)] + [V_2(A) - V_2(B)]_0 + \cdots .$$

We can simplify this if we write V as the total potential energy:

$$V = V_1 + V_2 \cdots .$$

Then

$$T(B) - T(A) = V(A) - V(B),$$

and further, rearranging terms, we have

$$T(B) + V(B) = T(A) + V(A). \tag{4.14}$$

Equation (4.14) states that the sum of kinetic energy and potential energy is a constant for a particle acted on by conservative forces. It is common to denote $T + V$ as the *total mechanical energy E*. Then (4.14) becomes

$$T + V = E = \text{const.} \tag{4.15}$$

Equation (4.15) now shows the reason for introducing the minus sign in (4.12).

Suppose now that the particle is acted on by *two* types of force, conservative and nonconservative. Let $\mathbf{F}_{con}$ be the vector sum of the conservative forces and $\mathbf{F}_{non}$ the vector sum of the nonconservative forces. Let V be the total potential energy of the conservative forces. The work–energy relation becomes

$$\int_A^B (\mathbf{F}_{non} + \mathbf{F}_{con}) \cdot d\mathbf{r} = T(B) - T(A).$$

But

$$\int_A^B \mathbf{F}_{con} \cdot d\mathbf{r} = V(A) - V(B).$$

Thus, denoting by W_{non} the work of the nonconservative forces, we have

$$(T + V)_B = (T + V)_A + W_{non} \tag{4.16}$$

or

$$E(B) - E(A) = W_{non}. \tag{4.17}$$

> The change in the total mechanical energy between point A and point B equals the work of the nonconservative forces.

If $E =$ const for a particle, we speak of the equation for E as a *first integral of the equation of motion* (the energy integral).

In the next section, we shall investigate the various forces we are likely to encounter in problems of dynamics. We shall indicate whether each force is conservative or nonconservative, and if it is conservative, we shall give the potential energy function for the force.

In order to illustrate (4.16) and (4.17), we get a bit ahead of ourselves by investigating the gravity force. Figure 4.11 shows a particle near the surface of the earth.

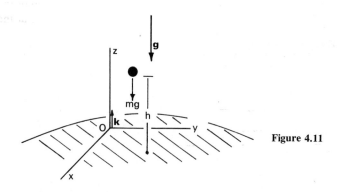

Figure 4.11

The force of gravity, the *weight,* can be written

$$\mathbf{W} = -mg\mathbf{k}. \tag{4.18}$$

If the force of gravity is conservative, a function $V = -\psi$ must exist so that (4.11) is valid. Noting (4.18), we must have

$$-mg\mathbf{k} = \frac{\partial \psi}{\partial x}\,\mathbf{i} + \frac{\partial \psi}{\partial y}\,\mathbf{j} + \frac{\partial \psi}{\partial z}\,\mathbf{k}$$

or

$$\frac{\partial \psi}{\partial x} = 0, \qquad \frac{\partial \psi}{\partial y} = 0, \qquad \text{and} \qquad \frac{\partial \psi}{\partial z} = -mg. \tag{a}$$

If we take $\psi = -mgz + C$, we see that the equations (a) are satisfied. And further, the potential energy V by (4.12) is

$$V = mgz + C.$$

The role of the *integration constant C* is interesting; its value is arbitrary, since

$$-\frac{\partial V}{\partial z} = -mg$$

no matter what C is. Thus the value of C can be assigned at our convenience. And this in turn means we can set the zero value of V at our convenience. Then

$$V = mgh,$$ (4.19)

where h is the vertical height above the zero level of V.

Example 8

In Fig. 4.12(a), a particle m slides down a smooth incline. Its speed at the top of the incline is zero, and the height of the incline is h. Find the speed of the particle at the bottom of the incline.

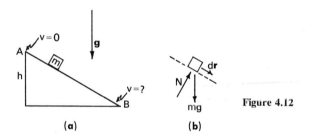

(a)   (b)   **Figure 4.12**

Solution

Figure 4.12(b) is a free-body diagram showing the forces acting on the particle m. Note that the normal force **N** is perpendicular to $d\mathbf{r}$. Thus **N** does no work. The weight of the particle is a conservative force.

Writing T and V at positions A and B gives

$$T(A) = 0, \qquad V(A) = mgh, \qquad T(B) = \tfrac{1}{2}mv^2, \qquad V(B) = 0.$$

Note that we have taken the zero level of V at the bottom of the incline. Applying (4.16) and noting that $W_{\text{non}} = 0$, we get

$$\tfrac{1}{2}mv^2 = mgh$$

or

$$v = \sqrt{2gh}.$$

Note that the speed of the particle at the bottom of the incline is independent of the mass of the particle or the slope of the incline.

Example 9

A simple pendulum of length l and mass m has speed v_0 at the bottom of its trajectory. Find its speed when it is at angle θ.

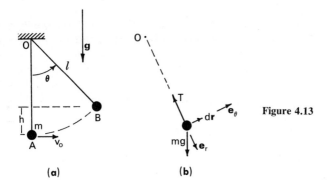

Figure 4.13

(a)   (b)

Solution

Figure 4.13(b) is a free-body diagram for m. The weight is a conservative force, and the tension in the string does no work. To show this, we note that the velocity of m is

$$\mathbf{v} = \dot{r}\,\mathbf{e}_r + r\dot{\theta}\,\mathbf{e}_\theta$$

but $r = l$ and thus $\dot{r} = 0$. Thus

$$\mathbf{v} = l\dot{\theta}\,\mathbf{e}_\theta.$$

And since $d\mathbf{r} = \mathbf{v}\,dt$, we get

$$d\mathbf{r} = l\,d\theta\mathbf{e}_\theta.$$

But the string tension $\mathbf{T} = -T\mathbf{e}_r$. Thus $\mathbf{T} \cdot d\mathbf{r} = 0$.
Writing the energies at A and B, we get

$$T(A) = \tfrac{1}{2}mv_0^2, \qquad V(A) = 0,$$

$$T(B) = \tfrac{1}{2}mv^2, \qquad V(B) = mgh = mgl(1 - \cos\theta)$$

and thus from (4.16) we get

$$\tfrac{1}{2}mv^2 + mgl(1 - \cos\theta) = \tfrac{1}{2}mv_0^2$$

or

$$v = \sqrt{v_0^2 - 2gl(1 - \cos\theta)}.$$

Again the result is independent of the mass m.

Example 10

Figure 4.14 shows three smooth curves, each of total height h. A bead slides on each, beginning with $v = 0$ at point A. Find the speed of the bead at the bottom in each case. Are the three motions identical?

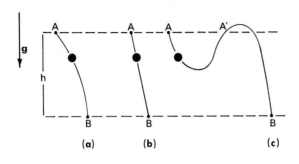

Figure 4.14

(a)   (b)   (c)

Solution

In each case, the system is conservative. Since each bead is at rest at point A, and since point B in each case has the same level, we can write for each case

$$T(A) = 0, \qquad T(B) = \tfrac{1}{2}mv^2,$$

$$V(A) = mgh, \qquad V(B) = 0,$$

and thus $v = \sqrt{2gh}$.

This analysis is valid in cases (a) and (b), but not in case (c). In case (c), the bead never gets beyond point A'. In fact, the motion in that case is an oscillation between points A and A'.

There is another difference between the motions in cases (a) and (b). Although the velocity of each bead at a given height is the same, the time it takes to get to that height depends on which curve the bead is on. Thus the time it takes for the bead to get from A to B is different in the two cases.

Example 11

A bead is restricted to slide on a smooth circular path (Fig. 4.15). A constant force P is applied to the bead so that the force remains horizontal. The bead is at rest at point A. Find the angle θ of the point B at which the bead again comes to rest.

Solution

We write

$$T(A) = 0, \qquad T(B) = 0,$$

$$V(A) = 0, \qquad V(B) = mgR(1 - \cos\theta).$$

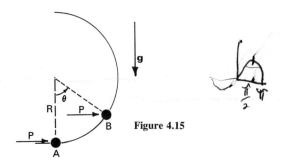

Figure 4.15

We can evaluate the work of P by multiplying P by its horizontal displacement:

$$W_P = PR \sin \theta, \qquad 0 \leqslant \theta \leqslant \pi \text{ radians.}$$

[Can you explain why this formula is valid for $(\pi/2) < \theta \leqslant \pi$ radians?] Thus, writing (4.17), we get

$$mgR(1 - \cos \theta) = PR \sin \theta \qquad \text{or} \qquad mg(1 - \cos \theta) = P\sqrt{1 - \cos^2 \theta}.$$

Squaring each side and rearranging the terms, we obtain

$$[(mg)^2 + P^2] \cos^2 \theta - 2(mg)^2 \cos \theta + [(mg)^2 - P^2] = 0.$$

Solving the quadratic equation for $\cos \theta$, we get

$$\cos \theta = \frac{(mg)^2 \pm P^2}{(mg)^2 + P^2}.$$

The plus root gives $\theta = 0$. The minus root is what we are after:

$$\theta = \cos^{-1} \left[\frac{1 - (P/mg)^2}{1 + (P/mg)^2} \right].$$

[Does it make sense that if $P = mg$, $\theta = (\pi/2)$ radians?]

6. FORCES IN DYNAMICS

A wide variety of forces can enter into a problem in dynamics. Some forces may be specified as functions of time, others as functions of position in space; still others may be defined by their effects. An example is the force required to constrain a mass to move on a given curve. Such forces, often called *forces of constraint,* can usually be determined only after the motion of the mass has been determined.

In this section, we shall briefly discuss gravitational forces, spring forces, the forces of sliding and viscous friction, and the general class of conservative forces. In each case, we shall indicate whether the force is conservative, and if so, we shall give its potential energy.

a) Gravitational Forces

On the basis of Kepler's three laws for the motion of the planets, Newton derived his law of universal gravitation.

Consider two homogeneous spherical bodies, of masses m_1 and m_2, with centers separated by a distance p. Let $\mathbf{e}_p$ be a unit vector along the line of centers (Fig. 4.16). Then the force of m_1 on m_2 is given by

$$\mathbf{F} = -\frac{\gamma m_1 m_2}{p^2} \mathbf{e}_p, \qquad (4.20)$$

where γ is called the *universal gravitational constant*. Its value is

$$3.45 \times 10^{-8} \text{ ft}^4/(\text{lb})(\text{sec})^4 \qquad \text{or} \qquad 6.68 \times 10^{-11} \text{ m}^3/(\text{kg})(\text{sec})^2.$$

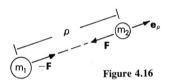

Figure 4.16

For a wide range of engineering problems, we can use an approximation to (4.20) to represent the force of gravity.

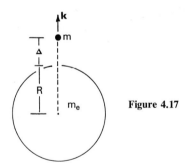

Figure 4.17

Let us consider the force which the earth exerts on a particle of mass m, which remains close to the surface of the earth. In Fig. 4.17, R is the radius of the earth and Δ is the distance of the particle from the earth's surface. Let the mass of the earth be m_e. Then if $\mathbf{k}$ is a unit vector along the local vertical direction:

$$\mathbf{F} = -\frac{\gamma m_e m}{(R + \Delta)^2} \mathbf{k}.$$

Suppose now that $(\Delta/R) \ll 1$. Then

$$\mathbf{F} = -m \frac{\gamma m_e}{R^2} \frac{1}{(1 + \Delta/R)^2} \mathbf{k} \approx -m \left(\frac{\gamma m_e}{R^2}\right) \mathbf{k}.$$

The quantity $(\gamma m_e/R^2)$ is usually abbreviated g and is called the *acceleration of gravity*. Its value is 9.8 m/sec² = 32.2 ft/sec².* Thus

$$\mathbf{F} = -mg\mathbf{k}. \tag{4.21}$$

> The force of gravity on a particle near the earth's surface is directed toward the center of the earth and has magnitude mg. This force is called the weight.

The force of gravity is conservative. The potential energy depends on which expression is used for gravity:

$$\mathbf{F} = -\frac{\gamma m_1 m_2}{\rho^2} \mathbf{e}_\rho, \qquad V = -\frac{\gamma m_1 m_2}{\rho}, \qquad \int \mathbf{F} \cdot d\mathbf{r} \tag{4.22}$$

or, for points close to the earth's surface,

$$\mathbf{F} = -mg\mathbf{k}, \qquad V = mgh. \tag{4.22'}$$

b) Spring Forces

Consider the spring in Fig. 4.18(a). We put various weights on the spring and then measure the deflection of its end from the equilibrium position. If we were to plot the magnitude of the force applied to the spring versus the deflection δ, we might get the plot shown in Fig. 4.18(b).

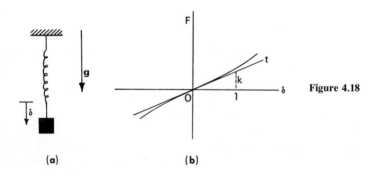

(a)   (b)   Figure 4.18

---

* Working in reverse, we can see that if $g = 32.2$ ft/sec², and if $R = 4000$ miles, $\gamma m_e = 1.255 \times 10^{12}$ mile³/hr².

If we pull on the spring, the deflection δ is positive. If we push, δ is negative. If we put a certain force on the spring and then let the force go to zero, we assume that δ will also go to zero. These assumptions may seem quite strict. One can easily think of springlike objects which do not respond exactly as we have assumed.

There is one further idealization which is commonly made with respect to springs. We assume that for small deformations we can write

$$F = k\delta,$$

where k is called the _spring constant_. This amounts to approximating the force–deflection curve by the tangent line through the origin.

In most problems in dynamics, we are concerned with springs as force producers. That is, we are interested in the force which the spring exerts on its environment. By Newton's third law, we have this force

$$f = -k\delta. \qquad\qquad (4.23)$$

> The force which a spring exerts on its environment has the same direction as the spring and the sense opposite to that of the spring's deflection from equilibrium.

Forces which are generated by elastic springs are conservative. Consider the spring in Fig. 4.19, which is linear, with spring constant k, and which has free length l_0.

Figure 4.19

We stretch the spring from its free length to position x. Then the force which the spring exerts is

$$\mathbf{f} = -k(x - l_0)\mathbf{i}.$$

For the potential energy, we seek a function V such that

$$-\frac{\partial V}{\partial x}\mathbf{i} - \frac{\partial V}{\partial y}\mathbf{j} - \frac{\partial V}{\partial z}\mathbf{k} = -k(x - l_0)\mathbf{i}$$

or

$$\frac{\partial V}{\partial x} = k(x - l_0), \qquad \frac{\partial V}{\partial y} = \frac{\partial V}{\partial z} = 0.$$

We get

$$V = \tfrac{1}{2}k(x - l_0)^2 + C.$$

It is reasonable, and convenient, to take the potential energy to be zero when the spring is at its free length, $x = l_0$. This means taking the constant $C = 0$:

$$V = \tfrac{1}{2}k(x - l_0)^2. \tag{4.24}$$

It is common to take δ to be the deflection of the spring from the free position. That is, $\delta = x - l_0$. Then (4.24) becomes

$$V = \tfrac{1}{2}k\delta^2. \tag{4.24'}$$

We note specifically that we should write (4.24') only if $\delta = 0$ is the unstretched or free position of the spring.

c) Sliding Friction

Suppose the block in Fig. 4.20 slides on the surface as shown. The theory of sliding friction is very complicated. It is based on experimental findings, whose results are simple,* even if somewhat surprising. They can be summarized as follows.

i) The force of sliding friction resists the relative motion of the adjacent surfaces and is tangent to the surfaces.

ii) The magnitude of the force of sliding friction is generally less than the magnitude of the maximum static friction force, and is independent of the area of contact of the surfaces, but is directly proportional to the normal force between the surfaces.

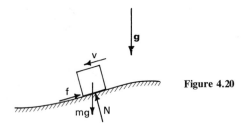

Figure 4.20

We model the force of sliding friction **f** to have the magnitude

$$f = \mu N. \tag{4.25}$$

---

* Part of the reason for the simplicity is that the results are not universally true. For example, these laws of sliding friction do not adequately describe the friction between an automobile tire and the road.

The parameter μ is called the *coefficient of sliding friction*. Engineering handbooks give various values of μ, generally between 0 and 1, but there is no theoretical reason to believe that they are bounded by 1. Sliding friction is a nonconservative force, and has no potential energy function.

d) Viscous Friction

In certain problems, the force of friction is not constant, as in the above model. The magnitude of the friction force might depend on the velocity of the body in some way. In some cases, air drag for example, the force on a body is proportional to the square of the velocity of the body in the surrounding medium. Of great importance in many problems is the *linear dashpot* shown schematically in Fig. 4.21.

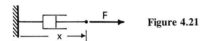

Figure 4.21

We note that, if a force F is applied to the dashpot, it will elongate at a rate proportional to the magnitude of F:

$$F = c\dot{x}.$$

Again we are interested in the force which the dashpot exerts on its environment:

$$f = -c\dot{x}. \tag{4.26}$$

This force is often referred to as *viscous friction*.

The forces of friction, both sliding and viscous, are nonconservative; they dissipate energy rather than store it, and have no potential energy function.

e) Conservative Forces: General Case*

We have said [see (4.11) and (4.12)] that a force is conservative if it can be written in the form

$$\mathbf{F} = -\nabla V,$$

where V, the potential energy, is a function of position in space only. Now let's reverse the question: Suppose $\mathbf{F}$ is given. How do we know whether the force is conservative or not? And further, if we determine that $\mathbf{F}$ is conservative, how do we determine the potential energy V? Fortunately, the answer to both these questions is straightforward.

---

* This material is not essential for an understanding of the remainder of the book.

First of all, suppose that

$$\mathbf{F} = -\nabla V,$$

and suppose we form

$$\nabla \times \mathbf{F} = -\nabla \times \nabla V.$$

In many instances, the del operator ∇ resembles a vector. In this particular case $\nabla \times \nabla$ resembles $\mathbf{a} \times \mathbf{a} = 0$. It is easily verified that $\nabla \times \nabla V = 0$, no matter what function V is. Thus, if $\mathbf{F}$ is conservative,

$$\nabla \times \mathbf{F} = 0.$$

The opposite of the above statement is also true.* Given a force $\mathbf{F}$, if $\nabla \times \mathbf{F} = 0$, there exists a potential energy function V such that

$$\mathbf{F} = -\nabla V.$$

Once we have determined that a force is conservative, we can determine V by means of successive partial integrations. This is best illustrated by means of examples.

Example 12

Recall Example 7. Determine whether $\mathbf{F}$ is conservative, and if it is, calculate its potential energy V. Find the work of $\mathbf{F}$ along any path between $(0,0,0)$ and $(a,a,0)$.

Solution

We had

$$\mathbf{F} = \frac{F_0}{a^3}\,(xy^2\mathbf{i} + x^2y\mathbf{j}).$$

Writing $\nabla \times \mathbf{F}$, we have†

$$\nabla \times \mathbf{F} = \begin{vmatrix} \mathbf{i} & \mathbf{j} & \mathbf{k} \\ \dfrac{\partial}{\partial x} & \dfrac{\partial}{\partial y} & \dfrac{\partial}{\partial z} \\ \left(\dfrac{F_0}{a^3}\,xy^2\right) & \left(\dfrac{F_0}{a^3}\,x^2y\right) & 0 \end{vmatrix}$$

$$= \mathbf{i}(0) - \mathbf{j}(0) + \mathbf{k}\left(\frac{F_0}{a^3}\,2xy - \frac{F_0}{a^3}\,x2y\right)$$

---

* Actually this statement is restricted to simply connected domains. See F. B. Hildebrand, *Advanced Calculus for Engineers*, Englewood Cliffs, N.J.: Prentice-Hall, 1949, page 309.

† See Appendix 1, Eq. (A1.20a).

Thus $\nabla \times \mathbf{F} = 0$, and we are assured that a function V exists.

Since V exists, it must satisfy $\mathbf{F} = -\nabla V$ or

$$F_x = \frac{F_0}{a^3}(xy^2) = -\frac{\partial V}{\partial x}, \tag{a}$$

$$F_y = \frac{F_0}{a^3}(x^2 y) = -\frac{\partial V}{\partial y}, \tag{b}$$

$$0 = -\frac{\partial V}{\partial z}. \tag{c}$$

We can start with any of the three equations. Suppose we take (a). Integrating partially with respect to x,* we have

$$-V = \frac{F_0}{2a^3}(x^2 y^2) + f(y,z). \tag{d}$$

The function $f(y,z)$ plays the role of an integration constant.

Moving on, we insert (d) into (b)

$$\frac{F_0}{a^3}(x^2 y) = \frac{F_0}{a^3}(x^2 y) + \frac{\partial f}{\partial y} \qquad \text{or} \qquad \frac{\partial f}{\partial y} = 0.$$

Noting that (d) also satisfies (c), we have

$$V = -\frac{F_0}{2a^3}(x^2 y^2) + C,$$

where C is an arbitrary constant. To compute the work between $A = (0,0,0)$ and $B = (a,a,0)$, we note (4.13),

$$W_{AB} = V(A) - V(B) = C - \left(-\frac{F_0}{2a^3}a^4 + C\right)$$

$$= \tfrac{1}{2}F_0 a.$$

Note that this agrees with the results of Example 7.

....

Example 13

Suppose that we have

$$\mathbf{F} = -\frac{1}{r}\mathbf{e}_r + \frac{1}{r}\tan\theta\,\mathbf{e}_\theta.$$

Determine whether $\mathbf{F}$ is conservative, and if so, determine V.

----

* That is, hold the variables y and z constant and integrate with respect to the variable x.

Solution

Refer to Appendix 1 for expressions for the gradient (∇) and the curl ($\nabla \times$) in cylindrical coordinates. We have

$$F_r = -\frac{1}{r}, \qquad F_\theta = +\frac{1}{r}\tan\theta, \qquad F_z = 0.$$

From (A1.20b), we have

$$\nabla \times \mathbf{F} = \left(\frac{1}{r}\right) \begin{vmatrix} \mathbf{e}_r & r\,\mathbf{e}_\theta & \mathbf{k} \\ \dfrac{\partial}{\partial r} & \dfrac{\partial}{\partial \theta} & \dfrac{\partial}{\partial z} \\ -\dfrac{1}{r} & \tan\theta & 0 \end{vmatrix}$$

$$= \frac{1}{r}\left[\mathbf{e}_r(0) - r\,\mathbf{e}_\theta(0) + \mathbf{k}(0)\right].$$

Thus $\mathbf{F}$ is a conservative force.

From (A1.17b) we have

$$-\frac{1}{r}\mathbf{e}_r + \frac{1}{r}\tan\theta\,\mathbf{e}_\theta = -\frac{\partial V}{\partial r}\mathbf{e}_r - \frac{1}{r}\frac{\partial V}{\partial \theta}\mathbf{e}_\theta - \frac{\partial V}{\partial z}\mathbf{k}$$

or

$$-\frac{1}{r} = -\frac{\partial V}{\partial r}, \tag{a}$$

$$+\frac{1}{r}\tan\theta = -\frac{1}{r}\frac{\partial V}{\partial \theta}, \tag{b}$$

$$0 = \frac{\partial V}{\partial z}. \tag{c}$$

Integrating (a) with respect to r, we obtain

$$V = \ln(r) + f(\theta, z).$$

Inserting this into (b) gives

$$\frac{1}{r}\tan\theta = -\frac{1}{r}\frac{\partial f}{\partial \theta}.$$

Thus $\partial f/\partial\theta = -\tan\theta$ and

$$f(\theta, z) = \ln(\cos\theta) + g(z).$$

Substituting this into (c) gives

$$0 = \frac{dg}{dz} \qquad \text{or} \qquad g = C.$$

And inserting g and f into V, we have

$$V = \ln (r) + \ln (\cos \theta) + C$$

or

$$V = \ln (r \cos \theta) + C.$$

There are many other forces which we have not described which enter into problems in dynamics: electrical forces, magnetic forces, etc. In addition, all the forces important in statics (see Section 7 of Chapter 2) are present in dynamics. Our point here is not to delineate all possible forces in problems of dynamics, but to explain how problems of dynamics are formulated mathematically and how they are solved. The forces we have described in this section are those which we shall encounter repeatedly in the rest of this book.

7. EXAMPLES

In this section, we shall present a variety of examples to illustrate the formulation of some typical problems in dynamics. Before launching into the examples, let us say a few words about the process of analysis.

There is such an enormous variety of possible problems in dynamics that students often have trouble translating a given problem into proper equations of motion. There is, however, a pattern to follow in all analyses of problems in dynamics.

I. Analyze the forces acting in the problem, using free-body diagram(s). All forces should have an identifiable physical origin.

II. Analyze the position, velocity, and acceleration of each particle of the problem, using some convenient coordinate system. In particular, if the motion of various parts of the system are related, write these kinematic relations.

III. Write the equation(s) of motion for the system:

a) $\mathbf{F} = \dot{\mathbf{p}}$

b) $\mathbf{M} = \dot{\mathbf{H}}$

c)   Work–energy

Steps I and II of this analysis scheme are relatively straightforward. And if you perform these steps methodically, chances are that step III will be easier. The problem is that step III is more than slightly ambiguous. The biggest difficulty in the analysis of a dynamics problem comes in deciding which of the equations of III to use, (a), (b), or (c); in fact, in many problems we shall use more than one.

In the examples, we shall give one solution to each problem. This should by no means be taken to mean that this is the only or even the best solution. If you think another approach might work or might be better than the one we have used, try it. You may well be right. In any event, you will learn something in the process.

Example 14

Figure 4.22(a) shows a vertical circle of radius R, which has two chords, AB and AB'. Two beads of mass m are released at A on the two chords at $t = 0$. Show that the two beads reach B and B' at the same time (given that the chords are frictionless).

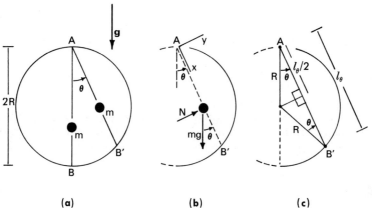

(a)   (b)   (c)   **Figure 4.22**

Solution

Consider the chord at angle θ in Fig. 4.22(b), which shows the free-body diagram for the particle.

Writing $\mathbf{F} = \dot{\mathbf{p}}$ in the x direction (along the incline), we have

$$F_x = ma_x: \qquad mg \cos \theta = m\ddot{x}.$$

Thus

$$\ddot{x} = g \cos \theta. \tag{a}$$

This is a case of constant-acceleration motion (recall Example 5 of Chapter 3). At $t = 0$:

$$x = \dot{x} = 0.$$

The solution to (a) is

$$x = \tfrac{1}{2}(g \cos \theta)t^2.$$ (b)

Now let the distance from A to B' be denoted by l_θ. Then the time to get from A to B' is, from (b),

$$t_{B'} = \left(\frac{2l_\theta}{g \cos \theta}\right)^{1/2}.$$ (c)

The last step in this solution is to determine the length l_θ. Consider Fig. 4.22(c). From this figure, we see that

$$\tfrac{1}{2}l_\theta = R \cos \theta \quad \text{or} \quad l_\theta = 2R \cos \theta.$$ (d)

Inserting (d) into (c) gives

$$t_{B'} = 2\sqrt{\frac{R}{g}}.$$ (e)

Finally, we note that the expression for $t_{B'}$ *does not depend on the angle θ*. This means that the time required to reach the end of any chord (including AB and AB') is the same: $t_{B'}$.

. . . .

Example 15

Figure 4.23(a) shows the problem we first talked about in Example 2 of Chapter 3: A particle slides along the frictionless curve, $y = x^2$, in the horizontal plane. Find the magnitude of the force normal to the curve at position x.

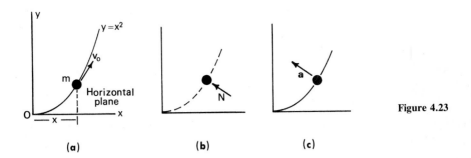

Figure 4.23

Solution

The only in-plane force on m is N, as shown in the free-body diagram in Fig. 4.23(b). Figure 4.23(c) shows the acceleration of m. Since

$$\dot{s} = v_0 \cdots \text{constant}, \quad \ddot{s} = 0.$$

Thus the acceleration is, by (3.9),

$$\mathbf{a} = \ddot{s}\, \mathbf{e}_t + \frac{(\dot{s})^2}{R}\, \mathbf{e}_n \qquad \text{or} \qquad \mathbf{a} = \frac{v_0^2}{R}\, \mathbf{e}_n.$$

The radius of curvature at x was determined in Example 2 of Chapter 3:

$$R = \tfrac{1}{2}(1 + 4x^2)^{3/2}.$$

Thus writing $\mathbf{F} = \dot{\mathbf{p}}$ in the normal direction, we have

$$F_n = ma_n, \qquad N = m\, \frac{2v_0^2}{(1 + 4x^2)^{3/2}}.$$

Note that as $x \to \infty$, $N \to 0$. What does this mean?

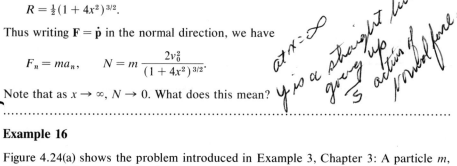

at $x = \infty$ y is a straight line going up ... action of normal force?

Example 16

Figure 4.24(a) shows the problem introduced in Example 3, Chapter 3: A particle m, which moves in a smooth horizontal plane, is attached to an inextensible string which wraps around a cylinder of radius a. In the first position shown, the free length of the string is l and the speed of m is v_0.

Find the speed of the particle and the tension in the string at the second position shown.

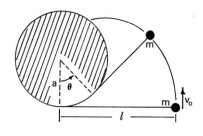

(a)

Figure 4.24

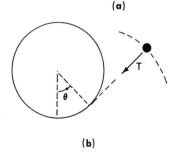

(b)

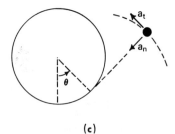

(c)

Solution

Figure 4.24(b) is the free-body diagram of the particle, indicating that the only in-plane force is the string tension T. Figure 4.24(c) shows the components of acceleration in the tangential and normal directions. Note that the normal direction is always aligned along the string (recall Example 3, Chapter 3).

Writing $\mathbf{F} = \dot{\mathbf{p}}$, we get

$$F_t = ma_t: \qquad 0 = m\ddot{s} \tag{a}$$

$$F_n = ma_n: \qquad T = m\left(\frac{\dot{s}^2}{R}\right) \tag{b}$$

Integrating equation (a), we obtain

$$m\dot{s} = C, \quad \ldots \text{ constant.} \tag{c}$$

But $\dot{s} = |\mathbf{v}|$. At the initial position, $\dot{s} = v_0$. Thus (c) says that $\dot{s} = v_0$ for all time:

$$\dot{s} = v_0, \quad \text{all } t. \tag{d}$$

From (b) we now can solve for the string tension T:

$$T = m\frac{v_0^2}{R}.$$

Again from Example 3 of Chapter 3, we can evaluate the radius of curvature, $R = l - a\theta$. Thus

$$T = m\frac{v_0^2}{l - a\theta}.$$

. . . .

Example 17

A particle falls vertically through the atmosphere. The drag force on it is proportional to the square of its velocity. Determine the terminal speed of the particle.

Solution

Figure 4.25 presents two free-body diagrams of the particle.

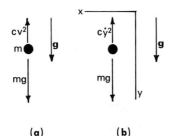

(a)   (b)

Figure 4.25

If the particle falls straight downward, the forces can be evaluated as in Fig. 4.25(b). Writing $\mathbf{F} = \dot{\mathbf{p}}$ in the y direction gives

$$F_y = ma_y, \qquad mg - c\dot{y}^2 = m\ddot{y}$$

or

$$\ddot{y} + \left(\frac{c}{m}\right)\dot{y}^2 = g.$$ (a)

Equation (a) is an example of a nonlinear differential equation. This particular non-linear problem can be solved, however, by using a slight trick: Let $v = dy/dt$. Then

$$\frac{dv}{dt} = \frac{d^2y}{dt^2} = \ddot{y}.$$

We write dy/dt in terms of the chain rule,

$$\frac{dv}{dt} = \frac{dv}{dy}\frac{dy}{dt} = \frac{dv}{dy}v.$$

And thus

$$\ddot{y} = v\,\frac{dv}{dy}.$$ (b)

Applying (b) to (a), we get

$$v\,\frac{dv}{dy} + \frac{c}{m}v^2 = g.$$

This can be rewritten

$$\frac{v\,dv}{g - (c/m)v^2} = dy.$$ (c)

Suppose that $v = 0$ at position $y = 0$. Then, integrating (c), we have

$$\int_0^v \frac{v\,dv}{g - (c/m)v^2} = \int_0^y dy.$$

The integrals can be evaluated* as

$$\frac{m}{2c}\ln\left[\frac{1}{1 - (c/mg)v^2}\right] = y.$$

We want to determine v when $y \to \infty$. The best way to do this is to invert the above relation. Multiplying each side of the equation by $(2c/m)$ and then raising e to the power indicated by each side of the equation, we obtain

$$\frac{1}{1 - (c/mg)v^2} = e^{(2c/m)y} \qquad \text{or} \qquad v^2 = \left(\frac{mg}{c}\right)(1 - e^{(-2c/m)y}).$$

And as $y \to \infty$, $v^2 \to (mg/c)$. Thus the terminal value of the speed is

$$v = \sqrt{\frac{mg}{c}}.$$

---

* For the integral on the left, use the substitution $u = g - (c/m)v^2$.

Note that the results of this problem are not irrelevant. If $c = 0$, we would be in real trouble if we were caught in a rainstorm.

....

Example 18

The system in Fig. 4.26(a) consists of a block and a rigid bar.

The block can slide on a frictionless surface, whereas the bar passes over a projection which is *not* frictionless. The coefficient of sliding friction is μ, x is the distance of the center of gravity of the bar from the projection, and a is the distance from the pin to the center of gravity of the bar. The speed of the system to the right is v_0 when $x = 0$, and the system comes to rest at $x = \frac{1}{2}a$. Determine the coefficient of sliding friction μ.

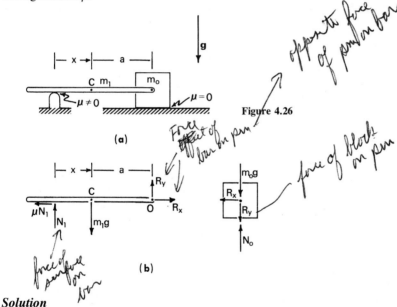

opposite force of pin on bar

Force of bar on pin

force of block on pin

Figure 4.26

(a)

(b)

force of pin on bar

Solution

Figure 4.26(b) gives free-body diagrams for the bar and block.

The entire system moves in the horizontal direction. Thus the acceleration of each point of the system can be written as $\ddot{x}$. Writing $\mathbf{F} = \dot{\mathbf{p}}$ in the horizontal direction for the bar and the block gives

Bar: $F_x = ma_x,$ $R_x - \mu N_1 = m_1 \ddot{x}.$ (a)

Block: $F_x = ma_x,$ $-R_x = m_0 \ddot{x}.$ (b)

If we add equations (a) and (b), we get

$$-\mu N_1 = (m_0 + m_1)\ddot{x}.$$ (c)

To complete the formulation of this problem, we must evaluate the normal force on the bar N_1.

Since the bar does not rotate, the moment of the forces on the bar about the center of gravity is zero (this is actually a result of Chapter 8). Noting Fig. 4.26(b), we take moments about the center of gravity.

Bar: $M_{Cz} = 0$: $R_y a - N_1 x = 0$,
          $F_y = 0$: $R_y - m_1 g + N_1 = 0$.

Thus

$$N_1 = m_1 g \left(\frac{a}{a + x} \right).$$ (d)

And thus (c) becomes

$$\ddot{x} + \mu \left(\frac{m_1 g}{m_0 + m_1} \right) \left(\frac{a}{a + x} \right) = 0.$$ (e)

To integrate Eq. (e), we employ the same trick used in the previous example. When we write

$$\dot{x} = v \qquad \text{and} \qquad \ddot{x} = \frac{dv}{dx} v,$$

(e) becomes [let $\mu m_1 g / (m_0 + m_1) = k$ for the present]

$$v \frac{dv}{dx} = -k \frac{a}{a + x},$$

or noting that at $x = 0$, $v = v_0$, we obtain

$$\int_{v_0}^{v} v \, dv = -\int_{0}^{x} \frac{ka}{a + x} \, dx.$$

Integrating gives

$$\tfrac{1}{2}(v^2 - v_0^2) = ka \ln \left[\frac{1}{1 + (x/a)} \right].$$

We can determine k (and have μ) if we note that $v = 0$ at $x = \tfrac{1}{2}a$. Thus

$$-\tfrac{1}{2}v_0^2 = ka \ln (\tfrac{2}{3}) = ka(-0.405) \qquad \text{or} \qquad k = (1.235)(v_0^2 / a).$$

And finally

$$\mu = (1.235) \left(\frac{m_0 + m_1}{m_1 g} \frac{v_0^2}{a} \right).$$

··

Example 19 (Projectile problem)

There are some circumstances in football in which it is reasonable to put the ball into play by means of a free kick.* Suppose the kicker is attempting to make a field goal.

─────────────────

* This situation occurred in 1962 in a game between the Green Bay Packers and Chicago Bears when Paul Hornung kicked a 52-yard field goal after a fair catch of a punt.

He would like to know what angle from the horizontal to kick the ball in order to have the best chance of making the field goal. Figure 4.27(a) shows this situation.

Solution

We model the football as a particle which is given an initial velocity v_0 at a distance L from the goal post, the crossbar of which is a height H above the ground.

If we assume that there is no air drag on the ball, the free-body diagram will have the weight $\mathbf{W}$ of the ball as the only force. Thus

$$\mathbf{W} = -mg\mathbf{j}.$$

The velocity and acceleration of the ball are

$$\mathbf{v} = \dot{x}\,\mathbf{i} + \dot{y}\,\mathbf{j}, \qquad \mathbf{a} = \ddot{x}\,\mathbf{i} + \ddot{y}\,\mathbf{j}.$$

Thus, writing $\mathbf{F} = \dot{\mathbf{p}}$, we get

$$F_x = ma_x, \qquad m\ddot{x} = 0, \qquad \text{or } \ddot{x} = 0,$$

$$F_y = ma_y, \qquad m\ddot{y} = -mg, \qquad \text{or } \ddot{y} = -g.$$

(a)

We integrate equations (a) twice to get

$$x = A_1 t + A_2, \qquad y = -\tfrac{1}{2}gt^2 + B_1 t + B_2,$$

(b)

where the constants A_1, A_2, B_1, B_2 are determined by the initial conditions. At $t = 0$,

$$x = 0, \qquad y = 0,$$
$$\dot{x} = v_0 \cos \theta, \qquad \dot{y} = v_0 \sin \theta.$$

Thus

$$x = (v_0 \cos \theta)t, \qquad y = -\tfrac{1}{2}gt^2 + (v_0 \sin \theta)t.$$

(c)

From one point of view, equations (c) represent the complete solution to the problem of finding the motion of the football. That is, for a given initial speed v_0 at inclination θ we have determined $x(t)$ and $y(t)$. The kicker is not, however, very interested in this information. He would like to know (i) the trajectory of the ball, (ii) whether the ball will clear the goal post, and (iii) for what angle θ he can make a field goal with minimum effort (i.e., minimum v_0).

We eliminate the time parameter t between the equations (c) to get the equation of the trajectory:

$$y = -\frac{1}{2}\frac{g}{(v_0 \cos \theta)^2} x^2 + (\tan \theta)x.$$

(d)

Equation (d) is linear in y and quadratic in x. Thus the trajectory is a parabola. (See Problem 4.91.)

From (d) we can determine the speed v_0 required at a given angle θ so that a field goal will just be made. That is, we require that

$$y(L) = H$$

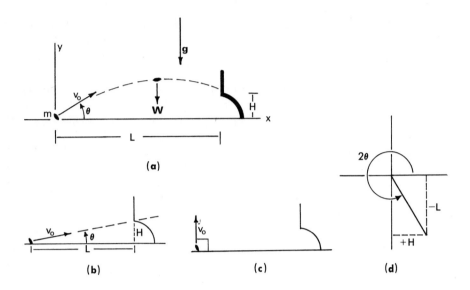

Figure 4.27

which gives the desired relation

$$v_0 = v_0(\theta).$$

We get

$$v_0 = \sqrt{\frac{gL^2}{2(L \tan \theta - H) \cos^2 \theta}}. \tag{e}$$

Now, considering (e), we can find, for given values of L and H, the angle θ which corresponds to a minimum value of v_0. First note that v_0 becomes infinite for two cases:

a) $L \tan \theta - H = 0$

b) $\cos \theta = 0, \qquad \theta = \dfrac{\pi}{2}$

Figures 4.27(b) and (c) diagram these cases, in which an infinite velocity is required in order to make the field goal. We thus seek a minimum of $v_0(\theta)$ between the limits:

$$\tan^{-1}\left(\frac{H}{L}\right) < \theta < \frac{\pi}{2}. \tag{f}$$

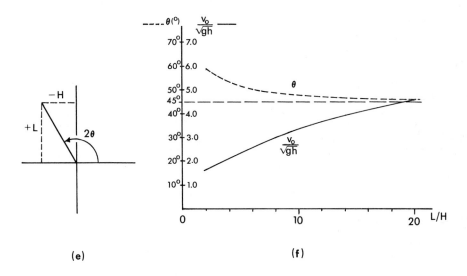

(e)   (f)

Figure 4.27 (cont.)

To obtain the minimum value of $v_0(\theta)$, we can consider the maximum value of (see Problem 4.94)

$$f(\theta) = (L \tan \theta - H) \cos^2 \theta.$$

From $f'(\theta) = 0$, we get

$$1 - 2 \sin^2 \theta + 2 \frac{H}{L} \sin \theta \cos \theta = 0.$$

But since

$$1 - 2 \sin^2 \theta = \cos 2\theta \qquad \text{and} \qquad 2 \sin \theta \cos \theta = \sin 2\theta,$$

we obtain the condition

$$\cos 2\theta + \frac{H}{L} \sin 2\theta = 0.$$

The minimum value of v_0 thus occurs at the angle

$$2\theta = \tan^{-1}\left(\frac{-L}{H}\right) \qquad \text{if } H \neq 0,$$

$$\theta = \frac{\pi}{4} \qquad \text{if } H = 0.$$

(g)

The result (g) requires some explanation. There are two roots to the equation for θ, one of which is extraneous. This arises from the two ways in which $-(L/H)$ can be expressed:

$$-\frac{L}{H} = \frac{-L}{H},$$

$$-\frac{L}{H} = \frac{L}{-H}.$$

In Fig. 4.27(d), the angle 2θ would occur in the fourth quadrant. Thus θ would be in the second quadrant. This is clearly the extraneous root.

In Fig. 4.27(e), the angle 2θ occurs in the second quadrant, and thus the angle θ occurs in the first quadrant, as required by (f).

Figure 4.27(f) shows a plot of the angle θ and the value of $v_0/\sqrt{gH}$ versus (L/H).

Example 20

A circular race track which has a perimeter of 5 miles is designed for cars which travel at 120 miles per hour. What banking angle should the track have?

Solution

Figures 4.28(a) and (b) show the layout of the track. Assuming the optimal condition that no sideward friction is necessary to keep the car on the track, we show the free-body diagram in Fig. 4.28(c). Figure 4.28(d) shows the acceleration of the particle (a constant speed around the track).

Writing $\mathbf{F} = \dot{\mathbf{p}}$ in the radial and tangential directions, we have

$$F_r = ma_r: \qquad -N \sin\theta = -m\frac{v_0^2}{R}, \tag{a}$$

$$F_z = ma_z: \qquad N\cos\theta - mg = 0. \tag{b}$$

Eliminating N between (a) and (b), we get

$$\tan\theta = \frac{v_0^2}{gR}. \tag{c}$$

Putting in the numerical values, we note that

$$60 \text{ mph} = 88 \text{ ft/sec.}$$

Thus

$$120 \text{ mph} = 176 \text{ ft/sec.}$$

Since the circumference of the track is 5 miles,

$$2\pi R = 5 \text{ miles}$$

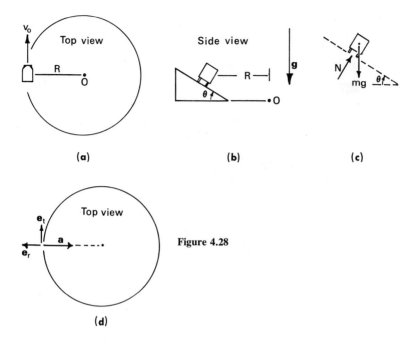

Figure 4.28

and $R = 0.796$ mile $= 4200$ ft. Thus from (c), we obtain

$$\tan \theta = \frac{(1.76 \times 10^2)^2}{(32.2)(4.2 \times 10^3)} = \frac{3.1 \times 10^4}{13.5 \times 10^4}.$$

Thus

$$\theta = 12.9°.$$

Example 21

The operator of an amusement park has an idea for a new thrill ride. Customers will ride in a powerless car on the path shown in Fig. 4.29(a). The straight portion of the curve becomes a vertical circle of radius R, and then continues again in a straight line. Find the minimum initial speed v_0 for which the car does not leave the track.

Solution

We begin by modeling the system. Suppose we consider the car and passengers to be a particle of mass m which passes over the indicated path with zero friction. In the actual problem, assumptions such as these must be evaluated carefully before any strong conclusions can be reached. This is particularly true when the safety of people is concerned.

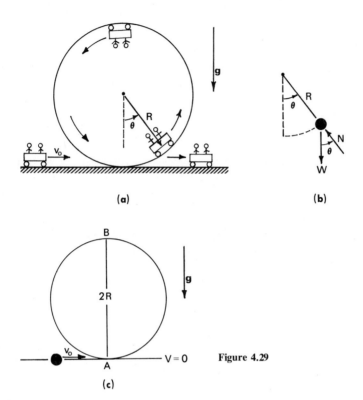

Figure 4.29

The free-body diagram (Fig. 4.29b) shows two forces on m: the weight W and the normal force N of the track on the car. We have

$$\mathbf{W} = mg(\cos\theta\ \mathbf{e}_r - \sin\theta\ \mathbf{e}_\theta)$$

$$\mathbf{N} = -N\mathbf{e}_r.$$

$$\text{(a)}$$

The velocity and acceleration of the particle (since $r = R = $ const) are

$$\mathbf{v} = R\dot\theta\ \mathbf{e}_\theta,$$

$$\mathbf{a} = -R\dot\theta^2\ \mathbf{e}_r + R\ddot\theta\ \mathbf{e}_\theta.$$

$$\text{(b)}$$

The point of greatest concern is the top of the circle. Let us assume that the car reaches this point. This will be true if at this point the force $N > 0$. If $N < 0$, the track would have to grab the car in order to keep it on the track (a condition somewhat disconcerting to the paying customers!). In this problem we would like to solve for N. When we write $\mathbf{F} = \dot{\mathbf{p}}$ we have

$$F_r = ma_r: \qquad -N + mg\cos\theta = -mR\dot\theta^2,$$

$$F_\theta = ma_\theta: \qquad -mg\sin\theta = mR\ddot\theta.$$

$$\text{(c)}$$

Thus

$$N = m(g \cos \theta + R\dot{\theta}^2)$$

and at $\theta = \pi$,

$$N(\pi) = m(-g + R\dot{\theta}^2).$$

Thus the condition that $N > 0$ implies

$$\dot{\theta}(\pi) > \sqrt{g/R} .\tag{d}$$

Note that (d) is independent of the mass m, which is important, since the number and dimensions of the passengers might vary considerably. Suppose $\dot{\theta}(\pi) = \sqrt{g/R}$. Can we be certain that the car will remain on the track at all other points? See Problem 4.108.

The next question concerns what speed v_0 should be given at the bottom of the path in order to achieve the inequality (d). Consideration of energy gives the simplest answer to this question. The system here is conservative, since **N** does no work and **W** is conservative. Thus at A (see Fig. 4.29c),

$$A: \quad T = \tfrac{1}{2}mv_0^2, \quad V = 0,$$

and at B,

$$B: \quad T = \tfrac{1}{2}m(R\dot{\theta})^2, \quad V = mg2R.$$

Equating $(T + V)$ at the two positions gives

$$\tfrac{1}{2}mv_0^2 = \tfrac{1}{2}m(R\dot{\theta})^2 + mg2R \quad \text{or} \quad v_0 = \sqrt{(R\dot{\theta})^2 + 4gR} ,$$

and from (d)

$$v_0 > \sqrt{5gR}.\tag{e}$$

Again the inequality for v_0 is independent of m. We should begin to ask ourselves at this point, "What does depend on m?" The answer to this question is contained in Problem 4.109.

Example 22

Figure 4.30(a) shows a particle m projected from rest at A by a linear spring. The unloaded spring stretches to the point B. The surface below the particle from A to B has a coefficient of sliding friction μ_1. Once the particle reaches point B, it slides inside a frictionless circular path (this path barely touches m before B).

Suppose $k = 10$ lb/in., $\mu_1 = 0.25$, $R = 8$ ft, $mg = 100$ lb.

i) Find the minimum deflection δ of the spring such that the particle will remain on the circular path at all points.

ii) Employing the value of δ determined in (i), find the normal force exerted by the curve on the particle at points $C-$ and $C+$, just before and just after leaving the circular portion of the path at the bottom.

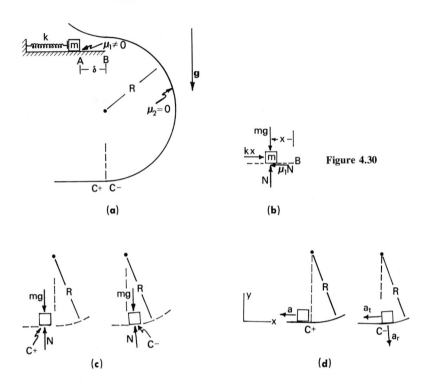

Figure 4.30

Solution

i) From the previous example, we know that the minimum speed required at point B is

$$R\dot\theta(\pi) = v_B = \sqrt{Rg}. \tag{a}$$

We wish to relate the value of δ to (a). Figure 4.30(b) is a free-body diagram of m before it reaches point B. We see that $N = mg$ in this case.

We write the energy equation between points A and B,

$$T(A) = 0, \qquad T(B) = \tfrac{1}{2}mv_B^2 = \tfrac{1}{2}mRg,$$

$$V(A) = \tfrac{1}{2}k\delta^2, \qquad V(B) = 0.$$

The work of sliding friction is

$$W_{AB} = -\mu_1 mg\delta.$$

Thus by (4.16) we get

$$\tfrac{1}{2}mRg = \tfrac{1}{2}k\delta^2 - \mu_1 mg\delta. \tag{b}$$

And therefore δ is determined as

$$\delta = \frac{\mu_1 mg}{k} \pm \sqrt{\left(\frac{\mu_1 mg}{k}\right)^2 + \frac{mgR}{k}}.$$

The minus sign gives an extraneous root. (It arises from the fact that (b) would seem to give positive friction work for $\delta < 0$. This, of course, is not the case.) Inserting the numerical values and noting that

$$k = 10 \text{ lb/in.} = 120 \text{ lb/ft,}$$

we get

$$\delta = \frac{(0.25)(100)}{120} + \sqrt{\left[\frac{(0.25)(100)}{120}\right]^2 + \frac{(100)(8)}{120}} \text{ ft.}$$

Thus $\delta = 2.84$ ft.

ii)   With the value of δ determined above, we have the speed at B,

$$v_B = \sqrt{gR}.$$

Writing the energies of the particle at B and C, we have

$$T(B) = \tfrac{1}{2}mv_B^2 = \tfrac{1}{2}mgR, \qquad V(B) = mg2R,$$

$$T(C) = \tfrac{1}{2}mv_C^2, \qquad V(C) = 0.$$

From (4.16), we obtain

$$\tfrac{1}{2}mv_C^2 = \tfrac{1}{2}mgR + 2mgR = \tfrac{5}{2}mgR.$$

Thus

$$v_C^2 = 5gR. \tag{c}$$

Figure 4.30(c) shows free-body diagrams at points $C-$ and $C+$. Figure 4.30(d) depicts the accelerations at $C-$ and $C+$.

Writing $\mathbf{F} = \dot{\mathbf{p}}$ at each point, we get:

At $C-$: $F_r = ma_r,$ $-N + mg = -m\dfrac{v_C^2}{R}.$ (d)

At $C+$: $F_y = ma_y,$ $N - mg = 0.$ (e)

From (c) and (d), we get

At $C-$: $N = mg + 5mg = 6mg,$

At $C+$: $N = mg.$

Thus

$$N(C-) = 600 \text{ lb,} \qquad N(C+) = 100 \text{ lb.}$$

Example 23 (The sliding girl)

A shy young coed is courted by two men: an arts major and an engineering student. The arts man has taken to reading *A l'Ombre des Jeunes Filles en Fleurs*—in the original. She understands neither the language nor the thoughts, but seems to be getting the message. To counteract this flowery foolishness, the engineer devises a plan of his own. He installs slippery seat covers in his car and removes the interior door handles from the right-hand side of his car. He then plots a course in his car which takes him on nothing but right-hand turns. You are asked to analyze the coed's motion.

Solution

Figure 4.31(a) indicates the geometry of the problem.

We begin by constructing a free-body diagram of the girl, here represented (with apologies) by a particle of mass m (Fig. 4.31b).

Here N_1 is the normal force of the back cushion against m, and f_1 is the friction due to the back cushion. N_2 is the normal force of the seat cushion and f_2 the corresponding friction force. Let us assume that the back cushion is vertical, the seat is horizontal, and the coefficient of sliding friction is μ. Then

$$f_1 = \mu N_1, \qquad f_2 = \mu N_2.$$

For the acceleration of m, we are advised to use polar coordinates (see Fig. 4.31c).

Now, writing $\mathbf{F} = \dot{\mathbf{p}}$, we get

$$\left.\begin{aligned}
F_r = ma_r: & \quad m(\ddot{r} - r\dot{\theta}^2) = -f_1 - f_2, \\
F_\theta = ma_\theta: & \quad m(r\ddot{\theta} + 2\dot{r}\dot{\theta}) = N_1, \\
\text{where } f_1 = \mu N_1, & \quad f_2 = \mu N_2
\end{aligned}\right\} \tag{a}$$

and, by vertical equilibrium,

$$N_2 - mg = 0.$$

Let us suppose that the car has a constant speed v_0. Then

$$\dot{\theta} = \frac{v_0}{R} = \text{constant} \quad \text{and} \quad \ddot{\theta} = 0. \tag{b}$$

From (a) and (b), we obtain finally

$$m\left[\ddot{r} - r\left(\frac{v_0}{R}\right)^2\right] = -2\mu m\left(\frac{v_0}{R}\right)\dot{r}$$

or

$$\ddot{r} + 2\mu\left(\frac{v_0}{R}\right)\dot{r} - \left(\frac{v_0}{R}\right)^2 r = -\mu g. \tag{c}$$

Equation (c) is the differential equation for the motion of the girl, $r(t)$. We specify initial conditions:

$$\text{At } t = 0, \qquad r = R_0, \qquad \dot{r} = 0, \qquad \theta = 0. \tag{d}$$

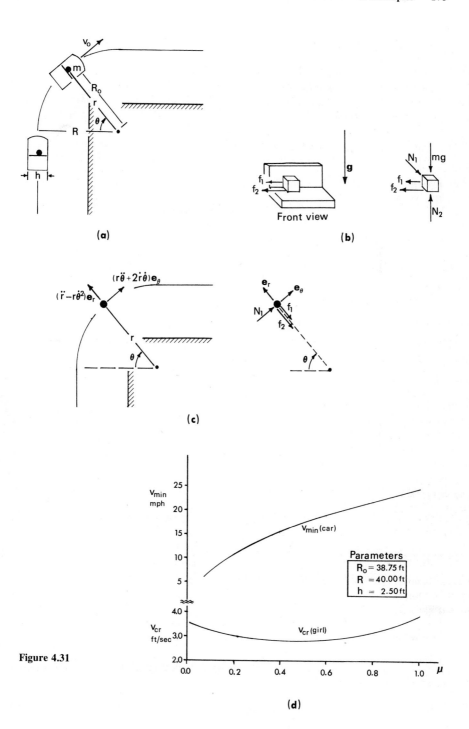

(a)

(b)

Front view

(c)

(d)

Figure 4.31

[Note that we assume that the damsel does not push off from her side of the car! Also note that neither (c) nor (d) in any way involves m. Thus the same analysis would apply to any other girl without regard to her tonnage, as long as the coefficient of sliding friction μ is the same.]

In Chapter 6 we shall talk in some detail about how to solve equations such as (c). Here we note only that

$$r(t) = (\mu g R^2/v_0^2) + e^{-[\mu(v_0/R)]t} \left(C_1 \cosh \frac{v_0}{R} \sqrt{1+\mu^2}\, t + C_2 \sinh \frac{v_0}{R} \sqrt{1+\mu^2}\, t \right) \quad \text{(e)}$$

is a solution to (c) for arbitrary values of C_1 and C_2. [To convince yourself that this is true, substitute (e) into (c).]

Requiring that (e) meet the initial conditions (d), we obtain

$$r(t) = (\mu g R^2/v_0^2) + e^{-\mu(v_0/R)t} (R_0 - \mu g R^2/v_0^2) \left(\cosh \frac{v_0}{R} \sqrt{1+\mu^2}\, t \right.$$

$$\left. + \frac{\mu v_0/R}{\sqrt{1+\mu^2}} \sinh \frac{v_0}{R} \sqrt{1+\mu^2}\, t \right). \quad \text{(f)}$$

For the case in which $\mu \to 0$, we get

$$r(t) = R_0 \cosh \frac{v_0}{R}\, t. \quad \text{(g)}$$

The above solutions (f) or (g) apply from

$$0 \leqslant t \leqslant \left(\frac{\pi R}{2 v_0} \right), \quad \text{(h)}$$

the time required to turn the corner. Will the engineer get the girl? If $\mu = 0$, it's just a matter of time. For once she is set in motion there is, so to speak, no stopping her. On the other hand, if $\mu \neq 0$, the question is more difficult to answer.

In fact, the main problem is not that which you might first suppose. If the girl slides at all, she will probably make it to the other side of the car well before the car gets around the corner. This is true primarily because $h \leqslant 3$ ft.

There is a minimum speed, $v_{\min}$, below which the girl will have no motion relative to the car. From (a), we see that the normal force N_1 is zero unless the girl is sliding ($\dot{r} \neq 0$). Thus the total static friction on the girl is

$$f_2 \leqslant \mu mg.$$

It is a simple exercise to show that in the static case

$$f_2 = m \frac{v^2}{R_0}.$$

Thus

$$m \frac{v^2}{R_0} \leqslant \mu mg \qquad \text{or} \qquad v \leqslant \sqrt{\mu R_0 g}.$$

In other words, the minimum speed for slipping is

$$v_{min} = \sqrt{\mu R_0 g}. \tag{i}$$

For a given corner R_0, the more slippery the seat covers, the more slowly the car can proceed around the corner. That is not the end of the problem, however. If the car is driven at v_{min}, the girl will reach the other side of the car. The question remains: What will her speed $\dot{r}$ be when she gets to the other side of the car? If this speed is too high, both girl and driver may find themselves on the pavement.

Figure 4.31(d) shows a plot of two curves: (i) v_{min} versus μ, (ii) v_{cr} versus μ, where v_{cr} is the value of $\dot{r}$ when $r = R_0 + h$, and is computed for the case in which the car is driven at speed v_{min}.

....

8. THE FREE-BODY DIAGRAM (AGAIN)

In this chapter, we have developed equations that apply to the description of the motion of a particle. The first step in analyzing a problem in dynamics is to construct a free-body diagram. Before we go on to other topics, let's reemphasize this step. Once again, we define a free-body diagram for a particle as follows.

> The free-body diagram of a particle includes all physically identifiable forces acting on the particle.

When we say that a force is "physically identifiable," we mean that we can identify its origin; for example, the force of gravity, a pin reaction, a friction force, a spring or dashpot force, etc. A force *should not* be postulated on the basis of its supposed effect. Let's look at an example which shows what we mean.

In Chapter 5, we shall discuss problems of central force motion. We give here a (mistaken) analysis which many students make of a problem involving motion caused by a central force.

Consider two particles in a plane, one fixed to the origin and the other free to move under the influence of the gravitational pull of the first (Fig. 4.32a).

We use polar coordinates to describe the motion. The acceleration of m_2 is then

$$\mathbf{a} = (\ddot{r} - r\dot{\theta}^2)\mathbf{e}_r + (r\ddot{\theta} + 2\dot{r}\dot{\theta})\mathbf{e}_\theta.$$

We now construct a free-body diagram (note that, in this analysis, we construct the free-body diagram *after* we discuss the acceleration of m_2). There is a force along $\mathbf{e}_r$ due to gravity. "But," says a casual analyst, "look at the acceleration component in the $\mathbf{e}_\theta$ direction:

$$a_\theta = (r\ddot{\theta} + 2\dot{r}\dot{\theta}). \tag{a}$$

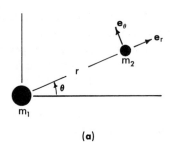

(a)

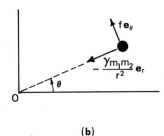

(b)

Figure 4.32

All those terms must come from somewhere! And besides, if m_2 is going around m_1, all the terms r, $\ddot{\theta}$, $\dot{r}$, and $\dot{\theta}$ can be nonzero. Let's call the force which caused this acceleration, $f\mathbf{e}_\theta$." The analyst mistakenly concludes that our free-body diagram must be as shown in Fig. 4.32(b).

The equations of motion for m_2 are now dictated by Fig. 4.32(b):

$$F_r = ma_r: \qquad m_2(\ddot{r} - r\dot{\theta}^2) = -\frac{\gamma m_1 m_2}{r^2} ,$$

$$F_\theta = ma_\theta: \qquad m_2(r\ddot{\theta} + 2\dot{r}\dot{\theta}) = f.$$

(b)

However, we thoughtful people know that there is something suspect about the above analysis. The casual analyst has defined the force f in terms of its supposed effect, namely that a_θ is nonzero; but he has not specified a physical identity for f.

In actual fact, as we shall see in Chapter 5,

$$a_\theta = 0.$$

That is, the parameters r, $\ddot{\theta}$, $\dot{r}$, and $\dot{\theta}$ arrange themselves in such a manner that

$$r\ddot{\theta} + 2\dot{r}\dot{\theta} \equiv 0.$$

And this is precisely because there is *no force component* in the tangential direction.

There is no great secret to the construction of the free-body diagram, yet time and again students make mistaken analyses in dynamics (and in strength of materials and fluid mechanics, etc.) because they can't construct a proper free-body diagram. Forces are pushes and pulls. Someone or something has to do the pushing and pulling. If you can't identify the source of the push or pull, chances are that it simply is not there!

Now in order to point out one other way in which students commonly make mistakes in constructing a free-body diagram, let us return to the example. Someone might say, "You have left out centrifugal force." What's the origin of this force? "The stuff that pulls m out along e_r — you know, the force which keeps water from falling out of a bucket when it's being swung around in a vertical circle."

Note that once again the "force" is defined in terms of its effect, not its physical origin. In this case, however, the student is on a bit safer ground. To be sure, there is no physical force corresponding to what people call "centrifugal force." There is, however, a way of viewing dynamics in which concepts such as centrifugal "force" are properly employed. This is commonly referred to as *d'Alembert's principle.**

From a pedagogical point of view, the problem lies in the fact that students have a hard time distinguishing between a nonforce like $f e_\theta$ of the example and a valid d'Alembert force. Until you have learned to identify physical forces in a wide variety of problems, don't try to use d'Alembert's principle.

9. SUMMARY

A) Equations of Particle Motion

i) $\mathbf{F} = \dot{\mathbf{p}}, \qquad \mathbf{p} = m\mathbf{v},$

$\mathbf{F}$ = total force on m.

$\mathbf{r}$ measured with respect to a fixed point of an inertial reference frame.

Thus $\mathbf{F} = m\ddot{\mathbf{r}}$.

ii) $\mathbf{M}_O = \dot{\mathbf{H}}_O, \qquad \mathbf{M}_O = \mathbf{r} \times \mathbf{F},$

$\mathbf{M}_O$ = moment of the total force about some fixed point O.

$\mathbf{H}_O = \mathbf{r} \times \mathbf{p} = \mathbf{r} \times m\mathbf{v}$

$\mathbf{H}_O$ = moment of the linear momentum about O.

iii) *Work–Energy Relation*
$E(B) = E(A) + W_{\text{non}},$

$E = T + V$ = total mechanical energy.

$$W_{\text{non}} = \int_A^B \mathbf{F}_{\text{non}} \cdot d\mathbf{r} = \text{work of nonconservative forces.}$$

---

* In actual fact, d'Alembert's principle is a variational statement.

B) Forces in Dynamics

Force		Potential energy
Gravity*	$\mathbf{F} = -\dfrac{\gamma m_1 m_2}{\rho^2}\,\mathbf{e}_\rho$	$V = -\dfrac{\gamma m_1 m_2}{\rho}$
Gravity	$\mathbf{F} = -mg\mathbf{k}$	$V = mgh$
Linear spring	$\mathbf{F} = -k\,\delta\,\mathbf{e}_\rho$	$V = \tfrac{1}{2}k\,\delta^2$
Sliding friction	$\mathbf{F} = -\mu N\,\mathbf{e}_t$	Nonconservative
Linear dashpot	$\mathbf{F} = -c\,\dot{\delta}\,\mathbf{e}_\rho$	Nonconservative
Conservative force	$\mathbf{F} = \nabla\psi, \qquad \nabla \times \mathbf{F} = 0$	$V = -\psi$

C) Free-Body Diagram

The free-body diagram of a particle includes all physically identifiable forces on the particle.

D) Mathematical Formulation of Dynamics Problems

The equations of motion for a particle are generally in the form of second-order differential equations, except in those cases in which a first integral to the problem (total energy, momentum, or moment of momentum) exists. Solutions of the equations involve arbitrary constants which must be specified in terms of initial position and velocity, or alternative, equivalent, quantities (if energy is conserved, one constant could be the total energy, E).

PROBLEMS

Section 4.2

Problem 4.1

The total force on a particle whose mass $m = 1.0$ slug is

$$\mathbf{F} = \sin t\,\mathbf{i} + \cos t\,\mathbf{j} \text{ lb.}$$

At $t = 0$, $\mathbf{r} = 0$ and $\mathbf{v} = 10\,\mathbf{k}$ ft/sec.

    a) Find $\mathbf{v}$ as a function of time.

    b) Note that v_z is constant. Explain why.

    c) Find $\mathbf{r}$ as a function of time.

Problem 4.2

The total force on a particle which weighs 8.0 lb is

$$\mathbf{F} = -5 \sin 2t\,\mathbf{i} - 8\,\mathbf{k} \text{ lb.}$$

---

* For the earth, $\gamma m_e = 1.255 \times 10^{12}$ miles^3/hr^2.

At time $t = 0$, $\mathbf{r} = 0$ and $\mathbf{v} = 0$.

    a) Find the velocity $\mathbf{v}$ at time t.

    b) Find the position $\mathbf{r}$ at time t.

Problem 4.3

A particle slides down a 45° incline (see figure). The coefficient of sliding friction is $\mu = 0.5$. Show that the time required for the particle to slide 20 ft is 1.88 sec.

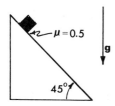

Figure 4.33

Problem 4.4

Consider the incline in Problem 4.3. Again the coefficient of sliding friction is $\mu = 0.5$. One particle is given an initial speed of 15.92 ft/sec down the incline. A second particle (of mass 1.0 slug), which is released from rest at the same position as the first particle, is subject to a 50-lb force (down the incline).

    a) Find the time required for the second particle to catch up with the first particle.

    b) Find the position at which the second particle catches up with the first particle.

Problem 4.5

Again consider the incline in Problem 4.3. Again the coefficient of sliding friction is $\mu = 0.5$. A particle which is released with initial speed v_0 slides down the 20-ft incline in 0.94 sec. Determine v_0.

Problem 4.6

A particle is released from rest at the top of the incline shown in the figure. The coefficient of sliding friction is μ.

    a) Find the distance x which the particle slides in t_0 seconds.

    b) Find the speed of the particle after it has slid a distance x.

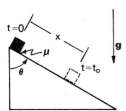

Figure 4.34

Problem 4.7

Suppose the speed $v = dx/dt$ is known as a given function of x (from energy, for example): $v = f(x)$. Given that $x = x_0$ at $t = t_0$, show that x can be determined as a function of time from

$$t = t_0 + \int_{x_0}^{x} \frac{dx}{f(x)}.$$

Problem 4.8

Suppose it is known that $v = -10x$ ft/sec, where $v = dx/dt$. Given that $x = 5$ ft at $t = 0$, find $x(t)$. [*Hint:* See Problem 4.7.]

Problem 4.9

Suppose that $v = x + \frac{1}{4}x^3$ ft/sec, where $v = dx/dt$, and that $x = 1$ ft when $t = 0$. Find $x(t)$. [*Hint:* Use Problem 4.7 and integrate, using the method of partial fractions.]

Problem 4.10

Suppose that $v = -x^3$ ft/sec, where $v = dx/dt$ and $x = 1$ at $t = 0$. Find $x(t)$. [*Hint:* Use Problem 4.7.]

Problem 4.11

Suppose that the acceleration $a = d^2x/dt^2$ is given as a known function of the speed $v = dx/dt$: $a = g(v)$. Show that the speed can be determined as a function of time from

$$t = t_0 + \int_{v_0}^{v} \frac{dv}{g(v)},$$

where v_0 is the speed at some instant t_0. Show further that if the above equation determines $v = h(t)$, then $x(t)$ can be found from

$$x(t) = x_0 + \int_{t_0}^{t} h(t) \, dt,$$

where $x = x_0$ at $t = t_0$.

Problem 4.12

Suppose it is given that $a = -2v$, where $a = \ddot{x}$ ft/sec^2, $v = \dot{x}$ ft/sec, and at $t = 0$, $\dot{x} = 10$ ft/sec and $x = -2$ ft. Find $\dot{x}(t)$ and $x(t)$. [*Hint:* Use Problem 4.11.]

Problem 4.13

Suppose that $a = -v^3$ ft/sec^2, where $a = \ddot{x}$ and $v = \dot{x}$. Given that $\dot{x} = 2$ ft/sec and $x = 10$ ft at $t = 0$, find $x(t)$. [*Hint:* Use Problem 4.11.]

Problem 4.14

Consider the pendulum in Fig. 4.35. Using x and y as coordinates, show that the equations of motion for the system are

$$m\ddot{x} = -N(x/l),$$
$$m\ddot{y} = -mg + N(1 - y/l),$$
$$(x/l)^2 + (1 - y/l)^2 - 1 = 0,$$

where N is the tension in the string.

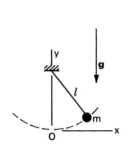

Figure 4.35

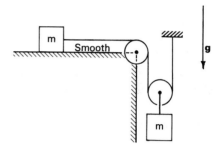

Figure 4.36

Problem 4.15

In the system shown in Fig. 4.36, the upper surface is frictionless and the pulleys are weightless. Each mass m is 491.7 slugs.

a) What is the acceleration of the upper mass?

b) What is the tension in the upper portion of the string?

Problem 4.16

For the system shown in Fig. 4.37, the mass $m_1 = 2$ slugs passes over a smooth rod at a constant speed of $v_0 = 20$ ft/sec under the influence of the force P. The mass m

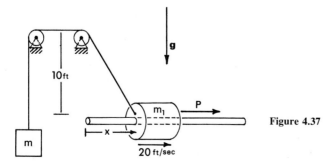

Figure 4.37

(which weighs 64 lb) is connected to m_1 by a weightless rope which passes over weightless pulleys. Find the force P when $x = 10$ ft.

Problem 4.17

A particle of mass $m = 1$ slug, which is on the end of a string attached to point O (see figure), is released from rest when it is at point A. Given that the string breaks under a tension of 50 lb, find the angle θ at which it breaks.

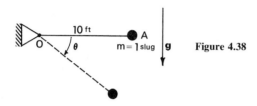

Figure 4.38

Problem 4.18

The x, y, z axes in Fig. 4.39 are fixed in space. The (x', z) plane rotates about the z axis at the constant angular rate $\dot{\theta} = \omega_0 = $ constant. A smooth curve which is inserted in the moving plane with one end through the origin has a shape defined by the condition that a bead which is placed at any point on the curve will remain where it was placed.

a) Find the curve $z = f(x')$.

b) Suppose that the bead is given an initial speed along the curve v_0 at the origin. Find the subsequent speed of the bead along the curve.

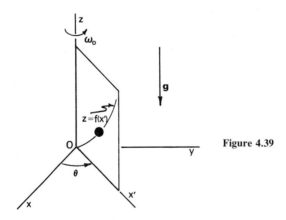

Figure 4.39

Problem 4.19

A pendulum of length l is mounted on a car which has a constant acceleration a_0 (see Fig. 4.40).

a) Show that the acceleration of the mass of the pendulum is

$$\mathbf{a} = (a_0 \sin \theta - l\dot{\theta}^2)\mathbf{e}_r + (l\ddot{\theta} + a_0 \cos \theta)\mathbf{e}_\theta.$$

b) Write a differential equation from which $\theta(t)$ could be determined.

c) Determine the angle about which the pendulum oscillates. [*Hint:* Find the angle at which $\ddot{\theta} = 0$.]

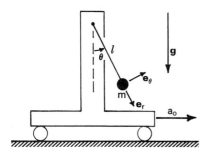

Figure 4.40

Section 4.3

Problem 4.20

The equation of motion of a simple pendulum is

$$\ddot{\theta} + (g/l) \sin \theta = 0.$$

a) Write $\omega = \dot{\theta}$ and then

$$\ddot{\theta} = \frac{d\omega}{dt} = \frac{d\omega}{d\theta}\frac{d\theta}{dt} = \omega\frac{d\omega}{d\theta}.$$

Assuming that $\dot{\theta} = \omega_0$ at $\theta = 0$, find ω as a function of θ.

b) Find the angle at which $\dot{\theta} = 0$.

Problem 4.21

Repeat Problem 4.20, using the approximation

$$\sin \theta \approx \theta.$$

Problem 4.22

The pendulum in Fig. 4.41 is modified by the barrier placed at a distance a from the support at O.

a) Write the equation of motion for $\theta > 0$.

b) Write the equation of motion for $\theta < 0$.

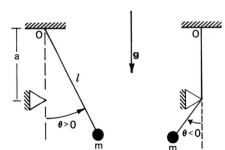

Figure 4.41

Problem 4.23

Consider the system in Problem 4.22. Suppose that the mass has the speed v_0 when the pendulum is vertical.

a) Find the maximum angular displacement if the mass swings to the right $(\theta > 0)$.

b) Find the maximum negative angular displacement if the mass swings to the left.

[*Hint:* Recall Problem 4.20.]

Problem 4.24

The system in Fig. 4.42 consists of a particle at the end of a string which passes through a hole in a smooth horizontal table. The string is pulled in at a constant rate $\dot{r} = -v_0 = \text{const.}$ It is given that $\dot{\theta} = 10$ rad/sec when $r = 10$ ft and that the particle weighs 16 lb.

a) Find the force P when $r = 10$ ft.

b) Find the force P when $r = 1$ ft.

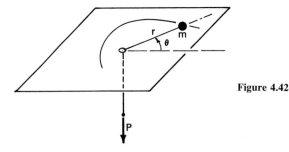

Figure 4.42

Problem 4.25

A particle m is restricted to move in the x,y plane, as shown in Fig. 4.43. The components of position and force on m are x,y and F_x, F_y.

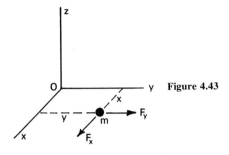

Figure 4.43

a) Write expressions for H_{Ox}, H_{Oy}, H_{Oz}.

b) Write expressions for M_{Ox}, M_{Oy}, M_{Oz}.

c) Write the equation $\mathbf{M}_O = \dot{\mathbf{H}}_O$ in component form.

Problem 4.26

The components of the position of a particle and of the forces on it are x, y, z and F_x, F_y, and F_z (see Fig. 4.44).

a) Write expressions for H_{Ox}, H_{Oy}, H_{Oz}.

b) Write expressions for M_{Ox}, M_{Oy}, M_{Oz}.

c) Write the equation $\mathbf{M}_O = \dot{\mathbf{H}}_O$ in component form.

d) Suppose that $M_{Oz} = 0$. Show that H_{Oz} is constant.

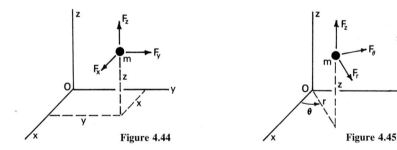

Figure 4.44   Figure 4.45

Problem 4.27

A particle has the components of position r and z (cylindrical coordinates). The force components applied to the particle are F_r, F_θ, and F_z (see Fig. 4.45).

a) Write expressions for the components H_{Or}, $H_{O\theta}$, H_{Oz}.

b) Write expressions for the components M_{Or}, $M_{O\theta}$, M_{Oz}.

c) Write the component equations for $\mathbf{M}_O = \dot{\mathbf{H}}_O$.

d) Suppose that $M_{Oz} = 0$. Show that $r^2 \dot{\theta} = \text{constant}$.

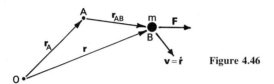

Figure 4.46

Problem 4.28

Suppose O is a fixed point and A is a moving point. We define $\mathbf{M}_A = \mathbf{r}_{AB} \times \mathbf{F}$ and $\mathbf{H}_A = \mathbf{r}_{AB} \times m\mathbf{v}$, where $\mathbf{v}$ is the velocity of m relative to O (see the figure). Determine the conditions under which we can write $\mathbf{M}_A = \dot{\mathbf{H}}_A$.

Problem 4.29

Suppose that $\mathbf{H}_{Ar}$ is defined to be the relative moment of momentum about A: $\mathbf{H}_{Ar} = \mathbf{r}_{AB} \times m\dot{\mathbf{r}}_{AB}$. (See the figure accompanying Problem 4.28.) Determine the conditions under which we can write $\mathbf{M}_A = \dot{\mathbf{H}}_{Ar}$.

Problem 4.30

Consider the conical pendulum shown in the figure.

    a) Write the components (cylindrical coordinates) of the equation $\mathbf{M}_O = \dot{\mathbf{H}}_O$.

    b) In particular, show that $r^2\dot{\theta} = (l \sin \phi)^2\dot{\theta}$ is constant.

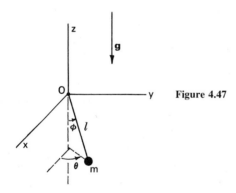

Figure 4.47

Problem 4.31

Suppose that the conical pendulum of Problem 4.30 has the trajectory $x^2 + y^2 = R^2$.

    a) Calculate the quantity $\dot{\theta}$ as a function of R, l, and g.

    b) Compute the amount of time it takes the pendulum to make one complete circuit about the trajectory.

Problem 4.32

Suppose that the length of the string attached to the conical pendulum in Problem 4.30 is 10 ft long, and that the mass of the pendulum travels in a circular path. One cycle around the path takes 0.5 sec. Find the radius of the circle. [*Hint:* See Problem 4.31.]

Section 4.4

Problem 4.33

A boy uses a force of 10 lb at an angle of 30° to pull the wagon in the figure a distance of 100 ft.

a) Compute the work done by the boy.

b) Suppose that the boy pulls the wagon at an angle of 10°. Compute the work done in moving the wagon 100 ft.

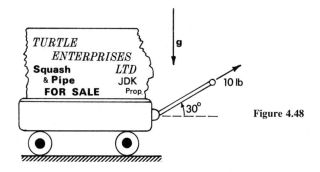

Figure 4.48

Problem 4.34

For the system in Problem 4.16, compute the work done by the force P in moving the mass m_1 from $x = 0$ to $x = 10$ ft. [*Hint:* Compute the change in energy of the system.]

Problem 4.35

Consider the system in Problem 4.24.

a) Compute the work done by the force P in moving the particle from $r = 10$ ft to $r = 1$ ft.

b) Evaluate the increase in kinetic energy of the particle in going from $r = 10$ ft to $r = 1$ ft.

Problem 4.36

Consider the force field

$$\mathbf{F} = \mathbf{i}(xy^2) + \mathbf{j}(x^2y + \sin y) \text{ lb.}$$

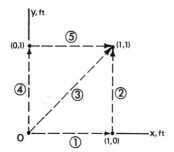

Figure 4.49

A particle moves along paths (1), (2), (3), (4), and (5) indicated in the figure. Compute the work done on the particle during its motion over each path.

Problem 4.37

A force field is given by

$$\mathbf{F} = \left(\frac{10}{x^2 + y^2}\right)(x\mathbf{i} + y\mathbf{j}) \text{ lb.}$$

Compute the work which the force field does on a particle which moves over paths (1), (2), and (3) shown in the figure.

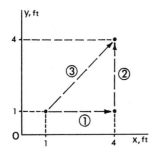

Figure 4.50

Problem 4.38

A force field is expressed in cylindrical coordinates

$$\mathbf{F} = F_r \, \mathbf{e}_r + F_\theta \, \mathbf{e}_\theta + F_z \, \mathbf{e}_z,$$

where F_r, F_θ, and F_z are functions of r, θ, and z. Show that the work done by the force field is

$$W_{AB} = \int_A^B [F_r \, dr + rF_\theta \, d\theta + F_z \, dz].$$

Problem 4.39

Compute the work done by the force field

$$\mathbf{F} = F_0 \left[\left(2x - \frac{y}{\sqrt{x^2 + y^2}} \right) \mathbf{i} + \left(2y + \frac{x}{\sqrt{x^2 + y^2}} \right) \mathbf{j} \right] \text{lb}$$

over the path from A to B indicated in the figure.

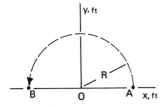

Figure 4.51

Problem 4.40

Suppose that a particle m moves in the x direction under the influence of the force F_x. Then $\mathbf{F} = \dot{\mathbf{p}}$ becomes

$$F_x = m\ddot{x}.$$

Write $v_x = \dot{x}$. Then show that

$$\ddot{x} = \frac{dv_x}{dt} = v_x \frac{dv_x}{dx}.$$

Given that the particle moves from x_0 to x, derive the relation

$$\int_{x_0}^{x} F_x \, dx = \tfrac{1}{2}mv_x^2 \Big|_{x_0}^{x}.$$

Problem 4.41

A particle m is restricted to move in the x direction. The forces on m are conservative, with total potential energy function $V(x)$. Let $v = \dot{x}$.

a) Let E be the total energy of the system. Sketch a representative plot of $\tfrac{1}{2}mv^2$ as a function of x.

b) From your plot in part (a), indicate (i) at what point the speed is a maximum, (ii) any bounds on the motion of the mass.

Problem 4.42

A particle has the following equation of motion:

$$m\ddot{x} + k_0 x + k_1 x^3 = 0.$$

Suppose that at $x = 0$, $v = \dot{x} = \sqrt{10}$ ft/sec. Let $m = 1$ slug and $k_0 = 10$ lb/ft. Using the methods of Problems 4.40 and 4.41, find the bounds on the motion of m if:

a) $k_1 = 4$ lb/ft³,

b) $k_1 = -4$ lb/ft³.

Problem 4.43

A particle whose mass $m = 2$ slugs is restricted to move in the horizontal x direction subject only to the spring force $f = -2x$ lb. Suppose that $\dot{x} = 2$ ft/sec when $x = 0$.

a) Let $v = \dot{x}$. Show that $v^2 = 4 - x^2$ (ft/sec)².

b) Plot v^2 from part (a) as a function of x.

c) Show that the maximum speed of the particle is 2 ft/sec, and show that the particle's motion is bounded by $x = \pm 2$ ft.

Problem 4.44

Suppose that a particle m is restricted to move in the x direction ($x \geqslant 1$ ft) under the influence of a force with potential energy function

$$V(x) = \frac{-4}{x} \text{ ft-lb.}$$

Using the technique of Problem 4.41, plot ($\frac{1}{2}mv^2$) as a function of x (where $v = \dot{x}$) for the following values of the total energy E:

a) $E = 0$ ft-lb

b) $E = +1$ ft-lb

c) $E = -2$ ft-lb

d) In each case, find (i) the maximum speed of the particle, (ii) any bound on the motion of the particle.

e) For the case in which the particle has no upper bound to its motion, it is said to "escape" from the force field represented by $V(x)$. Find the range of values of E for which the particle will escape.

Section 4.5

Problem 4.45

A force field has the potential energy function

$$V = 100(x^2 + xy + 2y) \text{ ft-lb}$$

a) Find the force components (i) F_x, and (ii) F_y as functions of x and y.

b) Determine the magnitude of the force field at $x = 1$ ft, $y = 2$ ft.

c) Determine the inclination of the force field at $x = 1$ ft, $y = 2$ ft. (That is, find the angle the force makes with the x axis.)

Problem 4.46

The potential energy of a force field is given by

$$V = \tfrac{1}{4}(x^4 - 4xy^3) \text{ ft-lb.}$$

a) Find the components of the force at the point (x,y). That is, find F_x and F_y.

b) Find the magnitude of the force at the point (x,y).

c) Find the orientation of the force (the angle the force makes with the x axis) at the points (i) $x = 0$, $y = -1$ ft, (ii) $x = 1$ ft, $y = 0$, (iii) $x = 1$ ft, $y = 2$ ft.

Problem 4.47

Suppose a potential energy function is given as

$$V = -\frac{2}{xy} \text{ ft-lb.}$$

a) Find the force components F_x and F_y.

b) Find the magnitude of the force at the point (x,y).

c) Find the angle the force at $x = 2$ ft, $y = 1$ ft makes with the positive x axis.

Problem 4.48

A force field has the following potential energy function expressed in cylindrical coordinates:

$$V = \tfrac{1}{2}r^2(1 + \cos^2 \theta) \text{ newton-m.}$$

a) Find the components of the force at the point (r,θ), F_r, and F_θ.

b) Find the magnitude of the force at the point (r,θ).

Problem 4.49

Consider the problem of Example 4.9. From the result

$$v = \sqrt{v_0^2 - 2gl(1 - \cos \theta)},$$

show that

$$t = \int_0^\theta \frac{l \, d\theta}{\sqrt{v_0^2 - 2gl(1 - \cos \theta)}},$$

where $\theta = 0$ when $t = 0$. Further, letting

$$\cos \theta \approx 1 - \tfrac{1}{2}\theta^2,$$

show that

$$\theta = \frac{v_0}{\sqrt{gl}} \sin \left(\sqrt{g/l}\, t\right).$$

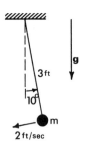

Figure 4.52

Problem 4.50

At $\theta = 10°$, the pendulum shown in the figure is given a speed of 2 ft/sec. The length of the pendulum is 3 ft and the mass m weighs 16 lb.

a) What is the maximum angular displacement of the pendulum?

b) Compute the force in the string when the pendulum passes through the lowest point of its trajectory.

Problem 4.51

The bead in Fig. 4.53 slides on the smooth curve shown. Determine the minimum speed v_0 at the point A so that the bead will eventually get to point B. Then, supposing that this is the speed at point A, find the speed at point B.

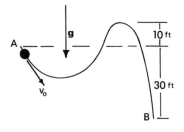

Figure 4.53

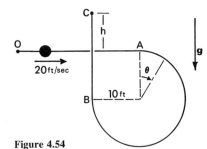

Figure 4.54

Problem 4.52

A bead which weighs 5 lb slides on the frictionless curve shown in Fig. 4.54. In the position shown, the speed of the bead is 20 ft/sec.

a) Find the normal force exerted by the curve on the bead on the following portions of the curve. (i) From 0 to A, (ii) at the angle θ between A and B, (iii) from B to C.

b) Find the height h above the original portion of the curve at which the speed of the bead becomes zero.

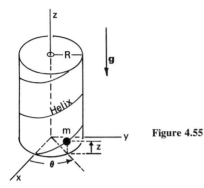

Figure 4.55

Problem 4.53

The bead shown in Fig. 4.55 slides down a frictionless helix,

$$r = R, \qquad z = k\theta.$$

The bead is released from rest at height $z = h$. Determine $\dot{z}$, the vertical component of the velocity, when $z = 0$.

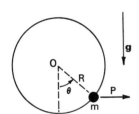

Figure 4.56

Problem 4.54

The bead m in Fig. 4.56 slides on the smooth circular wire, as a result of having a force P applied to it. This force P is *always horizontal* and has a constant magnitude (recall Example 11).

a) Write $\mathbf{M}_O = \dot{\mathbf{H}}_O$ for the bead. Determine the angle θ_2 at which $\ddot{\theta} = 0$.

b) Suppose that, at $\theta = 0$, $\dot{\theta} = 0$ (the situation in Example 11). Let

$$\theta_1 = \cos^{-1}\left[\frac{1 - (P/mg)^2}{1 + (P/mg)^2}\right].$$

Show that

$$\theta_1 = \tan^{-1}\left[\frac{2(P/mg)}{1 - (P/mg)^2}\right].$$

c) Show that $\theta_1 = 2\theta_2$. Interpret the result.

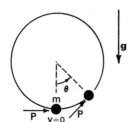

Figure 4.57

Problem 4.55

A particle slides on the smooth curve shown in Fig. 4.57. A force P, which is constant in magnitude and *always tangential to the curve*, is applied to the particle. The speed of the particle is zero at the bottom of the curve.

a) Find the magnitude of P so that the speed of the particle is zero again at $\theta = \frac{1}{2}\pi$ rad.

b) Assuming the value for P determined in part (a), find the angle θ at which $\ddot{\theta} = 0$.

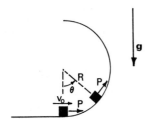

Figure 4.58

Problem 4.56

A particle moves along the frictionless path shown in Fig. 4.58. At point A, its speed is v_0. A force P, which has a constant magnitude and is *always tangential to the path*, is applied to the particle.

a) Compute the speed of the particle v as a function of the angle θ.

b) Suppose that $mg = 16$ lb and $R = 10$ ft. If $v_0 = 10$ ft/sec and $v = 0$ at $\theta = 90°$, determine the magnitude of the force P.

Problem 4.57

Modify Problem 4.56 in the following way: The force P is *always tangential to the path*, but the magnitude of P is an unspecified function of θ.

a) Determine the magnitude of P when the speed of the particle is constant along the entire trajectory.

b) Suppose that the speed of the particle is proportional to the angle θ, and that $v = 10\,\theta$ ft/sec. (In particular, $v_0 = 0$.) Given that $mg = 16$ lb and $R = 10$ ft, determine the magnitude of P as a function of θ.

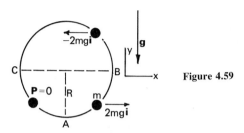

Figure 4.59

Problem 4.58

A bead of mass m moves on the smooth circular wire shown in Fig. 4.59, with a force P, which is *always horizontal*, applied to it. From A to B, $\mathbf{P} = 2mg\mathbf{i}$; from B to C, $\mathbf{P} = -2mg\mathbf{i}$; and from C to A, $\mathbf{P} = 0$. Given that the initial speed at A is $v = 0$, find the following:

a) the speed at B.

b) the speed at C.

c) the final speed at A.

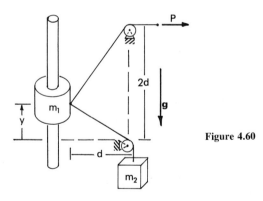

Figure 4.60

Problem 4.59

In the system shown in Fig. 4.60, the mass m_1 slides on a smooth, vertical rod. The pulleys and ropes are weightless.

a) For y, a given height of m_1 in the range $0 \leqslant y \leqslant d$, find the value of the force P required to maintain equilibrium. Assume $m_1 = m_2 = m$.

b) Suppose that the system is at rest when $y = 0$. Then a force $P = 30$ lb is applied. Given that $m_1 = m_2 = 0.3$ slug and $d = 10$ ft, find the speed of m_1 when $y = d$.

Problem 4.60

A particle which weighs 8 lb slides on the smooth curve shown in the figure. The curve has a parabolic shape, $y = x^2$. At the bottom of the curve, the particle has speed $v_0 = 8$ ft/sec.

a) Find the speed of the particle at the point (x,y) of the curve.

b) Find the x component of the velocity of the particle when it is at the position (x,y) of the curve.

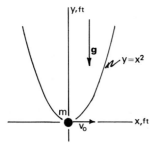

Figure 4.61

Problem 4.61

For the system in Problem 4.60, determine the normal force exerted by the curve on the particle at (a) $x = 0$, (b) $x = \frac{1}{2}\sqrt{3}$ ft.

Section 4.6

Problem 4.62

Consider the force of gravity

$$\mathbf{f} = -\frac{\gamma m m_e}{r^2}\,\mathbf{k}.$$

Let $r = R + h$, where R is the radius of the earth. Write the potential energy for $\mathbf{f}$, and show that $V \to mgh$ for small values of (h/R).

Problem 4.63

You are given the following values of g, γ, and R: $g = 9.80$ m/sec^2, $\gamma = 6.68 \times 10^{-11}$ m^3/(kg-sec^2), $R = 4 \times 10^3$ miles.

a) Compute the mass of the earth in kilograms.

b) Compute the mass of the earth in slugs.

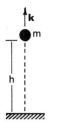

Figure 4.62

Problem 4.64

The particle shown in Fig. 4.62 is released from rest at a height h above the earth. Assuming that the force of gravity is $-mg\mathbf{k}$, find the speed of the particle when it reaches the surface of the earth.

Problem 4.65

A particle is released from rest at a height h above the earth. Assuming that the force of gravity is $-(\gamma mm_e/r^2)\mathbf{k}$, determine the speed of the particle when it reaches the surface of the earth. (See Fig. 4.62.) [*Hint:* At the initial instant $r = R + h$, where R is the radius of the earth.]

Problem 4.66

The speeds determined in Problems 4.64 and 4.65 were

$$v = (2gh)^{1/2} \quad \text{and} \quad v = \left(\frac{2\gamma m_e h}{R^2 + Rh}\right)^{1/2}.$$

Suppose that the second expression is taken to be exact. Show that the percentage error involved in using the first expression is approximately $50(h/R)\%$ for small values of h.

Problem 4.67

The bead m slides on the smooth wire in the shape $y = \sqrt{x}$ ft (see Fig. 4.63). A string which passes through the point A at $(2,0)$ exerts a force on m which is proportional to the distance between m and the point A ($f = -k\overline{AP}$). The motion takes place in the horizontal plane and the bead is released from rest at O.

a) Determine the speed of m as a function of x.

b) Compute the maximum speed and the point at which it occurs.

Problem 4.68

Consider again the system in Problem 4.67. Change A to the point $(0.4,0)$. Suppose that m is released from rest at the point $(1,1)$.

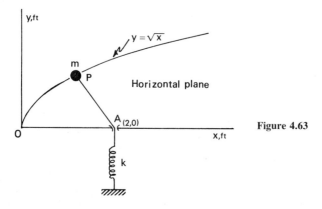

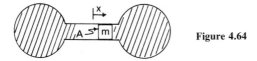

Figure 4.63

a) Find the point on the curve $y = \sqrt{x}$ at which the speed of m is maximum.

b) Suppose that m is released from rest at the point $(0,0)$. Describe the subsequent motion.

Problem 4.69

A particle m slides in a smooth tube (whose cross-sectional area is A) between two identical pressure vessels (see the figure). When the particle is at $x = 0$, the pressure and volume on both sides of it are p_0 and V_0. Suppose that both pressure vessels are isothermal, that is, that $pV = \text{constant}$.

Figure 4.64

a) Determine the total force on m when $x \neq 0$.

b) Determine a potential energy function for the total force on m.

Problem 4.70

Consider again the system in Problem 4.69. Suppose that the only change is that the two pressure vessels are insulated, so that pressure changes are adiabatic: $pV^\gamma = \text{constant}$, $\gamma > 1$.

a) Determine the total force on m when $x \neq 0$.

b) Determine a potential energy function for the total force on m.

Problem 4.71

The pressure vessel shown in Fig. 4.65 has smooth inner walls, and contains a gas which is isothermal. That is, the temperature of the gas is assumed to be a constant.

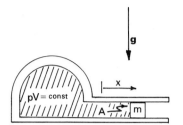

Figure 4.65

Thus

$$p_0 V_0 = pV.$$

Assume that the pressure outside the tube is zero. Consider the force exerted by the gas on the mass m.

a) Suppose that p_0 and V_0 are the pressure and volume of the gas when $x = 0$, and A is the cross-sectional area of the outlet tube. Write the expression for the force on m as a function of x.

b) Write a potential energy function for the force in (a).

c) Given that m has zero speed when $x = 0$, find the speed of m when $x = l$.

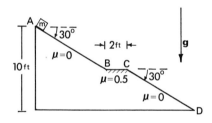

Figure 4.66   Figure 4.67

Problem 4.72

A particle m is released from rest at point A on the incline shown in Fig. 4.66. Determine the speed of the particle at point D. The paths from $A \rightarrow B$ and $C \rightarrow D$ are frictionless, and the coefficient of sliding friction on the path $B \rightarrow C$ is $\mu = 0.5$.

Problem 4.73

A particle m is released from rest at point A on the incline shown in Fig. 4.67. Determine the speed of the particle at point D. The paths from $A \rightarrow B$ and $C \rightarrow D$ have a coefficient of sliding friction $\mu = 0.5$. The path from $B \rightarrow C$ is frictionless.

Problem 4.74

A particle m slides down a frictionless incline, starting from rest at the top. The incline has height H and a variable length x (see Fig. 4.68). A spring at the end of the incline is designed to stop the particle in a deflection of δ_0 or less. Suppose that $H = 20$ ft, $mg = 10$ lb, $\delta_0 = 2$ ft, and $k = 105$ lb/ft. Determine the minimum length x.

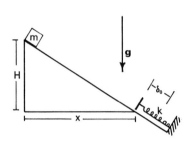
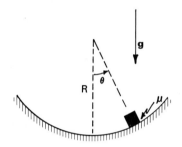

Figure 4.68   **Figure 4.69**

Problem 4.75

A particle slides on a vertical, circular path of radius R. The coefficient of sliding friction is μ (see Fig. 4.69). Write the equation of motion for the system, assuming that (a) $\dot{\theta} > 0$, (b) $\dot{\theta} < 0$.

Problem 4.76

Suppose that ψ is some function $\psi = \psi(x,y,z)$. Show that $\nabla \times (\nabla \psi) = 0$.

Problem 4.77

A force field is given by

$$\mathbf{f} = -\alpha(x^2 + y^2)(x\mathbf{i} + y\mathbf{j}) \text{ lb,}$$

where $\alpha = 10$ lb/ft³.

a) Determine whether the force field is conservative, and if so, determine the potential energy V.

b) A particle which weighs 32 lb is released from rest at the point $x = 1$ ft, $y = 1$ ft. Find the speed of the particle when it reaches the origin, $x = y = 0$.

c) Suppose that the particle in part (b) is released from rest at the point $x = 0$, $y = \sqrt{2}$ ft. Find the speed of the particle if it reaches the origin, $x = y = 0$.

Problem 4.78

Consider a plane central force field

$$\mathbf{F} = f(x,y)(x\mathbf{i} + y\mathbf{j}) \text{ lb.}$$

a) Show that the condition for the force $\mathbf{F}$ to be conservative is

$$y \frac{\partial f}{\partial x} - x \frac{\partial f}{\partial y} = 0.$$

b) Show that, if $f(x,y)$ is any function of $r = \sqrt{x^2 + y^2}$, $\mathbf{F}$ is a conservative force.

Problem 4.79

A force field is called a "central force field" if

$$\mathbf{F} = f(x,y,z)(x\mathbf{i} + y\mathbf{j} + z\mathbf{k}),$$

where $f(x,y,z)$ is some function of space.

a) Show that $\mathbf{F}$ always passes through the point $x = y = z = 0$.

b) Show that $\mathbf{F}$ is a conservative force field if

$$z \frac{\partial f}{\partial y} - y \frac{\partial f}{\partial z} = 0, \qquad x \frac{\partial f}{\partial z} - z \frac{\partial f}{\partial x} = 0, \qquad y \frac{\partial f}{\partial x} - x \frac{\partial f}{\partial y} = 0.$$

c) Show that if $f(x,y,z)$ is any function of $\rho = \sqrt{x^2 + y^2 + z^2}$, the central force $\mathbf{F}$ will be conservative.

Problem 4.80

Suppose that the potential energy function of a force field is expressed in cartesian coordinates:

$$\left.\begin{array}{ll}
\text{a) } V = \rho^n, & n \neq 0 \\
\text{b) } V = e^{-n\rho}, & n > 0 \\
\text{c) } V = e^{+n\rho}, & n > 0 \\
\text{d) } V = \sin(\rho^n), & n \neq 0.
\end{array}\right\} \qquad \text{where } \rho = \sqrt{x^2 + y^2 + z^2}$$

In each case, find the components of the force at the point (x,y,z); F_x, F_y, and F_z; show that the force lies on a line passing through the origin; and find the magnitude of the force as a function of ρ.

Problem 4.81

A central force field is expressed in cylindrical coordinates:

$$\mathbf{F} = f(r,\theta,z)(r\,\mathbf{e}_r + z\,\mathbf{k}).$$

a) Show that $\mathbf{F}$ will be a conservative force field if

$$\frac{\partial f}{\partial \theta} = 0 \qquad \text{and} \qquad z \frac{\partial f}{\partial r} - r \frac{\partial f}{\partial z} = 0.$$

b) Show that if $f(r,\theta,z)$ is any function of $\rho = \sqrt{r^2 + z^2}$, the force $\mathbf{F}$ is conservative.

Problem 4.82

Suppose the potential energy function of a force field is expressed in cylindrical coordinates:

a) $V = \rho^n$, $\quad n \neq 0$

b) $V = e^{-n\rho}$, $\quad n > 0$

c) $V = e^{+n\rho}$, $\quad n > 0$

d) $V = \cos(\rho^n)$, $\quad n \neq 0$.

where $\rho = \sqrt{r^2 + z^2}$

In each case, find the components of the force at the point (r,θ,z): F_r, F_θ, and F_z; show that the force lies on a line passing through the origin; and write the magnitude of the force as a function of ρ.

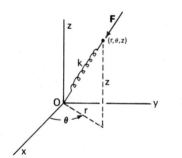

Figure 4.70

Problem 4.83

A spring k is stretched from an undeformed length l_0 to the position (r,θ,z) (see the figure).

a) Write the expression for the potential energy in the spring.

b) Derive the components of the force exerted by the spring: F_r, F_θ, and F_z.

[*Hint:* See Appendix A1, Eq. (A1.17b).]

Problem 4.84

A force field has the following components in spherical coordinates

$$\mathbf{F} = [-k(\rho - l_0) - mg \cos \phi]\, \mathbf{e}_\rho + mg \sin \phi \, \mathbf{e}_\phi,$$

where k, l_0, and mg are constants.

a) Determine whether $\mathbf{F}$ is a conservative force.

b) If $\mathbf{F}$ is conservative, find its potential energy function.

Section 4.7

Problem 4.85

The figure shows three smooth curves. In each case, a particle of mass m is released from a height h and allowed to slide down the curve (the mass is the same in each case). Find the normal force exerted by each curve on the mass at the bottom. In which case is the force greatest? Least?

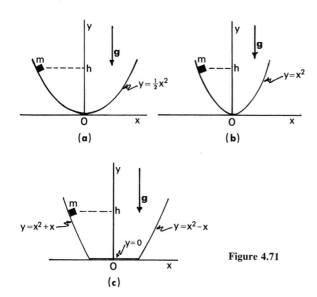

Figure 4.71

Problem 4.86

In the system shown in the figure, the string wraps around the rectangular box embedded in a smooth, horizontal plane. A particle of mass 1 slug is attached to the free end of the string of length 30 ft. In the position shown, the speed of the particle is 20 ft/sec. Find the tension in the string when it pivots about the following points: (a) A, (b) B, (c) C, (d) D.

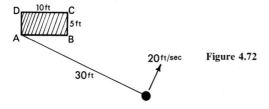

Figure 4.72

Problem 4.87

The body in the figure, which weighs mg lb, is released from rest at $t = 0$. Suppose that the drag force on the body is modeled as $\mathbf{f}_d = -c_0 \dot{y} \mathbf{j}$.

a) What is the speed of the body at time t?

b) What is the terminal speed of the body?

c) Show that the terminal speed can be found by seeking the speed at which the acceleration is zero.

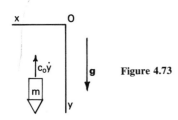

Figure 4.73

Problem 4.88

Consider the system in Problem 4.87. Model the drag force as $\mathbf{f}_d = -c_n \dot{y}^n \mathbf{j}$. Solve for the terminal speed of a body, using the condition that the terminal speed occurs when the acceleration is zero. Suppose that a body which weighs 1 oz has a terminal speed of 5 ft/sec.

a) Calculate c_1 for a drag force $\mathbf{f}_d = -c_1 \dot{y} \mathbf{j}$.

b) Calculate c_2 for a drag force $\mathbf{f}_d = -c_2 \dot{y}^2 \mathbf{j}$.

Problem 4.89

Consider the system in the figure (compare this system with that in Example 18). The coefficient of sliding friction between the rotating wheel and the bar is μ. Let x be the distance from the wheel to the point C of the bar. Let $\dot{x} = 0$ when $x = a$. Take $m_0 = 3m_1$, $\mu = 0.5$, and $a = 10$ ft.

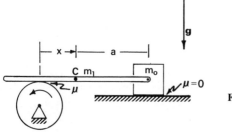

Figure 4.74

a) Compute $\dot{x}$ as a function of x.

b) Compute $\dot{x}$ when $x = 0$.

Problem 4.90

Consider the system of Problem 4.89 with the one change: Suppose that the rotation of the wheel is reversed (i.e., consider the rotation to be clockwise). Suppose that $\dot{x} = 0$ when $x = 0$.

a) Calculate $\dot{x}$ as a function of x.

b) Calculate $\dot{x}$ when $x = a$ $(a = 10$ ft$)$.

Problem 4.91

The mass in Example 19 has the trajectory

$$y = -\frac{1}{2}\frac{g}{(v_0 \cos\theta)^2} x^2 + (\tan\theta)x.$$

a) Show that the maximum height of the trajectory is

$$y_1 = \frac{1}{2}\left(\frac{v_0^2}{g}\right)\sin^2\theta,$$

which occurs at the position

$$x_1 = \left(\frac{v_0^2}{g}\right)\sin\theta\cos\theta.$$

b) Using the results of part (a), write

$$\eta = y_1 - y \qquad \text{and} \qquad \xi = x - x_1,$$

and determine the equation of the trajectory in terms of ξ and η. Does this show the parabolic nature of the trajectory?

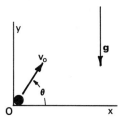

Figure 4.75

Problem 4.92

A projectile is fired into the air with a speed v_0 at the inclination θ (see figure).

a) Determine the point x, at which the trajectory is horizontal (that is, $dy/dx = 0$).

b) Find the velocity of the projectile at the point x determined in part (a).

Problem 4.93

A projectile is fired as in Problem 4.92.

    a) Determine the maximum height reached by the projectile.

    b) At what value of x does the height determined in part (a) occur?

Problem 4.94

    a) Suppose that $y_1 = f(x)$ has a relative minimum at $x = c$ and $f(c) > 0$. Show that $y_2 = [f(x)]^2$ has a relative minimum at $x = c$.

    b) Suppose that $y_1 = f(x)$ has a relative minimum at $x = c$ and $f(c) < 0$. Show that $y_2 = [f(x)]^2$ has a relative maximum at $x = c$.

    c) Suppose that $y_1 = f(x)$ has a relative minimum at $x = c$ and $f(c) \neq 0$. Show that $y_3 = 1/f(x)$ has a relative maximum at $x = c$.

    d) Consider the function $y = 1/\sqrt{f(x)}$, where $f(x) > 0$. Show that the conditions that $y(x)$ has a relative minimum at $x = c$ are that

$$\frac{df}{dx} = 0 \text{ at } c \quad \text{and} \quad \frac{d^2f}{dx^2} < 0 \text{ at } c.$$

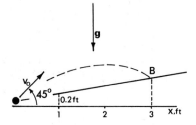

Figure 4.76

Problem 4.95

The projectile shown in the figure is fired at $45°$ to the horizontal with a speed v_0. It strikes the incline at point B.

    a) Determine v_0.

    b) Determine the speed of the projectile at impact.

Problem 4.96

Assume that the projectile in Problem 4.95 has the initial speed $v_0 = 20$ ft/sec. Determine the point of impact with the incline (i.e., determine the x value of the point of impact).

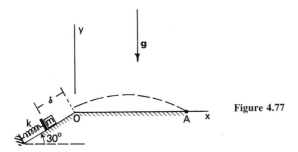

Figure 4.77

Problem 4.97

The spring shown in the figure is compressed an amount δ from its unstretched length. The particle is released from rest at the position shown. Assume that the surface is smooth, and let $\delta = 10$ ft, $k = 4.2$ lb/ft, and $m = 1$ slug.

a) Determine the trajectory of the particle after $x = 0$.

b) Determine the spot A at which the particle lands.

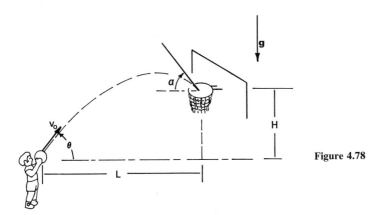

Figure 4.78

Problem 4.98

Suppose that, in order to make a free throw in basketball, a player wants the ball to enter the basket at angle α (see the figure). Show that the ball should be thrown at the angle θ given by

$$\tan \theta = \tan \alpha + 2 \frac{H}{L}.$$

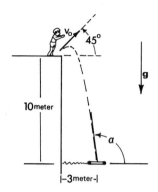

Figure 4.79

Problem 4.99

Dick Kimball dives through the ring shown in the figure, propelling himself with a speed v_0 at 45° to the horizontal.

   a) Calculate the speed v_0 such that he will pass through the ring.

   b) Calculate his speed at the instant he enters the water.

   c) Calculate the angle α from the horizontal at which he enters the water.

Problem 4.100

In the football problem (Example 19), we had $\dot{x} = v_0 \cos \theta = \text{constant}$. Use this fact to show that

$$\frac{dy}{dt} = (v_0 \cos \theta) \frac{dy}{dx}, \quad \text{or, in operator notation,} \quad \frac{d}{dt} = (v_0 \cos \theta) \frac{d}{dx}.$$

Use this result to rewrite the equation $\ddot{y} = -g$ in the form

$$\frac{d^2 y}{dx^2} = -\frac{g}{(v_0 \cos \theta)^2},$$

and thus solve directly for the ball's trajectory. (This procedure will be used in Chapter 5 to determine the trajectories of satellites.)

Problem 4.101

Use the results of Problem 4.100 to analyze the projectile problem (Example 19):

$$\dot{x} = v_0 \cos \theta, \quad \dot{y} = (v_0 \cos \theta) \frac{dy}{dx}.$$

Use the expression for the constant energy, $T + V$, to determine an equation for dy/dx. Integrate to determine the equation of the trajectory if $\theta = 45°$.

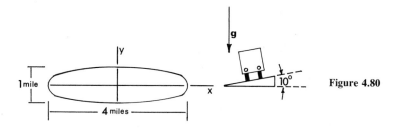

Figure 4.80

Problem 4.102

A racetrack which is built in the shape of the ellipse shown in the figure is banked 10°
at every point.

  a) Show that the radius of curvature at the point x of the ellipse can be expressed

   $R = \frac{1}{64}(64 - 15x^2)^{3/2}$ miles.

  b) Compute the ideal speed a car should have at point x so that it has no tendency
   to slide sideways. (Use $g = 7.9 \times 10^4$ miles/hr².)

  c) Compute the maximum and minimum ideal speeds on the track.

Problem 4.103

Suppose that the banking angle of the racetrack in Problem 4.102 is to be varied from
point to point so that the ideal speed for a car will be 250 mph at every point on the
track.

  a) Determine an expression for the banking angle as a function of the position x.

  b) Determine the maximum and minimum banking angles.

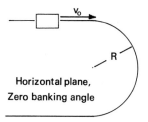

Figure 4.81

Problem 4.104

A car coasts around the track shown in the figure with a constant speed v_0.

  a) Given that the coefficient of static friction is μ, calculate the maximum speed
   v_0 before the car slides.

b) Suppose that $v_0 = 60$ mph and $\mu = 0.2$. Calculate the minimum radius R for the circular portion of the track before the car begins to slide.

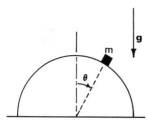

Figure 4.82

Problem 4.105

A particle m slides down a smooth cylindical surface as shown in the figure. It is released from rest at the top of the cylinder. Determine the angle θ at which the particle will leave the surface.

Problem 4.106

Suppose that in Problem 4.105 the particle is given a speed v_0 at the top of the cylinder.

a) Calculate v_0 such that m leaves the cylinder at $\theta = 30°$.

b) Calculate v_0 such that m leaves the cylinder at $\theta = 45°$.

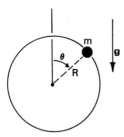

Figure 4.83

Problem 4.107

A bead of mass m which is restricted to slide on a frictionless ring of radius R (see Fig. 4.83) is released from rest at the top of the ring. What is the normal force exerted by the ring on the bead as a function of the angle θ?

Problem 4.108

The particle m shown in Fig. 4.84 has speed $v = \sqrt{gR}$ at the top point of the inside of a frictionless, circular track of radius R. Show that the particle will not leave the track

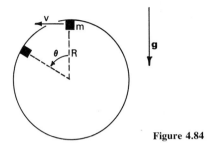

Figure 4.84

at any other point. That is, show that the normal force of the track on the particle is positive at any angle θ.

Problem 4.109

Refer to Example 21, which concerned the car in the amusement park (Fig. 4.85).

a) Show that the force exerted on the track is directly proportional to the mass of the car.

b) A constant force acts on the straight portion of the curve at the bottom to bring the car up to the speed v_0. Show that if the force F is to act over a fixed distance x_0, then F must be directly proportional to the mass of the car.

c) Suppose that the constant force F from part (b) is to act over a fixed time t_0 to bring the car up to speed. Show that F is again proportional to the mass of the car.

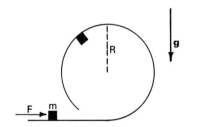

Figure 4.85

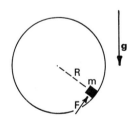

Figure 4.86

Problem 4.110

In Fig. 4.86, the particle m can slide inside the frictionless, circular curve. Suppose that the particle is at rest at the bottom of the curve, and that a force of constant magnitude which *remains tangential to the path* is applied to it. What is the minimum magnitude of F if the particle is to remain in contact with the curve at all points?

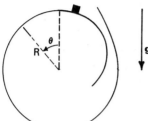

Figure 4.87

Problem 4.111

A particle which slides on the frictionless path shown in Fig. 4.87 is released from rest at the top of the path. Show that the particle leaves contact with the path at the angle θ given by

$$\theta = \cos^{-1}\left(\tfrac{2}{3}\right) = 48.2°.$$

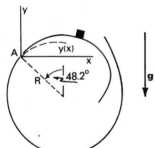

Figure 4.88

Problem 4.112

As in Problem 4.111, a particle slides on the frictionless path shown in Fig. 4.88. In Problem 4.111 it was shown that if the particle is released from rest at the top of the path, it leaves contact with the path at A:

$$\theta = \cos^{-1}\left(\tfrac{2}{3}\right) = 48.2°.$$

Determine the trajectory of the particle, $y = y(x)$, after it has left contact with the path.

Problem 4.113

The bead of mass m shown in Fig. 4.89 slides on a smooth wire bent in the shape of an ellipse:

$$\left(\frac{x}{a}\right)^2 + \left(\frac{y}{b}\right)^2 = 1.$$

The bead is released from rest at the top of the ellipse.

    a) What is the normal force exerted by the wire on the bead at point A?

    b) What is the normal force exerted by the wire on the bead at point B?

    [*Hint:* The radius of curvature at point B is $R = (a^2/b)$.]

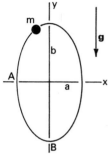

Figure 4.89

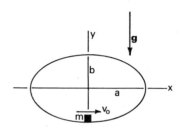

Figure 4.90

Problem 4.114

A particle m slides inside the frictionless, elliptical curve shown in Fig. 4.90.

    a) What is the minimum speed v_0 which m must have at the bottom of the curve in order that it stay in contact with the curve at the top? (Assume $a > b$.)

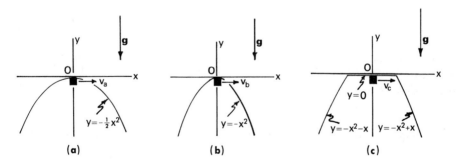

Figure 4.91

Problem 4.115

For each of the three smooth curves shown in the figure, determine the minimum speed a particle must have in order to maintain contact at the top of the curve.

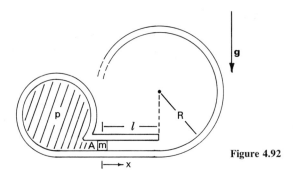

Figure 4.92

Problem 4.116

The pressure p of the gas in the blowgun in the figure is assumed to remain constant, and the mass m which slides over a smooth path is acted on by the blowgun for a distance l. Assume that the pressure outside the gun is zero.

a) Determine the minimum value of p in terms of m, g, R, l, and A (the cross-sectional area of the particle), so that m remains in contact with the entire path.

b) Given that $R = 10$ ft, $l = 1$ ft, $A = 12$ in², $mg = 25$ lb, calculate the value of the pressure p, in psi.

Problem 4.117

Consider the system in Problem 4.116. Suppose that the temperature of the gas in the blowgun is held constant so that pressure changes are isothermal:

$$pV = \text{constant},$$

where p and V are the pressure and volume of the gas. Again assume that the pressure outside the gun is zero.

a) Given that p_0 and V_0 are the pressure and volume of the gas when $x = 0$, and A is the cross-sectional area of the outlet tube, write the equation for the force on m as a function of x.

b) Write a potential energy function for the force in part (a).

c) Suppose that V_0, l, A, m, and R are given parameters for the system, and that m is released from rest at $x = 0$. Find the minimum pressure p_0 such that m will remain in contact with the entire path.

Problem 4.118

Again consider the system in Problem 4.116, and suppose now that the gas is thermally insulated, so that no heat enters or leaves the vessel. Then the pressure changes are adiabatic:

$$p_0 V_0^\gamma = pV^\gamma, \qquad \gamma > 1.$$

Again assume that the pressure outside the tube is zero.

a) Given that p_0 and V_0 are the pressure and volume of the vessel when $x = 0$ and A is the cross-sectional area of the outlet tube, write the expression for the force on m as a function of x.

b) Write a potential energy function for the force in (a).

c) Given that m has zero speed when $x = 0$, determine the value of p_0.

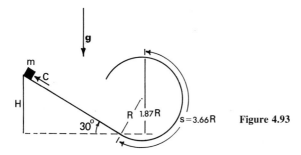

Figure 4.93

Problem 4.119

A particle m is released from rest on the track at the position shown in Fig. 4.93. There is a constant drag force, $f_d = -C$, acting on the particle. Find the minimum height H such that the particle will not leave the track at the top of the circular loop.

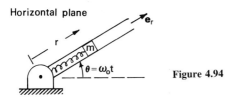

Figure 4.94

Problem 4.120

The particle of mass m shown in Fig. 4.94 is restricted to move in a frictionless tube which rotates in the *horizontal plane* so that $\theta = \omega_0 t, \ldots, \omega_0 = $ constant. A spring k exerts a force $\mathbf{f} = -kr\,\mathbf{e}_r$ on the particle.

a) Show that the differential equation for the motion of the particle is

$$\ddot{r} + \left(\frac{k}{m} - \omega_0^2\right) r = 0.$$

b) Show that the differential equation for the force exerted by the tube on the particle is

$$N = 2m\dot{r}\omega_0.$$

Problem 4.121

Suppose that, in Problem 4.120, $\dot{r} = v_0$ when $r = 0$.

   a) Determine $\dot{r}$ as a function of r.

   b) Determine the force N as a function of r.

   [*Hint:* Let $v = \dot{r}$; then note $\ddot{r} = v(dv/dr)$.]

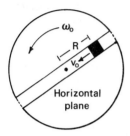

Horizontal
plane

Figure 4.95

Problem 4.122

The disk shown in the figure rotates at a constant angular rate of ω_0 rad/sec in the horizontal plane. A particle m, which can slide in a smooth track cut in the disk, is released at $r = R$ with a speed v_0 toward the center of the disk. Determine the minimum value of v_0 such that the particle will pass through $r = 0$. [*Hint:* Write $v = \dot{r}$ so that $\ddot{r} = v(dv/dr)$. Then find v_0 so that $v = 0$ at $r = 0$.]

Problem 4.123

Consider again the system in Problem 4.122. Show that the differential equation of motion for the particle,

$$m(\ddot{r} - \omega_0^2 r) = 0,$$

has the solution

$$r = C_1 e^{+\omega_0 t} + C_2 e^{-\omega_0 t},$$

where C_1 and C_2 are constants to be determined from the initial conditions $r(0) = r_0$, $\dot{r}(0) = v_0$.

   a) Determine the constants C_1 and C_2.

   b) Determine the condition under which $C_1 = 0$. Compare the result with that of Problem 4.122.

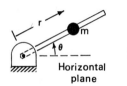

Figure 4.96

Horizontal
plane

Problem 4.124

The bead in the figure can slide without friction on the rod, which has the angular motion $\theta = \theta(t)$. It is found that the motion of the bead is

$$r = (5t + 2t^2) \text{ in.}$$

Determine $\dot{\theta}$ as a function of time. (Note that the motion takes place in the horizontal plane.)

5

CENTRAL FORCE MOTION

If some false principles led her astray,
how many admirable ones did she possess,
to which she always remained constant!

The Confessions of
Jean-Jacques Rousseau

1. INTRODUCTION

The motion of a particle under the influence of a force which always passes through a single point O, the origin of some inertial reference frame, is called *central force motion*. With the advent of man-made satellites and space vehicles, we have some dramatic examples of this sort of motion. Of course, other examples have been around for a long time: the motion of the planets about the sun, the motion of the moon about the earth, the motion of an electron, etc.

The motions of these systems have certain common characteristics:

i) The motion takes place in a single plane.

ii) The area swept out by the line segment from the center O to the particle is equal for equal times. We say that the *areal velocity* is constant in central force motion.

These two characteristics are common to all central force motion. If, in addition, we assume that the central force is of the special form

$$\mathbf{F} = -\frac{k}{r^2}\,\mathbf{e}_r, \qquad k = \text{const.}$$

(polar coordinates used here are defined in the single plane of motion), there are two additional results:

iii) The trajectory of the particle is a conic section.

220

iv) If the trajectory is an ellipse, the square of the time required for one complete orbit (the period) is proportional to the cube of the semi-major axis of the ellipse.

The properties described above are generally attributed to Johannes Kepler, who obtained them from the data of his teacher, Tycho Brahe. As we shall see, Isaac Newton later utilized these laws and the work of Kepler when he formulated his laws of dynamics in his *Philosophiae Naturalis Principia Mathematica* (1686).

2. GENERAL PROPERTIES OF CENTRAL FORCE MOTION

Consider a particle m under the influence of a force $\mathbf{F}$, which we assume always passes through the point O. This point will be called the *center* in our discussion (see Fig. 5.1).

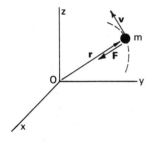

Figure 5.1

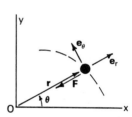

Figure 5.2

The magnitude of $\mathbf{F}$ may vary as m moves from point to point, but its direction is such that $\mathbf{F}$ has no moment about O. This leads us to write $\mathbf{M}_O = \dot{\mathbf{H}}_O$ for the particle. Since $\mathbf{M}_O = 0$ for all times, we have

$$\mathbf{H}_O = \mathbf{r} \times m\mathbf{v} = \text{constant vector.} \tag{5.1}$$

There are two important consequences of (5.1):

i) The direction of $\mathbf{H}_O$ is constant. Thus $\mathbf{r}$ and $\mathbf{v}$ are always perpendicular to $\mathbf{H}_O$, and the motion of m occurs in one plane.

ii) The magnitude of $\mathbf{H}_O$ is constant. Since m has plane motion, let us label that plane x,y and use polar coordinates to describe the position of m (Fig. 5.2).

The velocity of m in polar coordinates is

$$\mathbf{v} = \dot{r}\,\mathbf{e}_r + r\dot{\theta}\,\mathbf{e}_\theta.$$

Thus

$$\mathbf{H}_O = \mathbf{r} \times m\mathbf{v} = (r\ \mathbf{e}_r) \times m(\dot{r}\ \mathbf{e}_r + r\dot{\theta}\ \mathbf{e}_\theta)$$

$$= mr^2\dot{\theta}\ \mathbf{k} = \text{const.}$$

Since m is a constant, we can say that

$$r^2\dot{\theta} = C = \text{constant.} \tag{5.2}$$

This constant C, which applies to all problems of central force motion, will be taken to be one of the fundamental constants of the chapter.

Although we have in no way specified the magnitude of $\mathbf{F}$, we have determined that, in any central force motion, r and θ vary in such a way that the product $r^2\dot{\theta}$ remains constant throughout the motion.

We can interpret (5.2) in terms of the *areal velocity* of m. Consider the position of m at times t and $t + dt$ (Fig. 5.3).

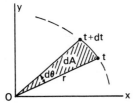

Figure 5.3

To the first order, the element of area dA swept out by the radius vector is

$$dA = \tfrac{1}{2}r^2\ d\theta.$$

Thus

$$\frac{dA}{dt} = \tfrac{1}{2}r^2\ \frac{d\theta}{dt} = \tfrac{1}{2}C.$$

Thus the areal velocity dA/dt is constant for any problem in central force motion, no matter what the force might be.

> The motion of a particle under the influence of a central force takes place in a single plane. The radius vector from the center sweeps out equal areas in equal time intervals. The center is that point through which the central force always passes.

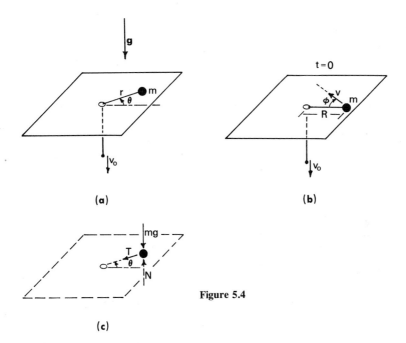

Figure 5.4

Example 1

Figure 5.4(a) shows a frictionless, horizontal table with a hole at its center. A particle m is attached to a string which passes through the hole. At $t = 0$, the position and velocity of the particle are as shown in Fig. 5.4(b). The string is pulled through the hole in the table at the constant rate v_0.

   i) Show that this is a case of central force motion.

   ii) Find $\theta = \theta(t)$.

Solution

The free-body diagram of the particle in Fig. 5.4(a) is shown in Fig. 5.4(c). We assume that the particle remains in the plane. Thus $z = 0$, and we write

$$F_z = ma_z, \qquad N - mg = 0.$$

Thus the normal force N and the weight mg cancel, and the net force on the particle is the string tension T, which always passes through the point O, at the hole. Thus this is a case of central force motion, and O is the center.

We must evaluate the constant C:

$$C = r^2\dot{\theta}.$$

The velocity of m is

$$\mathbf{v} = \dot{r}\,\mathbf{e}_r + r\dot{\theta}\,\mathbf{e}_\theta. \qquad (a)$$

At $t = 0$ (Fig. 5.4b), we have

$$\mathbf{v} = -v\cos\phi\,\mathbf{e}_r + v\sin\phi\,\mathbf{e}_\theta \qquad \text{and} \qquad r = R. \qquad (b)$$

But we also know that

$$\dot{r} = -v_0. \qquad (c)$$

Comparing (a), (b), and (c), we get $-v_0 = -v\cos\phi$. Thus

$$v = \frac{v_0}{\cos\phi}.$$

Comparing the $\mathbf{e}_\theta$ terms of (a) and (b) at $t = 0$, we have

$$R\dot{\theta} = v\sin\phi = v_0\tan\phi.$$

Thus, at $t = 0$,

$$C = r^2\dot{\theta} = Rv_0\tan\phi. \qquad (d)$$

Now note that $r = R - v_0 t$. Thus, by (d), we have

$$\frac{d\theta}{dt} = \frac{C}{(R - v_0 t)^2}.$$

Integrating, we obtain

$$\theta = \frac{C}{v_0}\frac{1}{R - v_0 t} + C_1.$$

If $\theta = 0$ at $t = 0$, the constant of integration C_1 can be determined. The final result is

$$\theta = \frac{Ct}{R(R - v_0 t)}.$$

Inserting the value of C from (d), we get

$$\theta = \frac{(v_0\tan\phi)t}{R - v_0 t}.$$

....

Example 2

For the setup described in Example 1, find the string tension T and the trajectory of m.

Solution

Noting the free-body diagram in Fig. 5.4(c), we write

$$F_r = ma_r, \qquad -T = m(\ddot{r} - r\dot{\theta}^2). \qquad (a)$$

We have $\dot{r} = -v_0$. Thus $\ddot{r} = 0$. And from $C = r^2\dot{\theta}$, we get

$$r\dot{\theta}^2 = \left(\frac{C^2}{r^3}\right).$$

But in Example 1, we determined that $r = R - v_0 t$ and $C = Rv_0 \tan\phi$. Thus (a) becomes

$$T = \frac{mR^2 v_0^2 \tan^2\phi}{r^3} = \frac{mR^2 v_0^2 \tan^2\phi}{(R - v_0 t)^3} \; .$$

To obtain the trajectory of m, we eliminate t between

$$r(t) = R - v_0 t, \qquad \theta(t) = \frac{(v_0 \tan\phi)t}{R - v_0 t}.$$

The result is

$$r(\theta) = \frac{R}{1 + (\cot\phi)\theta}.$$

....

Example 3

In this example, we compare two problems, (a) and (b), which seem to be very similar, but which are in fact quite different. In Figs. 5.5(a) and 5.5(a') we repeat Example 16 of Chapter 4.

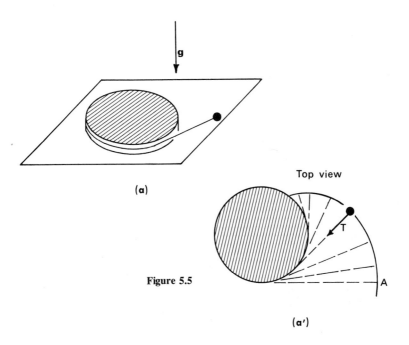

(a)

Figure 5.5

Top view

(a')

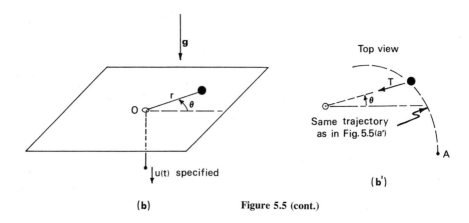

(**b**)   **Figure 5.5 (cont.)**

Figures 5.5(b) and 5.5(b′) show a system similar to that in Example 1 of this chapter, except that now we specify the displacement of the string in such fashion that the trajectory in the system is exactly the same as the trajectory in Fig. 5.5(a′).

At position A in each case, the speed of the particle is v_0. Compare the two systems from the following points of view:

H is (is not) constant
T, kinetic energy, is (is not) constant

Solution

Problem (a): In this case, the string tension **T** is perpendicular to the trajectory of the particle, and thus is perpendicular to $d\mathbf{r}$. No work is done on the particle; and since the potential energy of the particle is constant, the kinetic energy is constant. Thus

$T =$ kinetic energy $=$ constant.

The speed of the particle along its trajectory is constant. But since the orientation of the trajectory is changing at all times, there is no fixed point about which $\mathbf{r} \times m\mathbf{v}$ is constant. Thus **H** is not a constant.

Problem (b): Since the force on the particle always passes through O, the hole in the table, this is a case of central-force motion. Thus $\mathbf{H}_o =$ constant.

We know that in this problem the string tension does work. Since the force moves along its line of action, the work is positive when the string is pulled inward.

Thus the kinetic energy of the particle increases as the motion proceeds, and the speed of the particle increases. Therefore, the kinetic energy T is not constant.

One final word: The trajectories in these two problems are exactly the same, but that is about all that is the same. Just because the trajectories of two particles are the same, it does not follow that the causes of the motions are the same, or that the motions themselves are the same.

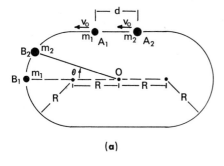

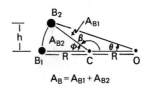

Figure 5.6

$A_B = A_{B1} + A_{B2}$

(b)

Example 4

Two particles m_1 and m_2 are in central-force motion about the center O (Fig. 5.6a). On the straight portion of the path, the particles have the same velocity.

When m_1 and m_2 have positions A_1 and A_2, they are separated by a distance d. When m_1 is at position B_1, m_2 is at B_2, which is identified by the angle θ. Find an equation by which the angle θ can be determined.

Solution

Since this is a case of central-force motion, the line from O to m_1 sweeps out equal areas in equal times. The same is true for the line from O to m_2. Since, on the straight portion of the trajectory, the velocities of m_1 and m_2 are the same, the areal velocities of the two particles are equal. Thus the enclosed area Om_1m_2 is constant throughout the motion. In particular, area OA_1A_2 and the area of the sector of the trajectory OB_1B_2 are equal. Denote these areas by A_A and A_B, respectively (Fig. 5.6b).

For area OA_1A_2, we have the area of a triangle:

$$A_A = \tfrac{1}{2} Rd. \tag{a}$$

The area A_B is not so simple to compute. We separate this area into two parts: A_{B1} and A_{B2}:

A_{B1} = area of isosceles triangle OCB_2,

A_{B2} = area of circular sector CB_1B_2.

For A_{B1}, we note that OB_2 has the length $OB_2 = 2R \cos \theta$. Thus the height of the triangle is $h = (2R \cos \theta) \sin \theta$. The base is R. Thus

$$A_{B1} = R^2 \cos \theta \sin \theta. \tag{b}$$

The area of the sector CB_1B_2 is easily computed when the angle ϕ is known. Note that

$$\phi + \beta = \pi, \tag{c}$$

and from the triangle OCB_2, we have

$$2\theta + \beta = \pi. \tag{d}$$

From (c) and (d), we get

$$\phi = 2\theta, \tag{e}$$

and thus

$$A_{B2} = \left(\frac{\phi}{2\pi}\right) \pi R^2 = R^2 \theta. \tag{f}$$

Equating areas A_A and A_B from (a), (b), and (f), we obtain

$$\tfrac{1}{2} Rd = R^2 \cos \theta \sin \theta + R^2 \theta \qquad \text{or} \qquad \sin 2\theta + 2\theta = \frac{d}{R}.$$

This transcendental equation can be used to determine θ.

....

Example 5

Figure 5.7(a) shows the top view of a frictionless horizontal plane. An elastic string, which has a particle m attached to one end, is relaxed when the particle is at O. The spring constant of the string is k lb/ft. Suppose that, at $t = 0$,

$$x = a \text{ ft}, \qquad y = 0,$$
$$\dot{x} = 0, \qquad \dot{y} = v_0 \text{ ft/sec.}$$

i) Calculate the trajectory of the motion.

ii) Verify that the areal velocity is constant.

Solution

Everything in this example leads us to consider the use of cylindrical coordinates. The string tension always passes through O, and every aspect of this problem would seem to indicate symmetry about the z axis through O. No one could complain if a student (or professor) started down that path of analysis. The fact is, however, that cylindrical coordinates are not very useful here, and—surprisingly enough—cartesian coordinates are extremely useful. So this situation has to be considered somewhat of a fluke.

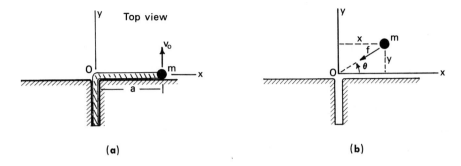

Figure 5.7

Figure 5.7(b) is a free-body diagram for the particle. The position of the particle is $(x,y,0)$. Thus the magnitude of the elastic string force is $|f| = k\sqrt{x^2 + y^2}$, and thus

$$f_x = -k\sqrt{x^2 + y^2}\ \cos\theta, \qquad f_y = -k\sqrt{x^2 + y^2}\ \sin\theta.$$

But

$$\cos\theta = \frac{x}{\sqrt{x^2 + y^2}}, \qquad \sin\theta = \frac{y}{\sqrt{x^2 + y^2}}$$

and hence

$$f_x = -kx, \qquad f_y = -ky.$$

Writing $\mathbf{F} = \dot{\mathbf{p}}$, we have then

$$F_x = ma_x, \qquad -kx = m\ddot{x}, \tag{a}$$

$$F_y = ma_y, \qquad -ky = m\ddot{y}, \tag{b}$$

or

$$\ddot{x} + (k/m)x = 0, \tag{a$'$}$$

$$\ddot{y} + (k/m)y = 0. \tag{b$'$}$$

After Chapter 6, solving (a$'$) and (b$'$) will be child's play. Here we note that

$$\left.\begin{array}{l} x = A\sin\omega_n t + B\cos\omega_n t \\[4pt] y = C\sin\omega_n t + D\cos\omega_n t \end{array}\right\} \qquad \text{where } \omega_n^2 = k/m, \tag{c}$$

are solutions to (a$'$) and (b$'$) for any values of the constants A, B, C, and D. These constants are determined from the initial conditions,

$$x(0) = a, \qquad \dot{x}(0) = 0, \tag{d}$$

$$y(0) = 0, \qquad \dot{y}(0) = v_0. \tag{e}$$

For example, from (c) and (d), we get

$$a = A \cdot 0 + B \cdot 1 = B,$$

$$0 = \omega_n A \cdot 1 - \omega_n B \cdot 0 = \omega_n A.$$

Thus $B = a$ and $A = 0$:

$$x = a \cos \omega_n t. \tag{f}$$

Similarly, we find that

$$y = \frac{v_0}{\omega_n} \sin \omega_n t. \tag{g}$$

To obtain the trajectory of m, we eliminate t between $x(t)$ and $y(t)$, that is, between (f) and (g). Note that

$$\frac{x}{a} = \cos \omega_n t, \qquad \frac{y}{v_0/\omega_n} = \sin \omega_n t,$$

and since

$$\cos^2 \omega_n t + \sin^2 \omega_n t = 1,$$

we obtain

$$\frac{x^2}{a^2} + \frac{y^2}{(v_0/\omega_n)^2} = 1.$$

Or, from (c), we get

$$\frac{x^2}{a^2} + \frac{y^2}{mv_0^2/k} = 1. \tag{h}$$

The trajectory is an ellipse with semimajor axis a and semiminor axis $(v_0 \sqrt{m/k}\,)$.

To check that the areal velocity is constant, we must see that, for our solution (f) and (g), the quantity $r^2 \dot{\theta} = \text{constant}$. We have

$$r^2 = x^2 + y^2 \tag{i}$$

and

$$\tan \theta = \frac{y}{x}. \tag{j}$$

From (f) and (g), we find that (j) becomes

$$\tan \theta = \left(\frac{v_0}{a\omega_n}\right) \tan \omega_n t.$$

Differentiating (using the chain rule on the left) with respect to time gives

$$\sec^2 \theta \, \frac{d\theta}{dt} = \left(\frac{v_0}{a\omega_n}\right) (\sec^2 \omega_n t)\omega_n = \left(\frac{v_0}{a}\right) \sec^2 \omega_n t.$$

Thus

$$\frac{d\theta}{dt} = \left(\frac{v_0}{a}\right) \cos^2 \theta \sec^2 \omega_n t.$$

Since

$$\cos \theta = \frac{x}{\sqrt{x^2 + y^2}} \qquad \text{and} \qquad \sec \omega_n t = \frac{a}{x},$$

we get

$$\frac{d\theta}{dt} = \left(\frac{v_0}{a}\right) \left(\frac{x^2}{x^2 + y^2}\right) \frac{a^2}{x^2} = \frac{v_0 a}{x^2 + y^2}.$$

Finally from (i), we obtain $r^2 \dot{\theta} = v_0 a$, which is constant.

....

3. THE GRAVITATIONAL FIELD

In this section, we consider the special case of the motion of one particle in the gravitational field of another particle. For the sake of simplicity, we assume that one particle has a fixed position in some inertial reference frame and that the second orbits the first,* as diagrammed in Fig. 5.8.

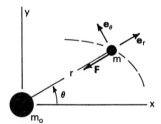

Figure 5.8

We assume that the force on the moving particle m is given by Newton's law of universal gravitation† (4.22):

$$\mathbf{F} = -\frac{\gamma m_0 m}{r^2} \mathbf{e}_r, \qquad V = -\frac{\gamma m_0 m}{r}. \tag{5.3}$$

Writing $\mathbf{F} = \dot{\mathbf{p}}$ for m, we now have

$$F_r = ma_r, \qquad m(\ddot{r} - r\dot{\theta}^2) = -\frac{\gamma m_0 m}{r^2}, \qquad \text{(a)}$$

$$F_\theta = ma_\theta, \qquad m(r\ddot{\theta} + 2\dot{r}\dot{\theta}) = 0. \qquad \text{(b)} \tag{5.4}$$

---

* This restriction can easily be relaxed. See Chapter 3 of H. Goldstein, *Classical Mechanics*, Reading, Mass.: Addison-Wesley, 1950.

† For the earth, γm_e has the value 1.255×10^{12} mi³/hr². For the computations involving satellites about the earth, this is a more useful constant than γ alone.

Writing $\mathbf{M} = \dot{\mathbf{H}}$ about the center, we obtain*

$$\mathbf{M}_0 = \dot{\mathbf{H}}_0, \qquad 0 = \frac{d}{dt}(mr^2\dot{\theta}\ \mathbf{k}),$$

or

$$r^2\dot{\theta} = C, \tag{5.5}$$

and the total energy of the system is a constant. We denote this by mE (for reasons which will soon be apparent):

$$\tfrac{1}{2}m(\dot{r}^2 + r^2\dot{\theta}^2) + \left(-\frac{\gamma m_0 m}{r}\right) = mE. \tag{5.6}$$

Let us emphasize that, in general, we do not expect central-force problems to be conservative. In general, we need not even know the nature of the central force. However, we are dealing here with a particular type of central-force motion: motion in a gravitational field. This problem *is* conservative.

We see that in each of the above three equations the mass m cancels out. Thus we have

$$\ddot{r} - r\dot{\theta}^2 + \frac{\gamma m_0}{r^2} = 0, \qquad \text{(i)}$$

$$r^2\dot{\theta} = C, \qquad \text{(ii)} \tag{5.7}$$

$$\tfrac{1}{2}(\dot{r}^2 + r^2\dot{\theta}^2) - \frac{\gamma m_0}{r} = E. \qquad \text{(iii)}$$

We would like to solve Eqs. (5.7) for the motion of m; that is, for $r(t)$ and $\theta(t)$. In actual fact we are probably more interested in finding the trajectory of m, $r = r(\theta)$. In order to accomplish this, we change the independent variable from time t to the angle θ by means of the chain rule and (5.7ii):

$$\frac{d}{dt} = \frac{d\theta}{dt}\frac{d}{d\theta} = \frac{C}{r^2}\frac{d}{d\theta}. \tag{5.8}$$

There are two ways we can derive $r(\theta)$: (a) from energy, or (b) from (5.7i). The first way is direct, but somewhat tedious. The second requires a transformation of variable, which is difficult to motivate.

---

* Equation (5.4b) is equivalent to (5.5). Differentiating (5.5) gives $2r\dot{r}\dot{\theta} + r^2\ddot{\theta} = 0$ or $r(r\ddot{\theta} + 2\dot{r}\dot{\theta}) = 0$.

a) Taking (5.7iii), eliminating $\dot{\theta}$ by (5.7ii), and changing to spatial derivatives (5.8), we get

$$\frac{C^2}{r^4}\left(\frac{dr}{d\theta}\right)^2 + \frac{C^2}{r^2} - \frac{2\gamma m_0}{r} = 2E.$$

Thus

$$\frac{dr}{d\theta} = \frac{r}{C}\sqrt{2Er^2 + 2\gamma m_0 r - C^2}$$

and

$$\int \frac{(C/r)\,dr}{\sqrt{2Er^2 + 2\gamma m_0 r - C^2}} = \theta + C_1,$$

where C_1 is a constant to be determined. This integral can be evaluated by most freshmen. For those of you more advanced in your education, a table of integrals might be helpful. We get

$$\theta + C_1 = \sin^{-1}\left[\frac{\gamma m_0 r - C^2}{r\sqrt{\gamma^2 m_0^2 + 2EC^2}}\right].$$

The integration constant C_1 is usually chosen so that

$$r = \frac{C^2/\gamma m_0}{1 - \sqrt{1 + 2EC^2/\gamma^2 m_0^2}\ \cos\theta}, \tag{5.9}$$

and (5.9) is generally written (in analytic geometry)

$$r = \frac{ep}{1 - e\cos\theta},$$

where

$$e = \sqrt{1 + (2EC^2/\gamma^2 m_0^2)}, \qquad p = \frac{C^2}{\sqrt{\gamma^2 m_0^2 + 2EC^2}} \tag{5.10}$$

and

$$ep = \frac{C^2}{\gamma m_0}.$$

Equation (5.10) gives the class of curves called *conic sections*. Here e is the eccentricity of the conic section and p is the distance between the focus of the conic and the directrix. (Don't panic; we'll define and review these quantities directly.) Bear in mind that E is not necessarily positive! [See (5.7iii).]

b) A simpler way of obtaining (5.10) amounts to using (5.7i). With (5.7ii) and (5.8), we get

$$\frac{C}{r^2}\frac{d}{d\theta}\left(\frac{C}{r^2}\frac{dr}{d\theta}\right) - r\left(\frac{C^2}{r^4}\right) + \frac{\gamma m_0}{r^2} = 0$$

or

$$\frac{d}{d\theta}\left(\frac{1}{r^2}\frac{dr}{d\theta}\right) - \frac{1}{r} = -\frac{\gamma m_0}{C^2}. \tag{5.11}$$

To solve this equation, we now transform the dependent variable r, writing

$$r = \frac{1}{u}; \quad \text{thus} \quad \frac{dr}{d\theta} = -\frac{1}{u^2}\frac{du}{d\theta} = -r^2\frac{du}{d\theta}.$$

Thus (5.11) becomes

$$\frac{d^2u}{d\theta^2} + u = \frac{\gamma m_0}{C^2}. \tag{5.12}$$

Equation (5.12) is one which we shall learn to solve in Chapter 6. For the time being, you should verify that

$$u = \frac{\gamma m_0}{C^2} + A_1 \sin\theta + A_2 \cos\theta$$

is a solution of (5.12) for arbitrary values of A_1 and A_2. Thus we see that

$$r = \frac{C^2/\gamma m_0}{1 + (C^2/\gamma m_0)(A_1 \sin\theta + A_2 \cos\theta)}. \tag{5.13}$$

Again by a proper choice of A_1 and A_2, we can write

$$r = \frac{ep}{1 - e\cos\theta},$$

where e and ep have the values given in (5.10).

Example 6

A particle m is shot from a gun with speed v_0 at a distance r_0 from the center of the earth, at an inclination α to the local vertical (Fig. 5.9). Calculate the trajectory of m.

Solution

We begin by evaluating the constants C and E. For C, we must evaluate r^2 and $\dot{\theta}$ at some instant (any convenient instant). Writing the velocity at the initial instant in polar coordinates, we have

$$\mathbf{v} = \dot{r}\mathbf{e}_r + r\dot{\theta}\,\mathbf{e}_\theta = (v_0 \cos\alpha)\mathbf{e}_r + (v_0 \sin\alpha)\mathbf{e}_\theta.$$

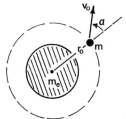

Figure 5.9

And since $r = r_0$ at this instant, we have, at $t = 0$,

$$r = r_0, \qquad \dot\theta = \left(\frac{v_0 \sin \alpha}{r_0}\right).$$

Thus

$$C = r_0 v_0 \sin \alpha. \tag{a}$$

For E, we look at the energy per unit mass at $t = 0$:

$$E = \tfrac{1}{2}v_0^2 - \frac{\gamma m_e}{r_0}. \tag{b}$$

Thus from (5.10) we obtain

$$e^2 = 1 + \left(v_0^2 - \frac{2\gamma m_e}{r_0}\right)\left(\frac{r_0 v_0}{\gamma m_e}\right)^2 \sin^2 \alpha,$$

$$ep = \frac{(r_0 v_0)^2}{\gamma m_e} \sin^2 \alpha, \tag{c}$$

and

$$r = \frac{ep}{1 - e \cos \theta}, \qquad \alpha \neq 0.$$

If $\alpha = 0$, $C = 0$. Thus $\dot\theta = 0$ for all t. The trajectory is a straight line through the center.

In order to obtain the trajectory of a particle in the $(1/r^2)$ gravitational field, we proceed as follows.

i) Obtain the constants C and E, the moment of momentum per unit mass, and the energy per unit mass.

ii) Obtain the geometric parameters e and ep from C and E, using (5.10). The trajectory is then

$$r = \frac{ep}{1 - e \cos \theta}.$$

iii) Interpret the results.

The final step requires a knowledge of the geometry of conic sections. This topic is taken up in the next section.

4. THE GEOMETRY OF A CONIC SECTION

For some reason, the mathematical training of engineers and scientists in recent years has tended to deemphasize geometry. This is regrettable, since many problems in engineering are directly related to geometry (and the geometry of the conic sections in particular). Notions from the geometry of conic sections pop up in such diverse areas as stress analysis, partial differential equations, linear algebra and matrix theory, etc. So let's give here a brief review of the properties of conic sections that are pertinent to our study of central-force motion in a gravitational field.* For an interesting and readable account of this and other areas of geometry, refer to the work of Hilbert and Cohn-Vossen.†

Conic sections, the ellipse, parabola, hyperbola (and the special limiting cases: circle, point, and straight line) can all be described analytically by (5.10) with a proper interpretation of the parameters e and p. We begin by considering a line l (called the *directrix*) and a point F (called the *focus*). (See Fig. 5.10.)

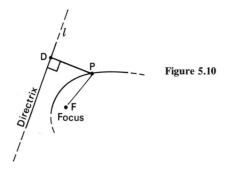

Figure 5.10

We now consider the locus of all points P in the plane of l and F such that the distance of P from F and from the line l have a constant ratio e, called the eccentricity:

$$\frac{PF}{PD} = e. \tag{5.14}$$

---

    * See, for example, G. B. Thomas, *Calculus and Analytic Geometry*, Reading, Mass.: Addison-Wesley, fourth edition, pages 354–55, 367–70.

    † D. Hilbert and S. Cohn-Vossen, *Geometry and the Imagination*, New York: Chelsea, 1952.

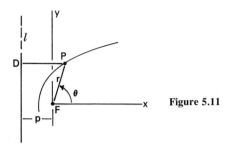

Figure 5.11

(The distance of a point from a line is taken to be the shortest distance of the point from any point on the line. This is, of course, the perpendicular distance of the point from the line.) In Fig. 5.11, we introduce polar coordinates to describe this situation.

Let the distance of F from l be denoted by p. Then we have

$$PD = p + r \cos \theta, \qquad PF = r,$$

and (5.14) becomes

$$r = \frac{pe}{1 - e \cos \theta} \qquad (5.15)$$

The various conic sections correspond to the following values of the eccentricity e.

Conic section	Eccentricity
Ellipse	$0 < e < 1$ ($e = 0 = $ circle)
Parabola	$e = 1$
Hyperbola	$e > 1$

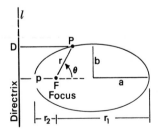

Figure 5.12

The parabola separates the conic sections into two categories: (a) ellipses — closed curves, (b) hyperbolas — open curves.

The case of $e = 0$ requires a separate explanation. Let us take the general case of the ellipse in Fig. 5.12. From

$$r = \frac{ep}{1 - e \cos \theta},$$

we consider

$$r_1 = r(0) = \frac{pe}{1 - e} \qquad \textit{apocentron}* \text{ (maximum distance from } F)$$

$$\tag{5.16}$$

$$r_2 = r(\pi) = \frac{pe}{1 + e} \qquad \textit{pericentron}* \text{ (minimum distance from } F)$$

The semimajor axis of the ellipse a is given by

$$2a = r_1 + r_2 = pe\left(\frac{1}{1 - e} + \frac{1}{1 + e}\right)$$

or

$$a = \frac{pe}{1 - e^2}. \tag{5.17a}$$

The semiminor axis b can be shown to be (see Problem 5.28)

$$b = a\sqrt{1 - e^2} = \frac{pe}{\sqrt{1 - e^2}}. \tag{5.17b}$$

From (5.17b) we see that if $a = b$, $e = 0$. This is the case of a circle. From (5.17a) we see that if we let $e \to 0$, we must let $pe \to a$. Then (5.10) gives $r = a$, a circle of radius a.

Finally, we review these geometric parameters in terms of the physical constants C and E. By (5.10) and (5.17), we have the following values.

Conic section, general

Eccentricity $\qquad\qquad\qquad\qquad\qquad\qquad e = \sqrt{1 + (2EC^2/\gamma^2 m_0^2)}$

Distance from focus to directrix $\qquad p = \dfrac{C^2}{\sqrt{\gamma^2 m_0^2 + 2EC^2}} \qquad$ (5.18)

$$ep = \frac{C^2}{\gamma m_0}$$

---

* If the ellipse is about the earth with the earth at the focus F, the apocentron and pericentron are called the *apogee* and *perigee*.

Elliptical orbit, $E < 0$

Apocentron* $\qquad r_1 = \dfrac{C^2/\gamma m_0}{1 - \sqrt{1 + (2EC^2/\gamma^2 m_0^2)}}$

Pericentron* $\qquad r_2 = \dfrac{C^2/\gamma m_0}{1 + \sqrt{1 + (2EC^2/\gamma^2 m_0^2)}}$

Semimajor axis $\qquad a = -\dfrac{\gamma m_0}{2E}$ $\qquad\qquad$ (5.19)

Semiminor axis $\qquad b = C\sqrt{\dfrac{-1}{2E}}$

Circular orbit (radius a) $\qquad C = \sqrt{a\gamma m_0}\,, \qquad E = -\dfrac{\gamma m_0}{2a}$

Example 7

A satellite is in a circular orbit of radius R about the earth. Evaluate the pertinent parameters and show that the eccentricity is zero.

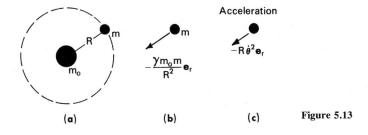

(a) $\qquad\qquad$ (b) $\qquad\qquad$ (c) $\qquad$ **Figure 5.13**

Solution

Consider the free-body diagram in Fig. 5.13(b). From $\mathbf{F} = \dot{\mathbf{p}}$, we have

$$F_r = ma_r, \qquad -\frac{\gamma m_0 m}{R^2} = -mR\dot{\theta}^2.$$

Thus

$$\dot{\theta}^2 = \frac{\gamma m_0}{R^3} \quad \text{and} \quad v^2 = \frac{\gamma m_0}{R}.$$

---

* The terms "apogee" and "perigee" refer to elliptical orbits about the earth. In that case, m_e replaces m_0.

The parameters E and C, (5.2) and (5.7iii), are

$$C^2 = R^4 \dot\theta^2 = R\gamma m_0,$$

$$E = \frac{1}{2}(R^2\dot\theta^2) - \frac{\gamma m_0}{R} = -\frac{1}{2}\frac{\gamma m_0}{R}.$$

From (5.10), the eccentricity e is

$$e^2 = 1 + 2EC^2/\gamma^2 m_0^2$$

$$= 1 + 2\left(-\frac{1}{2}\frac{\gamma m_0}{R}\right) R\gamma m_0/\gamma^2 m_0^2 = 0$$

as required.

....

Example 8

A satellite which is in an elliptical orbit about the earth has semimajor and semiminor axes: $a = 5100$ miles, $b = 5025$ miles. Find

   i) the apogee and perigee of the trajectory,

   ii) the closest that the satellite gets to the earth's surface,

   iii) the speed of the satellite at the perigee.

Solution

   i) From (5.17b), we have $b = a\sqrt{1 - e^2}$. Thus

$$\sqrt{1 - e^2} = \frac{5025}{5100} = 0.985,$$

and

$$e = 0.173. \tag{a}$$

From (5.17a), we get

$$pe = 4950 \text{ miles.} \tag{b}$$

From (5.16) we get: apogee, $r_1 = 5980$ miles, perigee, $r_2 = 4210$ miles.

   ii) Since the radius of the earth is 4000 miles and the focus of the ellipse is at the center of the earth, the closest the satellite gets to the surface of the earth is 210 miles.

   iii) From (5.18), we have

$$ep = \frac{C^2}{\gamma m_e}.$$

We know that $ep = 4950$ miles and

$\gamma m_e = 1.255 \times 10^{12}$ mi³/hr².

Thus

$C^2 = ep\gamma m_e = 6.21 \times 10^{15}$ mi⁴/hr²   and $C = 7.88 \times 10^7$ mi²/hr.

But $C = r^2\dot\theta$. And the speed $v = \sqrt{\dot r^2 + r^2\dot\theta^2}$. At perigee, $\dot r = 0$ (why?). Thus

$$v = r\dot\theta = \frac{C}{r}.$$

With $r = 4210$ miles at perigee, we get

$v = 1.87 \times 10^4$ mi/hr.

....

5. TRAJECTORIES IN THE GRAVITATIONAL FIELD

We can now classify the possible trajectories in terms of the physical parameters E and C.

Trajectory	Eccentricity	Condition	
Circle (radius a)	$e = 0$	$E = -\tfrac{1}{2}(\gamma m_0/C)^2,$ $C = \sqrt{a\gamma m_0},$	
		$E = -(\gamma m_0/2a)$	
Ellipse	$e < 1$	$E < 0$	
Parabola	$e = 1$	$E = 0$	(5.20)
Hyperbola	$e > 1$	$E > 0$	

If the trajectory is a circle or an ellipse, we can properly speak of an orbit. The parabolic and hyperbolic trajectories extend to infinity. We say that a particle with this type of trajectory "escapes" from the gravitational field of m_0. We say that a particle has "escape velocity" at the surface of m_0 if its velocity vanishes only at an infinite distance from m_0.

Figure 5.14 shows two positions of the particle: A, a point at the surface of m_0, and B, a point at infinity. The system is conservative. Thus we write

At A: $T = \tfrac{1}{2}mv_0^2,$ $V = -\dfrac{\gamma m m_0}{R}$.   At B: $T \geqslant 0,$ $V \to 0.$

Figure 5.14

Thus

$$\tfrac{1}{2}mv_0^2 - \frac{\gamma m m_0}{R} \geq 0.$$

Note that the left side of this inequality is mE. Thus, when $E > 0$, we have

$$v_0 > \sqrt{\frac{2\gamma m_0}{R}}. \tag{5.21}$$

The lowest value of $v_0 = \sqrt{2\gamma m_0/R}$ is referred to as the *escape velocity*. From the table (5.20), it is clear that the escape velocity corresponds to a parabolic trajectory. Larger values of v_0 correspond to hyperbolic trajectories.

....

Example 9

The mass of the moon is about one-eightieth of the mass of the earth. Its radius is about 1100 miles, compared to 4000 miles for the earth. Calculate:

   i) (γm_m) for the moon,

  ii) the escape velocity from the surface of the earth,

 iii) the escape velocity from the surface of the moon.

Solution

   i) We have $\gamma m_e = 1.255 \times 10^{12}$ mi^3/hr^2. Thus

$$\gamma m_m = \tfrac{1}{80}\gamma m_e = 1.57 \times 10^{10} \text{ mi}^3/\text{hr}^2.$$

  ii) $v_0 = \sqrt{\dfrac{2\gamma m_e}{R_e}} = \sqrt{\dfrac{2(1.255 \times 10^{12}) \text{ mi}^3/\text{hr}^2}{4.0 \times 10^3 \text{ mi}}}$

$$v_0 = 2.5 \times 10^4 \text{ mi/hr}.$$

 iii) $v_0 = \sqrt{\dfrac{2\gamma m_m}{R_m}} = \sqrt{\dfrac{2(1.57 \times 10^{10}) \text{ mi}^3/\text{hr}^2}{1.1 \times 10^3 \text{ mi}}},$

$$v_0 = 5.34 \times 10^3 \text{ mi/hr}.$$

....

Example 10

Consider Example 6, and determine the conditions on v_0 and α in order that the trajectory of m be (i) circular, (ii) elliptical, (iii) parabolic, and (iv) hyperbolic.

Solution

From (a) and (b) of Example 6, we had

$$C = r_0 v_0 \sin \alpha, \qquad E = \left(\tfrac{1}{2} v_0^2 - \frac{\gamma m_e}{r_0} \right).$$

i) For a circular orbit of radius r_0, we have, by (5.20),

$$C = \sqrt{r_0 \gamma m_e}, \qquad E = -\frac{\gamma m_e}{2 r_0}.$$

Comparing the expressions for E, we get

$$v_0 = \sqrt{\gamma m_e / r_0}. \tag{a}$$

Comparing the expressions for C, we get

$$\sin^2 \alpha = \frac{1}{v_0^2} \left(\frac{\gamma m_e}{r_0} \right).$$

But using (a), we get $\sin^2 \alpha = 1$. Thus

$$\alpha = \pi/2 \text{ rad.} \tag{b}$$

ii) For an elliptical orbit, we need $E < 0$ by (5.20). In addition, we need $C \neq 0$ if we are to avoid a straight-line degeneracy. Thus

$$0 < v_0 < \sqrt{\frac{2 \gamma m_e}{r_0}}, \qquad \alpha \neq 0. \tag{c}$$

iii) For a parabolic trajectory, we need $E = 0$ by (5.20). Again $\alpha = 0$ gives a straight-line degeneracy. Thus

$$v_0 = \sqrt{\frac{2 \gamma m_e}{r_0}}, \qquad \alpha \neq 0. \tag{d}$$

iv) For a hyperbolic trajectory, we need $E > 0$. Thus

$$v_0 > \sqrt{\frac{2 \gamma m_e}{r_0}}, \qquad \alpha \neq 0. \tag{e}$$

The values of v_0 determined in (d) and (e) are escape velocities.

....

6. KEPLER'S LAWS

We are now in a position to derive Kepler's three laws mathematically. Once again the laws are:

I) The planetary orbits are ellipses, with the sun at one focus.

II) The areal velocity of a planet is constant.

III) The square of the period of an orbit is proportional to the cube of the semimajor axis of the ellipse.

We have shown that the trajectories of motion in a gravitational field are conic sections. The only conic section which is closed is the ellipse. Hence the first law is verified. We have also seen that the areal velocity is constant in any central-force problem. Thus only the third law remains to be checked.

The area of an ellipse with semimajor axis a and semiminor axis b is

$$A = \pi ab.$$

Let τ be the period of the elliptic orbit; then, from the fact that $\frac{1}{2}C = \frac{1}{2}r^2\dot\theta$ is the constant areal velocity, we have

$$\frac{1}{2}C\tau = \pi ab \qquad \text{or} \qquad \tau = \frac{2\pi ab}{C}. \tag{a}$$

From (5.10), we have

$$ep = C^2/\gamma m_0, \tag{b}$$

and (5.17a, b) gives us

$$ep = b^2/a. \tag{c}$$

Thus

$$\tau^2 = \frac{4\pi^2 a^2 b^2}{C^2} = \frac{4\pi^2 a^2 b^2}{\gamma m_0 (b^2/a)} = \left(\frac{4\pi^2}{\gamma m_0}\right) a^3,$$

where m_0 is now the mass of the sun. This shows the proportionality expressed in the third law.

One further exercise would seem appropriate here. Suppose that we know Kepler's laws, but don't know the form of the law of gravitation. We can use the following procedure to derive the expression for the force of gravity.

The equation of motion for m is (5.7i),

$$m(\ddot r - r\dot\theta^2) + f = 0, \tag{a}$$

where $r^2\dot\theta = C$. We know that the orbit is an ellipse,

$$r = \frac{pe}{1 - e\cos\theta}, \qquad 0 \leqslant e < 1. \tag{b}$$

Transforming (a) into the spatial coordinate θ yields [see (5.11)]

$$\frac{d}{d\theta}\left(\frac{1}{r^2}\frac{dr}{d\theta}\right) - \frac{1}{r} = -\frac{fr^2}{mC^2}. \tag{c}$$

From (b), we get

$$\frac{dr}{d\theta} = \frac{-pe^2 \sin \theta}{(1 - e \cos \theta)^2} = \frac{-r^2}{p} \sin \theta.$$

Thus (c) becomes

$$\frac{d}{d\theta}\left(\frac{-1}{p} \sin \theta\right) - \frac{1}{r} = -\frac{fr^2}{mC^2}$$

or

$$\frac{-1}{p} \cos \theta - \frac{1}{r} = -\frac{fr^2}{mC^2}. \tag{d}$$

Solving for $\cos \theta$ from (b) and inserting in (d) gives

$$\cos \theta = \frac{1}{e} - \frac{p}{r}, \qquad \frac{-1}{pe} + \frac{1}{r} - \frac{1}{r} = -\frac{fr^2}{mC^2}$$

or

$$f = \left(\frac{mC^2}{pe}\right)\frac{1}{r^2}. \tag{e}$$

This shows that the force of gravity is proportional to $(1/r^2)$.

7. SOLUTION OF CENTRAL-FORCE PROBLEMS

The main problem involved in central-force motion problems is finding the trajectory. In the particular case of motion in a $(1/r^2)$ gravitational field, the trajectories can be shown to be conic sections. Once the parameters E and C are determined from some initial conditions, the particular trajectory can be determined as $r = r(\theta)$ using (5.10).

If we wish to know the velocity of m at a given position on the trajectory, we need only evaluate r at that position and find $\dot\theta$ from $\dot\theta = (C/r^2)$.

If we wish to know $r(t)$ and $\theta(t)$, the position of m as a function of time, we need only find $\theta(t)$ and then, from the trajectory equation,

$$r(t) = r[\theta(t)].$$

To obtain $\theta(t)$, we again use the (constant) areal velocity, $\frac{1}{2}C$. Since

$$dA = \tfrac{1}{2}r^2 \, d\theta = \tfrac{1}{2}C \, dt,$$

we obtain

$$\int_0^\theta dA = \frac{1}{2} \int_0^\theta r^2 \, d\theta = \frac{1}{2} \int_0^t C \, dt.$$

where t is the value of time which corresponds to the angle θ. Thus

$$t = \frac{1}{C} \int_0^\theta r^2 \, d\theta. \tag{5.22}$$

Equation (5.22) gives $t = t(\theta)$, which must be inverted to give $\theta = \theta(t)$, as desired. This inversion may well be difficult, and we might be able to accomplish the inversion most easily by graphical means.

Example 11

A particle m is in an elliptic trajectory about m_0:

$$r = \frac{pe}{1 - e \cos \theta}, \qquad 0 \le e < 1.$$

Find $r(t)$ and $\theta(t)$.

Solution

We have, by (5.22),

$$t = \frac{p^2 e^2}{C} \int_0^\theta \frac{d\theta}{(1 - e \cos \theta)^2}.$$

We can evaluate this integral by using the transformation* $z = \tan(\theta/2)$. We get

$$\cos \theta = \frac{1 - z^2}{1 + z^2}$$

and

$$d\theta = \frac{2dz}{1 + z^2},$$

and the integral becomes

$$\frac{2p^2 e^2}{(1 - e)^2 C} \int \frac{(1 + z^2)dz}{\left[1 + \left(\dfrac{1 + e}{1 - e} \right) z^2 \right]^2}. \tag{a}$$

---

* See G. B. Thomas, *op. cit.*, page 300.

After some manipulation, we can evaluate (a) as

$$t = \frac{p^2 e^2}{(1-e)^2 C} \left\{ \left[\left(\frac{1-e}{1+e}\right)^{1/2} + \left(\frac{1-e}{1+e}\right)^{3/2} \right] \tan^{-1} \left(\sqrt{\frac{1-e}{1+e}} \, \tan \frac{\theta}{2} \right) \right.$$

$$\left. + \left[\left(\frac{1-e}{1+e}\right)^{1/2} - \left(\frac{1-e}{1+e}\right)^{3/2} \right] \frac{(1-e)\tan\left(\frac{\theta}{2}\right)}{\left[1 + \left(\frac{1+e}{1-e}\right) \tan^2 \frac{\theta}{2} \right]} \right\}, \quad \text{(b)}$$

where $t = 0$ at $\theta = 0$. In theory, (b) can be inverted to yield $\theta = \theta(t)$.

To check to see that (b) is reasonable, we can take the particular case of a circle. Recall that we let $e \to 0$ as $ep \to a$, the radius of the circle. Then

$$\theta = 2 \tan^{-1} \left[\tan \frac{C}{2a^2} t \right] = 2 \frac{C}{2a^2} t = \frac{C}{a^2} t.$$

But in this case $C = a^2 \dot\theta$; thus $\dot\theta = \text{const}$, and the above gives

$$\theta = \dot\theta t.$$

....

8. SUMMARY

Central-force motion of a particle m: The force on m always passes through a point O, the origin of an inertial reference frame.

Central-force motion: (i) always occurs in a single plane, (ii) has a constant areal velocity

$$\frac{dA}{dt} = \tfrac{1}{2} r^2 \dot\theta = \tfrac{1}{2} C.$$

Central-force motion in the inverse-square, gravitational field:

$$\mathbf{F} = - \frac{\gamma m_0 m}{r^2} \, \mathbf{e}_r.*$$

The trajectories of m are conic sections,

$$r = \frac{pe}{1 - e \cos \theta},$$

where

$$pe = C^2 / \gamma m_0,* \qquad e = \sqrt{1 + 2EC^2/\gamma^2 m_0^2}.*$$

Here $C = r^2 \dot\theta$ (twice the areal velocity) and $E = [\tfrac{1}{2}(\dot r^2 + C^2/r^2) - \gamma m_0/r]$ (the total energy per mass).

If $0 \le e < 1$, the particle is in a closed (elliptic) orbit about m_0. If $e \ge 1$, the trajectory is either a parabola or a hyperbola, and we say that the particle

---

* The term γm_e (m_e being the mass of the earth) has the value 1.255×10^{12} mi^3/hr^2.

m escapes the gravitational field of m_0. The minimum velocity which m must have at the surface of m_0 in order to escape is called the escape velocity

$$v_0 = \sqrt{\frac{2\gamma m_0}{R}} \cdot *$$

The period τ of an elliptic orbit is given by

$$\tau = 2\pi \sqrt{a^3/\gamma m_0} \, ,*$$

where a is the semimajor axis of the ellipse.

PROBLEMS

Section 5.2

Problem 5.1

The particle m shown in the figure is in central-force motion about the point O. At the position shown, the speed of m is 100 ft/sec in the direction shown. Determine the constant $C = r^2\dot\theta$.

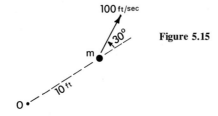

Figure 5.15

Problem 5.2

The particle m shown in the figure coasts along the line l until it reaches the point A. Then a central force of finite magnitude, passing through the point O, is turned on. Calculate the constant $C = r^2\dot\theta$.

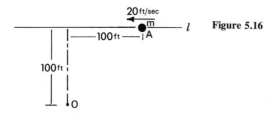

Figure 5.16

---

* The term γm_e (m_e being the mass of the earth) has the value 1.255×10^{12} mi³/hr².

Problem 5.3

The particles A and B shown in the figure are both in central-force motion about the point O. At $t = 0$, $r = 10$ ft for both particles, but the velocity of A is $v_A = 5e_\theta$ ft/sec and the velocity of B is $v_B = 4e_\theta$ ft/sec.

  a) Given that r for both particles is held at 10 ft, calculate the time required for particle A to "lap" particle B.

  b) Suppose the particles are pulled into a new radius of 5 ft. What would be the time required for particle A to "lap" particle B in the new configuration?

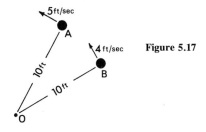

Figure 5.17

Problem 5.4

Two particles A and B are in central-force motion about the point O. In the position shown in the figure,

$$\dot{\theta}_A = \dot{\theta}_B = 60 \text{ rpm.}$$

Particle A is then drawn in to the radius of particle B. How long does it take for particle A to "lap" particle B?

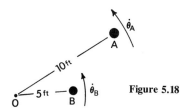

Figure 5.18

Problem 5.5

Suppose that the trajectory of a particle in central-force motion is given in the form $r = f(\theta)$, where the function $f(\theta)$ is given, and that the constant $C = r^2\dot{\theta}$ is also given. Take $\theta = \theta_0$ at $t = 0$.

  a) Show that the relation $t = t(\theta)$ can be represented as

$$t = \frac{1}{C} \int_{\theta_0}^{\theta} [f(\theta)]^2 \, d\theta.$$

b) Suppose that the relation of part (a) can be inverted to give $\theta = \theta(t)$. Show how $r = r(t)$ can be obtained.

Problem 5.6

A particle m is in central force motion about O. It is given that $\dot{\theta} = (10 + 5t)$ rad/sec. At $t = 0$, $r = 3$ ft.

    a) Calculate $r = r(t)$.

    b) Calculate the time at which $r = 2$ ft.

    c) Calculate the time at which $r = 1$ ft.

Problem 5.7

Again consider Problem 5.6.

    a) Compute $\theta = \theta(t)$. (Take $\theta = 0$ at $t = 0$.)

    b) Compute θ (rev) at $t = 2.5$ sec (i.e., when $r = 2$ ft).

    c) Compute θ (rev) at $t = 16$ sec (i.e., when $r = 1$ ft).

Problem 5.8

In Problems 5.6 and 5.7, we had

$$r = \left(\frac{90}{10 + 5t} \right)^{1/2} \text{ft}, \qquad \dot{\theta} = 10 + 5t \text{ rad/sec},$$

and $\theta = 0$ at $t = 0$. Given that the mass of the particle $m = 1$ slug, determine the central force:

    a) as a function of time t,

    b) as a function of r.

Problem 5.9

A particle in central-force motion has the trajectory $r = 2\theta$ ft (an Archimedean spiral). Suppose that at $t = 0$, $\theta = 90°$ and $\dot{\theta} = 3$ rad/sec.

    a) Find $\theta(t)$.   b) Find $r(t)$.

[*Hint:* Use Problem 5.5.]

Problem 5.10

A particle in central-force motion has the trajectory $r = e^{2\theta}$ ft, a logarithmic spiral. At $t = 0$, $\theta = 0$, $\dot{\theta} = 5$ rad/sec.

    a) Find $\theta(t)$.   b) Find $r(t)$.

[*Hint:* Use Problem 5.5.]

Problem 5.11

A particle m in central-force motion has the trajectory $r = 10/\theta$ ft, a hyperbolic spiral. At $t = 0$, $\theta = 2$ rad and $\dot{\theta} = 4$ rad/sec. Suppose that $m = 1$ slug.

    a) Show that $\theta(t) = \dfrac{2}{1 - 2t}$ rad.

    b) Show that $r(t) = 5(1 - 2t)$ ft.

    c) Show that the central force is $-\dfrac{10^4}{r^3}\, \mathbf{e}_r$ lb.

[*Hint:* Use Problem 5.5.]

Problem 5.12

Again consider Problem 5.11.

    a) Determine the kinetic energy at $\theta = 2$ rad.

    b) Determine the kinetic energy at $\theta = 5$ rad.

    c) Determine the work done by the central force during the motion from $\theta = 2$ rad to $\theta = 5$ rad. [*Hint:* Compute the change in kinetic energy.]

Problem 5.13

The central force in Problems 5.11 and 5.12 was

$$\mathbf{F} = -\frac{10^4}{r^3}\, \mathbf{e}_r.$$

    a) Write a potential energy function for the central force.

    b) Compute the difference in potential energy between $r = 5$ ft and $r = 2$ ft and confirm the computation of the work in Problem 5.12, part (c).

Problem 5.14

Suppose that in Example 5.5, $k = 0.2$ lb/in and $mg = 2$ lb. Suppose initially that

    $x(0) = 10$ in., $\dot{x}(0) = 0$,

    $y(0) = 0$, $\dot{y}(0) = 12.4$ in./sec.

    a) Find the positions at which the speed of m is
        i) maximum⎱
                Determine the speeds.
        ii) minimum⎰

    b) Find the positions at which the spring force is
        i) maximum⎱
                Determine the forces.
        ii) minimum⎰

Problem 5.15

For the system in Example 5.5, suppose that the particle weighs 6.4 lb and $k = 2$ lb/in. Given the initial conditions

$$x = 0, \qquad \dot{x} = 5 \text{ in./sec}$$
$$y = 10 \text{ in.}, \qquad \dot{y} = 3 \text{ in./sec},$$

calculate the equation of the trajectory. [*Hint:* Find $x(t)$ and $y(t)$, then eliminate time between the terms.]

Problem 5.16

We have written an element of area as

$$dA = \tfrac{1}{2}r^2 \, d\theta.$$

Write an expression for dA including the second-order terms involving dr and $d\theta$.

Problem 5.17

The figure shows a conical pendulum, whose string passes through the hole at O.

    a) Is this a case of central-force motion?

    b) Determine expressions for $\mathbf{H}_O$, $\mathbf{M}_O$ and, in particular, H_{Oz} and M_{Oz} in terms of r, θ, and z.

    c) Show that $C = r^2\dot{\theta}$ is constant.

    d) Suppose that the string is pulled through the hole in some arbitrary fashion. Is the system conservative?

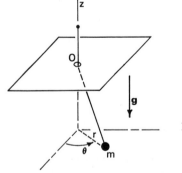

Figure 5.19

Problem 5.18

In Problem 5.17, assume that the length l of the string below O is held constant, l.

    a) Is the system conservative?

b) Let E be the energy per unit mass of the system and let $C = r^2\dot\theta$. Suppose that $\theta = \theta_0$ when $r = r_0$. Determine a definite integral from which the trajectory $r = r(\theta)$ could be determined.

Section 5.3

Problem 5.19

Give an example of a central-force problem in which the total energy $(T + V)$ is *not* constant.

Problem 5.20

The particle in Fig. 5.20 is given an initial velocity $\mathbf{v}_0$ as shown. The total force on the particle is $-f\mathbf{e}_r$, where f is constant.

a) Compute $C = r^2\dot\theta$.

b) Write a potential energy function for the force on the particle.

c) Determine $\dot r$ as a function of r. [*Hint:* Use energy considerations.]

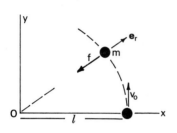

Figure 5.20

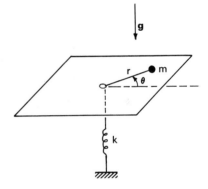

Figure 5.21

Problem 5.21

The particle m shown in Fig. 5.21 moves on a smooth, horizontal table. The string attached to it is connected to a linear spring of modulus k. The spring force is zero when $r = 0$. Initially $r = R$, $\dot r = 0$, and $\dot\theta = \omega$.

a) Write a differential equation from which $r(t)$ could be determined.

b) Determine $\dot r$ as a function of r. [*Hint:* Use energy considerations.]

Problem 5.22

A constant force P is applied to the string shown in Fig. 5.22, which is attached to a mass m which moves on a smooth, horizontal table.

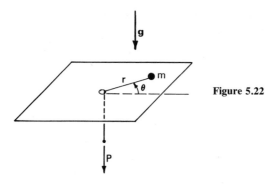

Figure 5.22

a) Is the force P conservative? If so, write its potential energy.

b) Suppose that when $r = r_0$, $\theta = \theta_0$. Denoting the total energy by mE, calculate a definite integral from which the trajectory $r = r(\theta)$ could be determined.

Problem 5.23

The two particles shown in the figure are connected by a string. Assume that the two particles have the same mass, and that the upper particle moves on a smooth, horizontal plane. At $t = 0$, $r = 2$ ft, $\dot{r} = 0$, and $\dot{\theta} = 10$ rad/sec.

a) Write $\mathbf{F} = \dot{\mathbf{p}}$ for each particle so that you get a differential equation from which $r(t)$ could be obtained.

b) Determine $\dot{r}$ as a function of r. [*Hint:* Use energy considerations.]

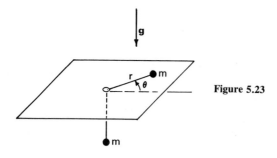

Figure 5.23

Problem 5.24

Show that $A \cos \theta + B \sin \theta = D \cos (\theta + \delta)$ if

$$D^2 = A^2 + B^2 \quad \text{and} \quad \delta = \tan^{-1}(-B/A).$$

Problem 5.25

The projectile shown in the figure is given the speed $v_0 = 2.0 \times 10^4$ mi/hr at the earth's surface. Given that the initial velocity is at 45° to the local vertical:

a) Find the parameters E and C.

b) Find the parameters e and pe.

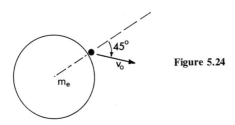

Figure 5.24

Problem 5.26

A body is given the speed of 1.5×10^4 mi/hr at the surface of the earth. The initial velocity occurs at an angle θ from the local vertical.

a) Calculate the parameters E and C as functions of θ.

b) Calculate the parameters e and ep as functions of θ.

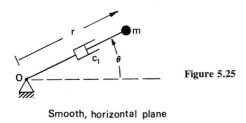

Figure 5.25

Smooth, horizontal plane

Problem 5.27

Consider the system shown in the figure.

a) Calculate the differential equation of motion for m.

b) Use Eq. (5.8) to determine a differential equation for the trajectory of m, $r = r(\theta)$.

Section 5.4

Problem 5.28

An ellipse is described in terms of the parameters p and e:

$$r = \frac{pe}{1 - e \cos \theta}.$$

Show that the semiminor axis b can be expressed as

$$b = \frac{pe}{\sqrt{1 - e^2}}.$$

[*Hint:* Consider the maximum of $y = r \sin \theta$.]

Problem 5.29

A particle has a parabolic trajectory $e = 1$:

$$r = \frac{p}{1 - \cos \theta}.$$

a) What is the minimum value of r?

b) Determine $r(\tfrac{1}{2}\pi)$ and $r(\tfrac{3}{2}\pi)$.

c) Sketch the trajectory $r = r(\theta)$.

Problem 5.30

A satellite is launched from the position shown in the figure. Find v_0 such that the trajectory is a circle.

a) Use the condition $e = 0$ to determine v_0.

b) Write $\mathbf{F} = \dot{\mathbf{p}}$ to determine v_0.

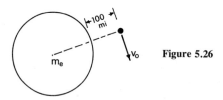

Figure 5.26

Problem 5.31

In Problem 5.30, find v_0 such that the trajectory of the satellite is a parabola. Find the equation of the trajectory.

Problem 5.32

The eccentricity of the trajectory of a satellite in motion about the earth is $e = 0.5$. The semimajor axis of the trajectory is $a = 8000$ miles.

a) What is the semiminor axis b?

b) Determine the apogee and perigee of the trajectory.

Problem 5.33

A satellite is in an *elliptical* trajectory about the earth. Show that the time required for one complete orbit can be written

$$\tau = \frac{\sqrt{2}}{2} \pi \frac{\gamma m_e}{(-E)^{3/2}}, \qquad E < 0.$$

[*Hint:* The area of an ellipse is $A = \pi ab$, where a and b are the semimajor and semiminor axes of the ellipse.]

Problem 5.34

The apogee and perigee of a satellite's trajectory are

$$r_1 = 6000 \text{ mi}, \qquad r_2 = 4500 \text{ mi}.$$

a) Determine the semimajor and semiminor axes of the trajectory.

b) Determine the time required for one complete orbit of the earth. [*Hint:* See Problem 5.33.]

Problem 5.35

A satellite is in a circular orbit 100 miles above the surface of the earth.

a) What is the speed of the satellite?

b) What is the time it requires for one complete orbit?

Problem 5.36

A satellite in a circular orbit about the earth has the same period of orbit as the period of the earth's rotation (i.e., the time it takes the satellite to make one complete orbit is 24 hr; thus the satellite appears to have a stationary position over the earth).

a) Determine the height of the satellite above the earth's surface.

b) Determine the speed of the satellite.

Problem 5.37

A bullet is shot from a gun at the surface of the earth ($R = 4000$ miles) with a muzzle velocity of 400 mi/hr at 45° to the local vertical.

a) Determine the parameters e and ep for the bullet's trajectory.

b) In Example 19 of Chapter 4, we found that the trajectory of a projectile is a parabola. Does this jibe with the result of part (a)?

Section 5.5

Problem 5.38

A satellite is in a circular trajectory of radius r_0 about the earth.

a) What is the areal velocity of the satellite?

b) Determine the period of the motion (the time required for one complete orbit of the earth).

c) Determine the periods of trajectories with radii (i) $r_0 = R + 100$ mi, (ii) $r_0 = R + 200$ mi, where R is the radius of the earth.

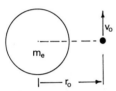

Figure 5.27

Problem 5.39

A satellite has the speed v_0 at the position shown in Fig. 5.27.

a) Determine the trajectory of the satellite.

b) Suppose that $e < 1$. Determine the apogee and perigee of the trajectory.

c) For what speed v_0 will the trajectory be a circle?

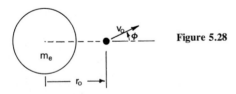

Figure 5.28

Problem 5.40

The satellite shown in Fig. 5.28 has speed v_0 at the inclination angle ϕ to the local vertical when $r = r_0$.

a) Determine the eccentricity of the trajectory of the satellite.

b) Under what conditions (on the parameters r_0, v_0, and ϕ) will the trajectory be a circle about the earth?

Problem 5.41

Suppose that, at the earth's surface, a bullet is fired from a gun, which is inclined at an angle α from the local vertical (see the figure). The speed of the bullet is the escape velocity.

a) Evaluate the parameters E and C.

b) Determine the parameters e and p which describe the trajectory of the bullet.

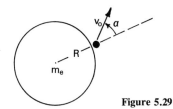

Figure 5.29

Problem 5.42

The satellite shown in the figure is in a circular trajectory about the earth at a radius r_0. Let v_0 be the speed of the satellite.

a) Determine v_0.

b) An increment in velocity Δv is given to the satellite at the angle α shown. Determine Δv so that the satellite will escape the earth's gravitational field.

c) Find the maximum and minimum values of $|\Delta v|$ in order that the satellite just escapes. Consider the range $0 \leqslant \alpha \leqslant \frac{1}{2}\pi$ rad.

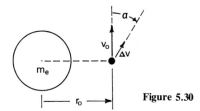

Figure 5.30

Problem 5.43

A rocket is in a circular trajectory about the earth at a radius $a = 4500$ mi.

a) What increment in tangential velocity is required in order that the rocket will escape the earth's gravitational field?

b) Suppose that the increment in velocity is in the radial direction. What is the minimum increment for escape?

Problem 5.44

A satellite is in a circular orbit 100 miles above the surface of the moon. (Refer to Example 9.)

   a) What is the speed of the satellite?

   b) How long does the satellite take to complete one orbit? (Compare the above results to those obtained in Problem 5.35.)

Problem 5.45

A rocket is in a circular trajectory about the *moon* at a radius of 1200 mi (refer to Example 9).

   a) What is the speed of the rocket?

   b) What is the minimum increment in tangential velocity required for the rocket to escape the moon's gravitational field?

   c) What is the minimum increment in radial velocity which the rocket requires for escape?

Section 5.6

Problem 5.46

Taking the semimajor axis of the earth's trajectory about the sun to be 9.3×10^7 mi and the period of the earth's orbit to be one year, show that the quantity γm_0 has the value 3.17×10^{25} $(\text{mi}^3/\text{yr}^2)$, where m_0 is the mass of the sun.

Problem 5.47

The planet Venus is in a trajectory about the sun with semimajor axis equal to 0.72 times that of the earth's trajectory. Find the period of Venus' orbit. [*Hint:* See Problem 5.46.]

Problem 5.48

The planet Pluto has a trajectory about the sun with a semimajor axis 39.5 times as long as that of the earth's trajectory. Determine the period of Pluto's orbit. [*Hint:* See Problem 5.46.]

★ Problem 5.49

Write the equation

$$mE = \tfrac{1}{2} m(\dot{r}^2 + r^2\dot{\theta}^2) + V(r) = \text{const},$$

and show that if the trajectory $r = r(\theta)$ is a conic section, then the potential energy is proportional to $-1/r$. (This is analogous to showing that the central force is proportional to $-1/r^2$.)

   ★ Starred problems are those which are more difficult than average.

Problem 5.50

A particle in central-force motion about the point O has the trajectory $r = e^{a\theta}$ ft, $a =$ constant. Show that the (inward) central force is

$$f = mC^2(1 + a^2)(1/r^3).$$

Section 5.7

Problem 5.51

A particle in central-force motion has the trajectory

$$r = \frac{10}{\theta} \text{ mi.}$$

At $t = 0$, $\theta = 1$ rad, and the speed of the particle is $v = \sqrt{2}$ mi/min, with the particle moving inward.

    a) Determine the constant $C = r^2\dot{\theta}$.

    b) Determine $\theta = \theta(t)$ and $r = r(t)$.

Problem 5.52

A particle in central-force motion has the trajectory

$$r = 2\theta \text{ ft.}$$

At $t = 0$, $\theta = 3$ rad and $\dot{\theta} = 1$ rad/sec.

    a) Determine the constant $C = r^2\dot{\theta}$.

    b) Determine $\theta = \theta(t)$ and $r = r(t)$.

Problem 5.53

A particle in central-force motion has the trajectory

$$r = 10e^{-2\theta} \text{ ft.}$$

At $t = 0$, $\theta = 0$, and the speed of the particle is $v = \sqrt{5}$ ft/sec, with the particle moving inward.

    a) Calculate the constant $C = r^2\dot{\theta}$.

    b) Calculate the relation $t = t(\theta)$.

    c) Calculate the relation $\theta = \theta(t)$, and hence determine $r = r(t)$.

★ Problem 5.54

A body is in a parabolic trajectory about the earth (see the figure):

$$r = \frac{p}{1 - \cos\theta}.$$

    a) Compute the areal velocity of the body.

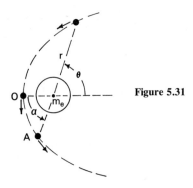

Figure 5.31

b) Suppose that, at $t = 0$, the body is at the point O. How long will it take the body to arrive at the point A?

c) Let $p = 12,000$ mi and $\alpha = \frac{1}{2}\pi$ rad. How long does it take the body to get from O to A?

6
ELEMENTARY VIBRATION THEORY

Turandot:

Ice which gives you fire
and which your fire freezes
still more!
Lily-white and dark,
if it allows you your freedom,
it makes you a slave;
if it accepts you as a slave,
it makes you a King!

Hurry, stranger, you turn pale
from fear!
You know you are lost!
Hurry, stranger, ice which gives
you fire,
what is it?

Turandot
Giaccomo Puccini—Giaccomo Adami

1. INTRODUCTION

N. W. McLachlan began his book* on vibrations with the following statement: "Vibration is ubiquitous."

Vibrations occur in every facet of life. They are present in one form or another in nearly every engineering project. The theory of vibrations forms the heart of the theory of wave motion and control theory. And if one were to name the one most important topic in dynamics, it would probably be vibrations.

Vibratory motion is often undesirable or even dangerous. Therefore much of the study of vibrations is concerned with means of eliminating or at least minimizing this type of motion. However, before we can learn to control

---

* N. W. McLachlan, *Theory of Vibrations*, New York: Dover, 1951.

or isolate vibrations, we must first learn to analyze the motions. Let us first examine the most important aspect of elementary vibration theory: the motion of a linear, one-degree-of-freedom oscillator. It is fortunate (and remarkable) that a complete analysis of this rather simple problem contains the essential details of the vibrations of much more complicated systems: systems with more than one degree of freedom, and continuous systems such as beams and structures.

We shall proceed with our study of vibrations in the following sequence:

a)   free motion of an undamped system

b)   harmonically forced motion of an undamped system

c)   free motion of a damped system

d)   harmonically forced motion of a damped system

e)   arbitrarily forced motion of a damped system

The systems we shall study in this chapter are relatively simple. We investigate them in order to give a clear idea of the fundamental concepts of vibrations: natural frequency, forcing frequency, amplitude, phase, damping ratio, etc. This is not to say that only simple systems vibrate. In fact all systems—complicated and simple alike—vibrate, and the notion of a natural frequency has the same significance for all systems.

In Chapter 8 we shall give several examples which involve the vibrations of rigid bodies. You will find that the mathematical analysis of these problems is remarkably similar to the analysis of the simple problems in this chapter.

Before we get going, let's say a few words about the mathematics of this chapter. We shall be studying one basic differential equation:

$$\ddot{x} + 2\zeta\omega_n\dot{x} + \omega_n^2 x = F_0(t),$$

where $2\zeta\omega_n$ and ω_n^2 are constants and $F_0(t)$ is usually taken to be a harmonic function ($\sin \omega t$ or $\cos \omega t$). If you have not had a course in differential equations, you may feel unprepared for this material, but this is not true. We shall be deriving the solutions to this differential equation, often on an ad hoc basis. A mathematician might obtain these solutions with a bit more flair and eloquence, but he would obtain *the same solutions*. In any event, it is *not* your job to derive solutions; we have done that! Your job is to recognize a vibration problem when it occurs and to obtain the parameters which describe the problem (natural frequency, damping ratio, etc.). The final step of the analysis is to *plug* these parameters into the solutions we have provided. You *do not* derive the solution.

Thus — although a knowledge of differential equations is useful in understanding much of the text of this chapter — it is not necessary for understanding the examples or doing the problems.

2. EXAMPLES OF SIMPLE VIBRATION PROBLEMS

Example 1

Our first example of a vibration problem is given by the simple pendulum which we analyzed in Chapter 4 (Example 2). By writing $\mathbf{M}_O = \dot{\mathbf{H}}_O$, we had (see Fig. 6.1)

$$\mathbf{M}_O = \dot{\mathbf{H}}_O: -mgl \sin \theta \, \mathbf{k} = ml^2 \ddot{\theta} \, \mathbf{k} \qquad \text{or} \qquad \ddot{\theta} + \frac{g}{l} \sin \theta = 0.$$

As this equation stands, it is "nonlinear." We shall define this term shortly. For the time being, we point out that if we restrict ourselves to small angles such that

$$\sin \theta \approx \theta,$$

the governing equation of motion can be approximated by

$$\ddot{\theta} + \left(\frac{g}{l}\right) \theta = 0. \tag{a}$$

Figure 6.1

Example 2

Consider the so-called mass–spring oscillator shown in Fig. 6.2. The particle m can move without friction in the horizontal direction. The displacement x is taken relative

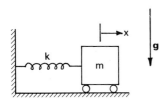

Figure 6.2

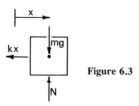

Figure 6.3

to the undeformed position of a linear spring of spring constant k. The free-body diagram for m in a deformed position is shown in Fig. 6.3. The weight mg and the normal force N cancel. Thus when we write $\mathbf{F} = \dot{\mathbf{p}}$, we get

$$F_x = ma_x: -kx = m\ddot{x} \quad \text{or} \quad \ddot{x} + \left(\frac{k}{m}\right)x = 0. \tag{a}$$

From a mathematical point of view, equations (a) of Example 1 and (a) of Example 2 are identical. We have the second derivative of some quantity plus a (positive) constant times the quantity itself equal to zero. If we can solve (a) of Example 1, then we can obtain the solution of (a) of Example 2 by making the analogy

$$\theta \to x, \quad \left(\frac{g}{l}\right) \to \frac{k}{m}.$$

....

Example 3

Figure 6.4 shows a third example of a vibration problem. A large mass m is attached to a wall by means of a linear spring of modulus k. Attached to the mass m is an unbalanced mass m_0, which rotates at a constant rate ω:

$$\dot{\theta} = \omega = \text{const.}$$

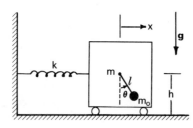

Figure 6.4

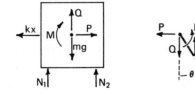

Figure 6.5

We measure the displacement of m by means of x measured from the equilibrium position of m. Figure 6.5 shows free-body diagrams for m, the weightless bar linking m_0 to m, and for m_0. (Can you justify the forces and moments shown on the bar?)

Let us assume that the mass m has no vertical motion. Then we can write the positions of m and m_0:

m: $\quad \mathbf{r} = x\,\mathbf{i}$,

m_0: $\quad \mathbf{r} = (x + l\sin\theta)\mathbf{i} + (h - l\cos\theta)\mathbf{j}$.

Thus we obtain the accelerations of the masses by differentiating the position vectors:

m: $\quad \ddot{\mathbf{r}} = \ddot{x}\,\mathbf{i}$,

m_0: $\quad \ddot{\mathbf{r}} = (\ddot{x} - l\omega^2\sin\omega t)\mathbf{i} + (l\omega^2\cos\omega t)\mathbf{j}$,

(recall that $\dot\theta = \omega = \text{const.}$) Writing $\mathbf{F} = \dot{\mathbf{p}}$ in the horizontal direction for each particle, we get:

Main mass, m	$F_x = ma_x$:	$P - kx = m\ddot{x}$	(a)
Unbalanced mass, m_0	$F_x = ma_x$:	$-P = m_0(\ddot{x} - l\omega^2\sin\omega t)$	(b)

If we add (a) to (b), the force P drops out, and we get

$$(m_0 + m)\ddot{x} + kx = m_0 l\omega^2 \sin\omega t \tag{c}$$

or

$$\ddot{x} + \left(\frac{k}{m_0 + m}\right)x = \left(\frac{m_0 l\omega^2}{m_0 + m}\right)\sin\omega t. \tag{d}$$

The left-hand side of (d) has the appearance of (a) of Examples 1 and 2. However there is a harmonic term on the right of (d). We shall speak of (d) as an example of *harmonically forced motion of an undamped system.*

If we had inserted a dashpot c in parallel with the spring in this problem, we would have had (prove it)

$$(m_0 + m)\ddot{x} + c\dot{x} + kx = m_0 l\omega^2 \sin\omega t \tag{c'}$$

in place of (c), or

$$\ddot{x} + \left(\frac{c}{m_0 + m}\right)\dot{x} + \left(\frac{k}{m_0 + m}\right)x = \frac{m_0 l\omega^2}{m_0 + m}\sin\omega t, \tag{d'}$$

instead of (d). This would have been an example of *harmonically forced motion of a damped system.*

In each of the final equations of motion for the three examples, we have written the unknown parameter on the left and the known parameter on the right. Each of these equations is an example of what is called a second-order, linear, differential equation with constant coefficients. The most important of these properties is the linearity.

Linearity in an equation can be recognized by the fact that the unknown and its derivatives (say x, $\dot{x}$, $\ddot{x}$) occur only to the first power. For example, a linear equation would *not* have terms such as

$$x\dot{x}, \qquad x^2, \qquad \dot{x}^2, \qquad \sin ax, \qquad \text{etc.}$$

However, an equation

$$\ddot{x} + t\dot{x} + t^2 x = f(t)$$

would be linear, even though some of its coefficients are now variable.

The fundamental ramification of linearity in a differential equation is that we can construct its solution in steps by adding (or superimposing) partial solutions.

Those of you who have yet to encounter a course in differential equations may be a bit mystified by this discussion (those of you who *have* studied differential equations should be bored!). We shall not assume that you know anything about differential equations, however. In this chapter we are concerned with the definition of various classes of vibration problems. We shall define those classes in terms of several standard differential equations.

In the course of this chapter and, in particular, in the chapter summary, we provide complete solutions to these differential equations. Thus your problem becomes:

  i) defining the differential equation for the problem at hand,

 ii) identifying the parameters of the equation with the parameters of the standard differential equations,

iii) writing the solution to the standard differential equation.

You do *not* derive the solution of the differential equation. To do so would be redundant, since we have provided the solutions. In fact, as we have said, this chapter does not depend on a previous course on differential equations; rather, it could serve as an introduction to such a course.

3. FREE MOTION OF AN UNDAMPED SYSTEM

Consider the mass–spring system in Fig. 6.2. We had

$$m\ddot{x} + kx = 0.$$

There are two ways to solve this equation. First we might note that the system is conservative, and the total energy*

$$E = \tfrac{1}{2}m\dot{x}^2 + \tfrac{1}{2}kx^2$$

is a constant. Thus solving for $\dot{x}$:

$$\frac{dx}{dt} = \sqrt{\frac{1}{m}(2E - kx^2)},$$

and thus

$$t + C = \int^x \frac{dx}{\sqrt{(1/m)(2E - kx^2)}} = \sqrt{\frac{m}{k}} \sin^{-1}\left(\frac{x}{\sqrt{2E/k}}\right)$$

or

$$x = \sqrt{\frac{2E}{k}} \sin\left[\sqrt{\frac{k}{m}}(t + C)\right]. \tag{6.1}$$

The value of E and C are determined from initial conditions, that is, the values of x and $\dot{x}$ at some particular time.

The above method is direct, well motivated, and always works for the free motion of an undamped system. For forced motion or for the motion of a damped system, energy is not conserved. Hence we shall now use another ad hoc approach to the solution of the equations.

Suppose we standardize the form of the differential equation. In general, we consider

$$\ddot{x} + \omega_n^2 x = 0. \tag{6.2}$$

Here ω_n is called the *natural circular frequency* of the system. In the mass-spring system, its value is $\sqrt{k/m}$.

To solve (6.2), we need a function which is very similar to its own second derivative. We might guess sine functions, cosine functions, or exponentials. Suppose we try

$$x(t) = Ce^{st}. \tag{6.3}$$

Inserting (6.3) into (6.2) yields

$$(s^2 + \omega_n^2)Ce^{st} = 0.$$

Thus $x = Ce^{st}$ will be a solution if either

(i) $(s^2 + \omega_n^2) = 0$ or   (ii) $Ce^{st} = 0$.

---

* In this chapter and subsequently, we shall use E to denote the total energy, as in Chapter 4. In Chapter 5, E had the special meaning of total energy per unit mass.

The second case is not very interesting, since it means $x = 0$. That is, no motion. From (i), we get

$$s = \pm i\omega_n, \qquad \text{where } i^2 = -1.$$

This says that either

$$x = C_1 e^{+i\omega_n t} \qquad \text{or} \qquad x = C_2 e^{-i\omega_n t}$$

is a solution. In fact, because of the linearity of (6.2), the combination

$$x = C_1 e^{+i\omega_n t} + C_2 e^{-i\omega_n t} \tag{6.4}$$

is a solution for arbitrary values of C_1 and C_2. The solution as presented here is a bit clumsy to use, since we are interested only in *real* solutions. From Euler's theorem,*

$$e^{i\theta} = \cos \theta + i \sin \theta, \qquad e^{-i\theta} = \cos \theta - i \sin \theta$$

Thus (6.4) becomes

$$x = i(C_1 - C_2) \sin \omega_n t + (C_1 + C_2) \cos \omega_n t.$$

Now for x to be real, we must have the imaginary parts of the constants multiplying sine and cosine be zero:

$$\text{Im}\{i(C_1 - C_2)\} = \text{Im}\{C_1 + C_2\} = 0.$$

If we let $C_1 = a_1 + ib_1$, etc., this shows that C_1 and C_2 must be complex conjugates. The point is that

$$x(t) = A \sin \omega_n t + B \cos \omega_n t \tag{6.5}$$

is a solution for arbitrary values of the *real* constants A and B.

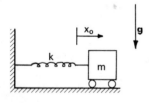

Figure 6.6

Example 4

Consider the mass–spring oscillator. In the first case (Fig. 6.6), the mass is displaced from equilibrium to a position x_0 and then released from rest at $t = 0$. In the second case (Fig. 6.7), the mass is hit with a hammer in the equilibrium position $x = 0$, giving the mass a speed v_0 at $t = 0$. Find the motion in each case.

---

* See Appendix 1, Eq. (A1.26).

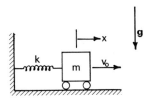

Figure 6.7

Solution

i)   The initial conditions are (Fig. 6.6)

$$x(0) = x_0, \qquad \dot{x}(0) = 0.$$

From (6.5), we have

$$x(0) = B = x_0, \qquad \dot{x}(0) = \omega_n A = 0.$$

Thus $x = x_0 \cos \omega_n t$.

ii)   The initial conditions are (Fig. 6.7)

$$x(0) = 0, \qquad \dot{x}(0) = v_0.$$

Thus, from (6.5), we have

$$x(0) = B = 0, \qquad \dot{x}(0) = \omega_n A = v_0.$$

Thus $x = \left(\dfrac{v_0}{\omega_n}\right) \sin \omega_n t$.

· ·

Suppose that at $t = 0$ we had the following initial conditions for the system described by (6.2):

$$x(0) = x_0 \qquad \text{and} \qquad \dot{x}(0) = v_0.$$

Then the solution (6.5) would become

$$x(t) = \left(\frac{v_0}{\omega_n}\right) \sin \omega_n t + x_0 \cos \omega_n t.$$

There is an alternative form of (6.5) which is useful. Since

$$\sin (\alpha - \beta) = \sin \alpha \cos \beta - \cos \alpha \sin \beta,$$

we can try to write

$$A \sin \omega_n t + B \cos \omega_n t = A_0 \sin (\omega_n t - \phi)$$

$$= A_0 (\sin \omega_n t) \cos \phi - A_0 (\cos \omega_n t) \sin \phi.$$

Equating the factors of $\sin \omega_n t$ and $\cos \omega_n t$, we get

$$A = A_0 \cos \phi, \qquad B = -A_0 \sin \phi,$$

which must be satisfied by A_0 and ϕ if the representation is to be valid. To solve these equations, we note that

$$A^2 + B^2 = A_0^2 \cos^2 \phi + A_0^2 \sin^2 \phi = A_0^2, \qquad -\frac{B}{A} = \tan \phi.$$

Thus

$$A_0 = \sqrt{A^2 + B^2}, \qquad \phi = \tan^{-1}\left(-\frac{B}{A}\right).$$

Thus (6.5) can be written

$$x(t) = A_0 \sin (\omega_n t - \phi). \tag{6.6}$$

Figure 6.8 shows a plot of (6.6).

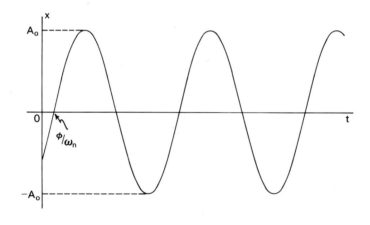

Figure 6.8

In practice, representing the solution by (6.5) may well be as useful as representing it by (6.6). However, (6.6) shows that no matter what the initial conditions might be, the free motion of an undamped vibrator is a sinusoidal function with amplitude A_0. In the present case, the quantity ϕ, called the *phase angle*, has limited importance, since we could eliminate it from the equations by shifting the origin of time. The phase angle will play a more important role in the case of the forced motion of a damped system. There we shall want to measure the difference in phase between the force and the response.

Suppose now that we are confronted with an undamped vibratory system in free motion. Our problem consists of two parts:

i)   Find the undamped natural frequency ω_n.

ii) Using the initial conditions, find the motion, Eq. (6.5).

How do we find ω_n? There are two useful techniques.

First we can write the differential equation of motion and compare it with (6.2). For the pendulum in Example 7, we had

$$\ddot{\theta} + \left(\frac{g}{l}\right)\theta = 0,$$

which we compare with (6.2),

$$\ddot{x} + \omega_n^2 x = 0.$$

Thus we have the comparison

$$x \to \theta, \qquad \omega_n^2 \to \left(\frac{g}{l}\right),$$

and the solution (6.5) now becomes

$$\theta = A \sin \sqrt{g/l}\ t + B \cos \sqrt{g/l}\ t.$$

Example 5

A simple pendulum of length l (Fig. 6.1) has the initial conditions at

$$t = 0: \qquad \theta = \theta_0, \qquad \dot{\theta} = 0.$$

Find the solution, $\theta = \theta(t)$.

Solution

From (6.5) we write

$$\theta = A \sin \omega_n t + B \cos \omega_n t$$

where ω_n has been determined to be $\omega_n = \sqrt{g/l}$. At $t = 0$,

$$\theta = B = \theta_0, \qquad B = \theta_0,$$

and

$$\dot{\theta} = \omega_n A = 0, \qquad A = 0.$$

Thus

$$\theta = \theta_0 \cos \sqrt{\frac{g}{l}} \; t.$$

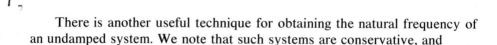

There is another useful technique for obtaining the natural frequency of an undamped system. We note that such systems are conservative, and

$$T + V = E = \text{constant.} \tag{a}$$

Consider the mass–spring system in Fig. 6.2:

$$\tfrac{1}{2}m\dot{x}^2 + \tfrac{1}{2}kx^2 = E. \tag{b}$$

Now suppose that

$$x = A \sin \omega_n t. \tag{c}$$

We assume that the system is in free motion at its natural frequency. We do not assume that we know the expression for ω_n, however. Inserting (c) into (b), we obtain

$$\tfrac{1}{2}m\omega_n^2 A^2 \cos^2 \omega_n t + \tfrac{1}{2}kA^2 \sin^2 \omega_n t = E. \tag{d}$$

Consider the two times

$$\text{(i)} \quad \omega_n t_1 = 0 \quad \text{and} \quad \text{(ii)} \quad \omega_n t_2 = \frac{\pi}{2}.$$

For (i), (d) becomes

$$\tfrac{1}{2}m\omega_n^2 A^2 = E \tag{e}$$

and for (ii)

$$\tfrac{1}{2}kA^2 = E. \tag{f}$$

Since E, the total energy, is constant, we have from (e) and (f) that

$$\tfrac{1}{2}m\omega_n^2 A^2 = \tfrac{1}{2}kA^2$$

or

$$\omega_n^2 = \frac{k}{m}.$$

Thus the parameter ω_n is determined.

In effect, what we have done above is the following: We note that the energy in an undamped system oscillates from kinetic to potential energy. This means that at some time, the kinetic energy attains its maximum value, and at other times, all the energy is potential energy. Suppose that when

$$T = T_{max}, \qquad V = V_0,$$

and when

$$T = 0, \qquad V = V_1.$$

Then, from (a), we get

$$T_{max} + V_0 = E, \qquad 0 + V_1 = E$$

and we have

$$T_{max} = V_{max}, \tag{6.7}$$

where we have taken

$$V_{max} = V_1 - V_0.$$

To use (6.7) to determine ω_n, we proceed as follows.

i) Write the energy functions T and V.

ii) Assume a harmonic solution (for example, $x = A \sin \omega_n t$, $\theta = \theta_0 \sin \omega_n t$, etc.).

iii) Put the harmonic solution into T and V and apply (6.7). The above procedure is called *Rayleigh's method.**

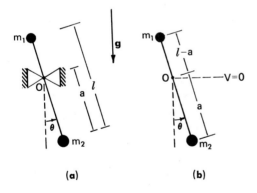

(a)   (b)   **Figure 6.9**

Example 6

Figure 6.9(a) shows two particles m_1 and m_2 connected by a rigid, weightless bar of length l. The bar is pinned to a support at a distance a from the lower mass m_2.

---

   * See S. H. Crandall, *Engineering Analysis*, New York: McGraw-Hill, 1956, pages 83–84.

Find the undamped natural frequency ω_n, and find the value of a so that ω_n is maximum and is minimum for the case $m_2 = 2m_1$.

Solution

Figure 6.9(b) shows the pendulum at an inclination angle θ.

Let the pin O be the zero level of potential energy. Let θ_1 be the maximum angular displacement. Then

$$V_1 = m_1 g(l - a) \cos \theta_1 - m_2 ga \cos \theta_1.$$

The kinetic energy is maximum at $\theta = 0$. Thus

$$V_0 = m_1 g(l - a) - m_2 ga.$$

And therefore

$$V_{max} = -m_1 g(l - a)(1 - \cos \theta_1) + m_2 ga(1 - \cos \theta_1). \tag{a}$$

We are interested in small motions of the system. Thus we take*

$$\cos \theta_1 \approx 1 - \frac{\theta_1^2}{2}.$$

Thus we have

$$V = \tfrac{1}{2}[m_2 a - m_1(l - a)]g\theta^2. \tag{b}$$

For the kinetic energy, suppose that the bar rotates at rate $\dot{\theta}$. The speed of m_1 is then $(l - a)\dot{\theta}$ and the speed of m_2 is $a\dot{\theta}$. Thus

$$T = \tfrac{1}{2}[m_1(l - a)^2 + m_2 a^2]\dot{\theta}^2. \tag{c}$$

Let $\theta = A \sin \omega_n t$. Equating the maximum values of (b) and (c) gives (6.7):

$$\tfrac{1}{2}[m_1(l - a)^2 + m_2 a^2]\omega_n^2 A^2 = \tfrac{1}{2}[m_2 a - m_1(l - a)]gA^2$$

or

$$\omega_n^2 = \frac{[m_2 a - m_1(l - a)]g}{m_2 a^2 + m_1(l - a)^2}. \tag{d}$$

The minimum value of ω_n is 0. This value occurs if

$$m_2 a - m_1(l - a) = 0 \quad \text{or} \quad a = \frac{m_1 l}{m_1 + m_2}.$$

If $m_2 = 2m_1$, we get

$$a = \tfrac{1}{3}l.$$

---

* It is important in writing energy expressions that the cosine approximation contain at least two terms. If $\cos(\theta) = 1$ were the approximation, V would be a constant (which could be disregarded!).

This is the center of gravity of the system. If the bar is pinned here, there is no oscillation. For the maximum value of ω_n, we note that, with $m_2 = 2m_1$, (d) becomes

$$\omega_n^2 = \left[\frac{3a - l}{3a^2 - 2al + l^2} \right] g. \tag{d'}$$

Taking $d(\omega_n^2)/da = 0$, we get

$$a = \tfrac{1}{3}l(1 \pm \sqrt{2}).$$

The minus root is extraneous. Thus

$$a = \tfrac{1}{3}l(1 + \sqrt{2}) = l(0.805),$$

and from (d')

$$\omega_{n\,\text{max}}^2 = 1.06 \left(\frac{g}{l} \right).$$

Example 7

A liquid in the U-tube in Fig. 6.10 is in free oscillation. Calculate the natural frequency of the liquid, neglecting frictional effects.

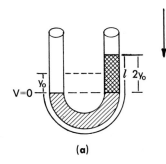
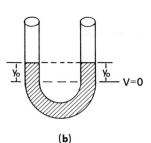

(a)   (b)   **Figure 6.10**

Solution

Let the maximum amplitude of the motion be y_0 (Fig. 6.10a). If we take the level of zero potential energy as shown, the potential energy in the first case is

$$V_1 = (\rho A 2y_0)gy_0,$$

where ρ is the density of the liquid and A is the cross-sectional area of the U-tube. The position of maximum kinetic energy is shown in Fig. 6.10(b). Here

$$V_0 = 2(\rho A y_0)g^{\frac{1}{2}}y_0.$$

Thus

$$V_{\text{max}} = V_1 - V_0 = (\rho A g)y_0^2. \tag{a}$$

Suppose the length of the fluid is l. Then the total mass is $m = \rho A l$ and the kinetic energy is

$$T = \tfrac{1}{2}(\rho A l)\dot{y}^2.$$ $\alpha \alpha p \, 274$ (b)

Assuming the harmonic solution

$$y = y_0 \sin \omega_n t$$

and applying (6.7), we get

$$\tfrac{1}{2}\rho A l y_0^2\,\omega_n^2 = \rho A g y_0^2 \qquad \text{or} \qquad \omega_n^2 = \left(\frac{2g}{l}\right).$$

The energy approach has much to recommend it, particularly when the energy functions are easy to express. The next example shows a situation in which this is not the case. Here it is probably a better idea to write $\mathbf{F} = \dot{\mathbf{p}}$ and compare the equation of motion to (6.2) to obtain ω_n.

Example 8

Figure 6.11(a) shows a particle on a smooth horizontal plane suspended by a string which has an initial tension T_0. Suppose that, for small motions of m, the tension remains essentially T_0. Find the natural frequency of the system.

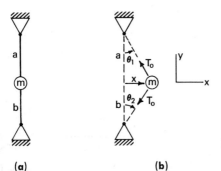

Figure 6.11

(a)   (b)

Solution

Figure 6.11(b) shows the free-body diagram. The force on m in the x direction is

$$F_x = -T_0 \sin \theta_1 - T_0 \sin \theta_2$$

where

$$\sin \theta_1 = \frac{x}{\sqrt{a^2 + x^2}} \qquad \text{and} \qquad \sin \theta_2 = \frac{x}{\sqrt{b^2 + x^2}}.$$

For $x \ll a$ and $x \ll b$, we approximate

$$\sin \theta_1 \approx \frac{x}{a}, \qquad \sin \theta_2 \approx \frac{x}{b}$$

and thus

$$F_x = -T_0 \left(\frac{1}{a} + \frac{1}{b} \right) x.$$

And writing $\mathbf{F} = \dot{\mathbf{p}}$ in the x direction, we get

$$F_x = ma_x : \qquad -T_0 \left(\frac{1}{a} + \frac{1}{b} \right) x = m\ddot{x}$$

or

$$\ddot{x} + \frac{T_0}{m} \left(\frac{a+b}{ab} \right) x = 0.$$

Thus

$$\omega_n^2 = \frac{T_0}{m} \left(\frac{a+b}{ab} \right).$$

....

Before we leave the subject of free motion of undamped systems, let us make one more point. Many authors, noting that

$$T + V = E = \text{constant},$$

analyze this sort of system by writing

$$\frac{d}{dt} (T + V) = 0.$$

This is supposed to yield the differential equations of motion. The method works for very simple, one-degree-of-freedom systems (problems in which the dynamics can be written entirely in terms of one variable). However, it is not valid for more general systems.* For this reason, we have not presented this method even for simple systems.

4. PERIODIC FUNCTIONS OF TIME

It is clear from the above section that in studying vibration problems, we shall be dealing a great deal with sinusoidal functions. Sinusoidal functions are one

---

* See S. H. Crandall, unpublished memorandum, 1956, also see L. A. Pars, *A Treatise on Analytical Dynamics*, London: Heinemann, 1965, page 86.

example of functions which repeat themselves after a given interval of time. Such functions are called *periodic*.

Consider the function represented in Fig. 6.12.

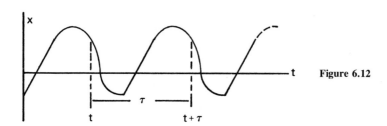

Figure 6.12

We see that the curve repeats itself after an interval of time τ. This interval is called the *period* of the motion. The motion which occurs during one period of time is called one *cycle*.

The number of cycles which occur in one second is called the *frequency**
of the periodic motion. Since one cycle takes τ seconds, the frequency f is

$$f = 1/\tau. \tag{6.8}$$

The frequency is measured in hertz (abbreviated Hz; one hertz equals one cycle per second).

Let us now discuss the special case of sinusoidal motion

$$x = A_0 \sin(\omega t - \phi).$$

Figure 6.13 is a plot of this function.

The sine function repeats when its argument increases by 2π radians. Thus

$$[\omega(t + \tau) - \phi] - [\omega t - \phi] = 2\pi \text{ rad}$$

or

$$\tau = (2\pi/\omega) \text{ sec} \tag{6.9a}$$

and the frequency f is

$$f = 1/\tau = \omega/2\pi \text{ hertz.} \tag{6.9b}$$

The quantity ω, which will play an important role in the equations we shall write, is called the *circular frequency*. The circular frequency ω is usually

---

* The *frequency* f (cycles per second) of a harmonic function is related to the circular frequency ω by $\omega = 2\pi f$.

circular frequency   natural frequency ω_n natural circular frequency

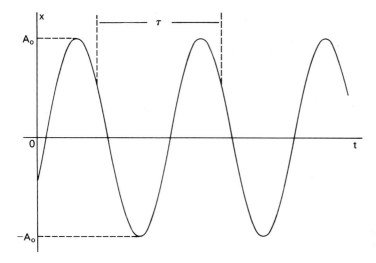

Figure 6.13

specified in units of radians per second. We should emphasize here that we shall want to use ω in most of the equations we write. However, in describing a vibration, we ordinarily speak in terms of so many cycles per second. That is, we talk in terms of the frequency f. (Does the electric company describe its current as 377 radians per second or as 60 cycles per second?)

In the case of the mass–spring oscillator, the natural, circular frequency is ω_n:

$$\omega_n = \sqrt{\frac{k}{m}} \quad \text{(radians/second)},$$

the natural frequency is

$$f = \frac{1}{2\pi} \sqrt{\frac{k}{m}} \quad \text{(hertz)},$$

and the period of the motion is

$$\tau = 2\pi \sqrt{\frac{m}{k}} \quad \text{(seconds)}.$$

We give one final example of free, undamped motion. Our point here is that it may not be obvious that a system will exhibit this sort of motion until we obtain the differential equation of motion. If the equation has the form of (6.2), no matter what the system is, the motion will have the form of (6.5) or (6.6).

Example 9

Figure 6.14 shows a uniform bar resting on two counter-rotating cylinders. The coefficient of sliding friction between the bar and the cylinders is μ. The analysis of this problem requires a knowledge of some ideas from Chapter 8. However, the problem is relevant in the present context.

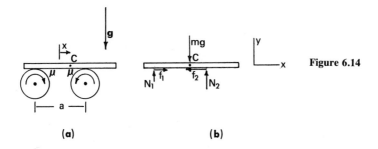

Figure 6.14

(a)   **(b)**

Figure 6.14(b) is a free-body diagram of the bar. The only motion of this bar is in the horizontal direction. Thus the forces on it must be equivalent to a single force in the horizontal direction. The net force in the vertical direction is zero, and there is no net moment about the center of mass:

$$F_y = 0: \qquad N_1 + N_2 - mg = 0, \tag{a}$$

$$\mathbf{M}_C = 0: \qquad \left[-N_1 \left(\frac{a}{2} + x \right) + N_2 \left(\frac{a}{2} - x \right) \right] \mathbf{k} = 0. \tag{b}$$

The moment equation (b) is written about the center of gravity of the bar. Solving (a) and (b) for N_1 and N_2, we get

$$N_1 = \frac{1}{2} mg \left(\frac{a - 2x}{a} \right), \qquad N_2 = \frac{1}{2} mg \left(\frac{a + 2x}{a} \right).$$

Thus the forces of friction are

$$f_1 = \frac{1}{2} \mu mg \left(\frac{a - 2x}{a} \right), \qquad f_2 = \frac{1}{2} \mu mg \left(\frac{a + 2x}{a} \right).$$

And finally, writing $\mathbf{F} = \dot{\mathbf{p}}$ for the bar, we get

$$F_x = ma_x: \qquad \frac{1}{2} \mu mg \left[\left(\frac{a - 2x}{a} \right) - \left(\frac{a + 2x}{a} \right) \right] = m\ddot{x}.$$

Putting this equation into standard form, we have

$$\ddot{x} + \left(\frac{2\mu g}{a}\right) x = 0.$$

Thus the net effect of the forces of friction is a linear restoring force. In this case, sliding friction results in a linear restoring force. In this regard, refer also to Example 23 in Chapter 4, in which sliding friction has an effect resembling that of a dashpot.

In the present problem the natural circular frequency ω_n is

$$\omega_n = \sqrt{\frac{2\mu g}{a}} \ \text{rad/sec.}$$

The natural frequency and period of the motion are

$$f = \frac{1}{\pi} \sqrt{\frac{\mu g}{2a}} \ \text{Hz,} \qquad \tau = \pi \sqrt{\frac{2a}{\mu g}} \ \text{sec.}$$

....

5. HARMONICALLY FORCED MOTION OF AN UNDAMPED SYSTEM

Consider the motion of a system described by the equation

$$\ddot{x} + \omega_n^2 x = F_0 \sin \omega t, \qquad F_0 = \text{const.} \tag{6.10}$$

There are two cases:

i) $\omega \neq \omega_n$, ω = forcing frequency,
ii) $\omega = \omega_n$, ω_n = natural frequency.

The second case indicates the most important ramification of the natural frequency of an undamped system.

i) $\omega \neq \omega_n$: We wish to have a function x which, when substituted into the left side of (6.10), gives $\sin \omega t$ as a result. It is reasonable to try, as a trial solution,

$$x = C \sin \omega t,$$

where C is to be determined. Inserting this into (6.10) yields

$$(-\omega^2 + \omega_n^2)C \sin \omega t = F_0 \sin \omega t.$$

Thus if C is chosen to be

$$C = \frac{F_0}{\omega_n^2 - \omega^2},$$

a solution to (6.10) will be

$$x = \left(\frac{F_0}{\omega_n^2 - \omega^2}\right) \sin \omega t. \tag{6.11}$$

The only problem with (6.11) is that there is no way to satisfy initial conditions on x and $\dot{x}$. Suppose we add the free solution (6.5) to (6.11). We can do this because (6.10) is linear. We have then

$$x = \left(\frac{F_0}{\omega_n^2 - \omega^2}\right) \sin \omega t + A \sin \omega_n t + B \cos \omega_n t. \tag{6.12}$$

The free solution gives zero when substituted into the left side of (6.10), but now we have two arbitrary constants, A and B, which can be used in (6.12) to meet the initial conditions. If

$$x(0) = x_0, \qquad \dot{x}(0) = v_0,$$

the solution is (see Problem 6.67):

$$x = \left(\frac{F_0}{\omega_n^2 - \omega^2}\right) \sin \omega t + \frac{1}{\omega_n}\left(v_0 - \frac{\omega F_0}{\omega_n^2 - \omega^2}\right) \sin \omega_n t + x_0 \cos \omega_n t. \tag{6.13}$$

We shall commonly speak of *forced* and *free solutions* (the terms *particular* and *complementary* or *homogeneous* are commonly used in mathematics texts):

Forcing function $F_0 \sin \omega t$

Forced solution $\left(\dfrac{F_0}{\omega_n^2 - \omega^2}\right) \sin \omega t$

Free solution $A \sin \omega_n t + B \cos \omega_n t$

We shall refer to (6.13) as the *complete solution.*

Note that the forced solution is merely a magnification of the forcing function if $\omega < \omega_n$. We say that the forcing and response functions are *in phase*. If $\omega > \omega_n$, the functions are minus one another. We say that they are (180°) *out of phase*. We shall explain this more fully when we discuss Fig. 6.24.

It is useful to plot the amplitude of the response versus (ω/ω_n). To nondimensionalize the plot, we introduce the magnification factor μ:

$$\mu = \left|\frac{F_0/(\omega_n^2 - \omega^2)}{x_{st}}\right|,$$

where x_{st} is the static deflection (see Problem 6.64):

$$x_{st} = F_0/\omega_n^2.$$

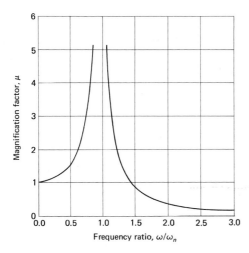

Figure 6.15

Thus

$$\mu = \left| \frac{1}{1 - (\omega/\omega_n)^2} \right|.$$

Figure 6.15 shows a plot of this.

ii) $\omega = \omega_n$: If we try to guess a solution to

$$\ddot{x} + \omega_n^2 x = F_0 \sin \omega_n t, \tag{6.14}$$

it would be

$$x = C \sin \omega_n t.$$

But this is a free solution. Thus the technique used above fails in the case $\omega = \omega_n$.

There are well-defined methods to determine the forced solution (e.g., variation of parameters or the superposition integral; see Example 16). Suppose we guess that the solution will have the frequency ω_n, but an amplitude that increases with time. We might guess $Ct \sin \omega_n t$, for example. Too bad! It's close, but no cigar! If we try a solution of this form, it will be clear that we should have tried

$$x = Ct \cos \omega_n t; \tag{6.15}$$

then

$$\ddot{x} = C(-2\omega_n \sin \omega_n t - t\omega_n^2 \cos \omega_n t).$$

If we insert this into the left-hand side of (6.14), we get

$$C(-2\omega_n \sin \omega_n t - t\omega_n^2 \cos \omega_n t) + \omega_n^2 Ct \cos \omega_n t = -2C\omega_n \sin \omega_n t,$$

and equating this to

$$F_0 \sin \omega_n t,$$

we see that if

$$C = -\left(\frac{F_0}{2\omega_n}\right),$$

(6.15) will be a (forced) solution to (6.14):

$$x = -\left(\frac{F_0}{2\omega_n}\right) t \cos \omega_n t. \tag{6.16}$$

Figure 6.16 gives a representative plot of (6.16).

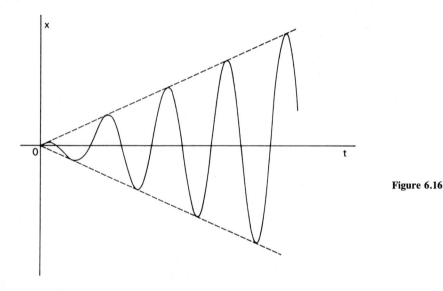

Figure 6.16

We see that the amplitude of x increases linearly with time. Obviously this is a dangerous condition, in that the component parts of the vibratory system will eventually fly apart.

In many systems described by (6.10), the operating frequency ω is greater than ω_n. But, as in the case of a rotary turbine, in which ω represents the turbine angular rate, the frequency ω must go from zero to the operating value. Thus ω must pass through ω_n. The machine will not fly apart if ω passes through ω_n fast enough (so the amplitude does not have a chance to build up, as in Fig. 6.16). This example illustrates why it is often desirable to know the natural frequencies of a system.

Example 10

Figure 6.17(a) shows a pendulum whose support has a sinusoidal motion:

$$x_0(t) = A_0 \sin \omega t, \qquad \omega \neq \sqrt{\frac{g}{l}}.$$

What is the motion $\theta(t)$, given that m is at rest at $t = 0$?

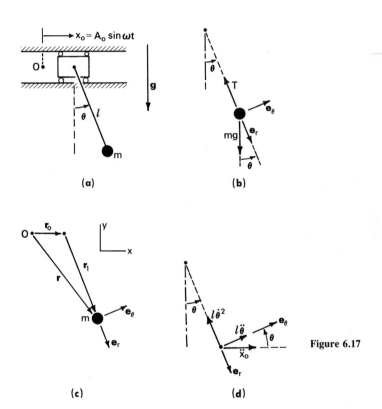

Figure 6.17

Solution

Figure 6.17(b) presents the free-body diagram for m. Figure 6.17(c) shows the position vector for m:

$$\mathbf{r} = \mathbf{r}_0 + \mathbf{r}_1. \tag{a}$$

From (a), we have

$$\ddot{\mathbf{r}} = \ddot{\mathbf{r}}_0 + \ddot{\mathbf{r}}_1. \tag{b}$$

Since $\mathbf{r}_0 = x_0(t)\mathbf{i}$, we have

$$\ddot{\mathbf{r}}_0 = \ddot{x}_0\mathbf{i}. \tag{c}$$

The term $\ddot{\mathbf{r}}_1$ is most easily written in polar coordinates,

$$\ddot{\mathbf{r}}_1 = -l\dot{\theta}^2\,\mathbf{e}_r + l\ddot{\theta}\,\mathbf{e}_\theta. \tag{d}$$

The terms (b) and (c) must be written in one coordinate system. (See Fig. 6.17d). We use the polar coordinates, $\mathbf{e}_r$ and $\mathbf{e}_\theta$. Equation (b) becomes

$$\mathbf{a} = \ddot{\mathbf{r}} = (\ddot{x}_0 \sin \theta - l\dot{\theta}^2)\mathbf{e}_r + (\ddot{x}_0 \cos \theta + l\ddot{\theta})\mathbf{e}_\theta,$$

and from $x_0 = A_0 \sin \omega t$, we get finally

$$\mathbf{a} = -(\omega^2 A_0 \sin \omega t \sin \theta + l\dot{\theta}^2)\mathbf{e}_r + (-\omega^2 A_0 \sin \omega t \cos \theta + l\ddot{\theta})\mathbf{e}_\theta.$$

Noting the free-body diagram (Fig. 6.17b), we write $\mathbf{F} = \dot{\mathbf{p}}$:

$$F_r = ma_r: \quad -T + mg \cos \theta = -m(\omega^2 A_0 \sin \omega t \sin \theta + l\dot{\theta}^2),$$

$$F_\theta = ma_\theta: \quad -mg \sin \theta = m(-\omega^2 A_0 \sin \omega t \cos \theta + l\ddot{\theta}).$$

To determine the motion $\theta(t)$, we need only the second equation, which can be written

$$\ddot{\theta} + \left(\frac{g}{l}\right) \sin \theta = \frac{\omega^2 A_0}{l} \sin \omega t \cos \theta.$$

We linearize this problem by considering θ small enough so that*

$$\sin \theta \approx \theta, \quad \cos \theta \approx 1.$$

We obtain finally

$$\ddot{\theta} + \left(\frac{g}{l}\right) \theta = \left(\frac{\omega^2 A_0}{l}\right) \sin \omega t.$$

We compare this equation to (6.10),

$$\ddot{x} + \omega_n^2 x = F_0 \sin \omega t,$$

and obtain the following correspondence

$$x \to \theta \qquad \text{variable}$$

$$\omega_n^2 \to \left(\frac{g}{l}\right) \qquad \text{natural frequency}$$

$$F_0 \to \frac{\omega^2 A_0}{l} \qquad \text{force amplitude}$$

$$\omega \to \omega \qquad \text{force frequency}$$

---

* When writing energy expressions, we must take $\cos \theta \approx 1 - \theta^2/2$. In the differential equation of motion, one can ordinarily use the approximation employed here.

Equation (6.12) or (6.13) gives the solution for the initial conditions of the problem:

at $t = 0$ $\begin{cases} \theta = 0 \\ \dot{\theta} = 0 \end{cases}$ (at rest at $t = 0$).

Thus

$$\theta(t) = \frac{\omega^2 A_0/l}{(g/l) - \omega^2} \sin \omega t - \sqrt{\frac{l}{g}} \left(\frac{\omega^3 A_0/l}{(g/l) - \omega^2} \right) \sin \sqrt{\frac{g}{l}} \, t.$$

Example 11

Figure 6.18(a) shows a particle m attached to a spring k, the upper end of which is assumed to move

$$y_A = R \sin \omega t. \tag{a}$$

i) Determine the forced motion of m, $y(t)$, where y is measured from the equilibrium position.

ii) Determine the range of frequencies for which the string in Fig. 6.18(a) is in tension (and for which the assumption $y_A = R \sin \omega t$ is valid).

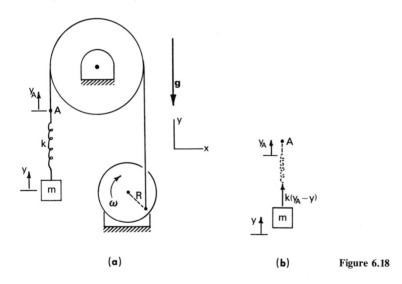

(a)   (b)   Figure 6.18

Solution

At the equilibrium position, $y = 0$, the forces of the weight and spring cancel. The free-body diagram for the system can then be given as in Fig. 6.18(b).

Writing $\mathbf{F} = \dot{\mathbf{p}}$ in the vertical direction

$$F_y = ma_y: \qquad k(y_A - y) = m\ddot{y} \qquad \text{or} \qquad \ddot{y} + \left(\frac{k}{m}\right) y = \frac{ky_A}{m} \, ,$$

or, from (a),

$$\ddot{y} + \left(\frac{k}{m}\right) y = \frac{kR}{m} \sin \omega t.$$

Thus

$$\omega_n^2 = \frac{k}{m}, \qquad F_0 = \frac{kR}{m}, \qquad \omega = \omega.$$

The forced solution, (6.11), is

$$y(t) = \frac{kR/m}{(k/m) - \omega^2} \sin \omega t, \qquad \omega \neq \sqrt{\frac{k}{m}} \tag{b}$$

If the string is to be in tension, the spring force must be in tension. The total force in the spring is

$$f = mg + k(y_A - y). \tag{c}$$

(Note that if $y_A = y = 0$, $f = mg$, which is just opposite to the weight. This is the equilibrium position, as was assumed.) We must have $f \geq 0$. With (a) and (b), this condition becomes

$$mg + k\left(R \sin \omega t - \frac{kR/m}{(k/m) - \omega^2} \sin \omega t\right) \geq 0. \tag{d}$$

Suppose that $\omega^2 < (k/m)$; then

$$\frac{g}{R} \geq \frac{\omega^2}{1 - (m/k)\omega^2} \sin \omega t. \tag{e}$$

Now $-1 \leq \sin \omega t \leq +1$. The inequality becomes violated first when $\sin \omega t = +1$ and (e) is an equality:

$$\omega_{\max}^2 = \frac{k}{m}\left[1 - \frac{1}{(mg/Rk) + 1}\right]$$

Thus

$$0 \leq \omega < \sqrt{\frac{k}{m}\left[1 - \frac{1}{(mg/Rk) + 1}\right]}.$$

Now suppose that $\omega^2 > (k/m)$. Returning to (d) and repeating the argument, we obtain the results:

if $\qquad \dfrac{mg}{Rk} \leq 1 \qquad$ no upper operating region if $\dfrac{mg}{Rk} > 1$ $\sqrt{\dfrac{k}{m}}\left[1 + \dfrac{1}{(mg/Rk) - 1}\right] < \omega$ gives the upper operating region.

....

L   R   C

6. FREE MOTION OF A DAMPED SYSTEM

Consider the system in Fig. 6.19(a): We have now added a linear dashpot to the mass–spring oscillator, and here we show the free-body diagram. Writing $\mathbf{F} = \dot{\mathbf{p}}$ gives

$$F_x = ma_x: \qquad -c\dot{x} - kx = m\ddot{x} \qquad \text{or} \qquad m\ddot{x} + c\dot{x} + kx = 0.$$

dashpot spring

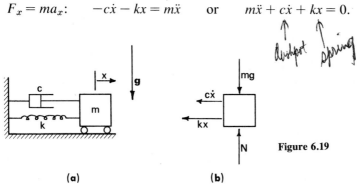

Figure 6.19

(a)   (b)

We put this equation in standard form,

$$\ddot{x} + 2\zeta\omega_n\dot{x} + \omega_n^2 x = 0, \tag{6.17}$$

where

$$\omega_n = \sqrt{\dfrac{k}{m}}, \qquad \zeta = \dfrac{c}{2\sqrt{mk}}.$$

circular

The term ω_n is the *undamped natural frequency* (as before) and ζ^* is called the *damping ratio.* The rationale for this name will become apparent shortly.

To solve (6.17), we again attempt exponential solutions of the form

$$x = Ce^{st}.$$

Inserting this into (6.17) yields

$$(s^2 + 2\zeta\omega_n s + \omega_n^2)Ce^{st} = 0.$$

_____

* For the mass–spring–dashpot system in Fig. 6.19(a), ζ is sometimes defined to be $\zeta = c/c_{\text{cr}}$, where c_{cr}, the critical dashpot damping, has the value $2\sqrt{mk}$.

Taking the expression inside the parentheses to be zero gives

$$s = \omega_n(-\zeta \pm \sqrt{\zeta^2 - 1}).$$

There are three cases of the solution, depending on the sign of $(\zeta^2 - 1)$:

i) $0 \leqslant \zeta < 1$ underdamped

ii) $\zeta = 1$ critically damped

iii) $1 < \zeta$ overdamped

Case (i): *Underdamped*

If $0 \leqslant \zeta < 1$, $\zeta^2 - 1 < 0$ and

$$\sqrt{\zeta^2 - 1} = i\sqrt{1 - \zeta^2}, \qquad i^2 = -1.$$

The solutions to (6.17) are then

$$x = C_1 e^{\omega_n(-\zeta + i\sqrt{1-\zeta^2})t} + C_2 e^{\omega_n(-\zeta - i\sqrt{1-\zeta^2})t}$$
$$= e^{-\omega_n \zeta t}(C_1 e^{i\omega_n \sqrt{1-\zeta^2}t} + C_2 e^{-i\omega_n\sqrt{1-\zeta^2}t}).$$

The terms inside the parentheses can be re-expressed [as we did in (6.4) and (6.5)] in terms of sine and cosine functions,

$$x = e^{-\omega_n \zeta t}(A \sin \omega_n \sqrt{1 - \zeta^2}\, t + B \cos \omega_n \sqrt{1 - \zeta^2}\, t). \tag{6.18}$$

If we have the following conditions on x and $\dot{x}$:

$$x(0) = x_0, \qquad \dot{x}(0) = v_0,$$

then the solution (6.18) becomes

$$x = e^{-\omega_n \zeta t}\left(\frac{v_0 + \omega_n \zeta x_0}{\omega_n \sqrt{1 - \zeta^2}} \sin \omega_n \sqrt{1 - \zeta^2}t + x_0 \cos \omega_n \sqrt{1 - \zeta^2}\, t\right). \tag{6.19}$$

The terms inside the parentheses of (6.18) are equivalent to a single sine function with amplitude $A_0 = (A^2 + B^2)^{1/2}$ and circular frequency $[\omega_n \sqrt{1 - \zeta^2}]$. Thus we can write (6.18) in the equivalent form

$$x = A_0 e^{-\omega_n \zeta t}(\sin [\omega_n \sqrt{1 - \zeta^2}\, t - \phi]), \tag{6.20}$$
$$A_0 = \sqrt{A^2 + B^2}, \qquad \phi = \tan^{-1}(-B/A).$$

The sinusoidal part of the solution varies between $+1$ and -1. Thus $x(t)$ has the form of a sinusoid with exponentially decreasing amplitude $A_0 e^{-\omega_n \zeta t}$. Figure 6.20 shows this.

We note that the decreasing amplitude of this motion can be directly attributed to the damping term, since, if $\zeta = 0$, there is no exponential decay. In

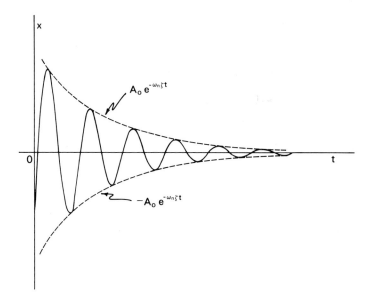

$A_0 e^{-\omega_n \zeta t}$

$-A_0 e^{-\omega_n \zeta t}$

Figure 6.20

fact, we can get some idea of the damping present by looking at any two successive peaks of the motion.

We begin by noting that although the motion (6.20) is not periodic in the sense we have defined in Section 4, the time between successive zeros is constant. Thus we could speak of (6.18) as *time*-periodic *motion*, with "period"

$$\tau = \frac{2\pi}{\omega_n \sqrt{1 - \zeta^2}}. \tag{6.21}$$

Suppose we consider the amplitude at time t and $t + \tau$ (Fig. 6.20). We have

$$A(t) = A_0 e^{-\omega_n \zeta t} [\sin (\omega_n \sqrt{1 - \zeta^2}\, t - \phi)],$$

$$A(t + \tau) = A_0 e^{-\omega_n \zeta(t+\tau)} [\sin (\omega_n \sqrt{1 - \zeta^2}\, t - \phi)].$$

(Note that the sinusoidal parts are the same, since the sine function exactly repeats from t to $t + \tau$.) Consider the ratio

$$\frac{A(t)}{A(t + \tau)} = e^{\omega_n \zeta \tau}. \tag{6.22}$$

By (6.21) we can write

$$\omega_n \zeta \tau = \frac{2\pi \zeta}{\sqrt{1 - \zeta^2}} = \delta. \tag{6.23}$$

logarithmic decrement

The quantity δ is called the *logarithmic decrement* because, by (6.22), we have

$$\frac{A(t)}{A(t+\tau)} = e^\delta \quad \text{or} \quad \delta = \ln\left[\frac{A(t)}{A(t+\tau)}\right].$$

If we let $A(t+\tau) = x$, then

$$A(t) = x + \Delta x$$

[recall that $A(t)$ is decreasing with time]. Then

$$\delta = \ln\left(\frac{x+\Delta x}{x}\right) = \ln\left(1 + \frac{\Delta x}{x}\right)$$

$$= \frac{\Delta x}{x} - \frac{1}{2}\left(\frac{\Delta x}{x}\right)^2 + \frac{1}{3}\left(\frac{\Delta x}{x}\right)^3 \cdots$$

If $\Delta x << x$, we can truncate this series at the first term to give an approximate formula for the logarithmic decrement:

$$\delta \approx \frac{\Delta x}{x}, \tag{6.24}$$

and thus, from (6.23), we have

$$\zeta \approx \frac{\Delta x/x}{\sqrt{4\pi^2 + (\Delta x/x)^2}} \approx \frac{\Delta x/x}{2\pi}, \tag{6.25}$$

where $A(t) = x + \Delta x$ and $A(t+\tau) = x$.

Case (ii): *Critically damped*

If $\zeta = 1$, then $\zeta^2 - 1 = 0$, and the two exponential solutions found above,

$$C_1 e^{\omega_n(-\zeta+\sqrt{\zeta^2-1})t} \quad \text{and} \quad C_2 e^{\omega_n(-\zeta-\sqrt{\zeta^2-1})t},$$

both become multiples of $e^{-\omega_n t}$.

As in Section 5, we are led to try

$$x = Cte^{-\omega_n t} \tag{6.26}$$

for the second solution to

$$\ddot{x} + 2\omega_n\dot{x} + \omega_n^2 x = 0, \quad \zeta = 1. \tag{6.27}$$

It is a trivial exercise to show that (6.26) is a solution to (6.27) for arbitrary values of C. Thus we have the free solution for the critically damped case,

$$x = (C_1 + C_2 t)e^{-\omega_n t}. \tag{6.28}$$

Students occasionally worry about the term $te^{-\omega_n t}$ becoming large as $t \to \infty$. However, for all cases in which $\omega_n > 0$, it can be shown that

$$t^m e^{-\omega_n t} \to 0 \qquad \text{as} \qquad t \to \infty,$$

where m is any finite, positive number. (See Problem 6.89.)

For the initial conditions

$$x(0) = x_0 \qquad \text{and} \qquad \dot{x}(0) = v_0,$$

we then have (6.28),

$$x = [x_0 + (v_0 + \omega_n x_0)t]e^{-\omega_n t}. \tag{6.29}$$

Suppose that this system is released from some position x_0 at some velocity v_0. We can ask if there is any finite time when $x = 0$. Since $e^{-\omega_n t} \to 0$ only as $t \to \infty$, if $x = 0$ at some finite value of time $t = t_1$, we must have

$$x_0 + (v_0 + \omega_n x_0)t_1 = 0 \qquad \text{or} \qquad t_1 = \frac{-x_0}{v_0 + \omega_n x_0}.$$

We are interested only in positive times t_1. Since $\omega_n > 0$, we have the following two cases:

a) $x_0 > 0,$ then $v_0 < -(\omega_n x_0),$

b) $x_0 < 0,$ then $v_0 > -\omega_n x_0.$

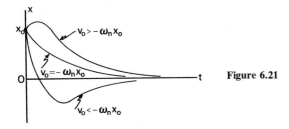

Figure 6.21

In these cases x moves from x_0 across $x = 0$ and then becomes 0 again only at $t = \infty$. Any other values of v_0 than indicated in (a) and (b) will result in $x = 0$ only at $t = \infty$. Thus we have the two possibilities shown in Fig. 6.21.

We shall see that the overdamped case resembles the critically damped case in that the free motion has essentially no oscillatory characteristic.

Case (iii): Overdamped

If $\zeta > 1$, then $\zeta^2 - 1 > 0$, and

$$s_1 = \omega_n(-\zeta + \sqrt{\zeta^2 - 1}), \qquad s_2 = \omega_n(-\zeta - \sqrt{\zeta^2 - 1}).$$

Thus the free solution in the overdamped case is

$$x = A e^{\omega_n(-\zeta + \sqrt{\zeta^2 - 1})t} + B e^{\omega_n(-\zeta - \sqrt{\zeta^2 - 1})t}. \tag{6.30}$$

We can rewrite (6.30) in terms of hyperbolic functions,

$$x = e^{-\omega_n \zeta t}(A \sinh \omega_n \sqrt{\zeta^2 - 1}\, t + B \cosh \omega_n \sqrt{\zeta^2 - 1}\, t), \tag{6.31}$$

where the A, B in (6.30) are *not* the same as those in (6.31). The appearance of Eq. (6.31) reminds us of the equation for the free motion of an under-damped system, (6.18). However, the similarity is a matter of form alone.
   If we have initial conditions

$$x(0) = x_0 \qquad \text{and} \qquad \dot{x}(0) = v_0,$$

the solution (6.31) becomes

$$x = e^{-\omega_n \zeta t}\left(\frac{v_0 + \omega_n \zeta x_0}{\omega_n \sqrt{\zeta^2 - 1}} \sinh \omega_n \sqrt{\zeta^2 - 1}\, t + x_0 \cosh \omega_n \sqrt{\zeta^2 - 1}\, t \right). \tag{6.32a}$$

[Recall the form of (6.19).]
   We can again ask if there are any finite values of t such that $x = 0$. Again $e^{-\omega_n \zeta t} \to 0$ only as $t \to \infty$. Thus, if $x = 0$ at $t = t_1$, we must have

$$\frac{v_0 + \omega_n \zeta x_0}{x_0 \omega_n \sqrt{\zeta^2 - 1}} \tanh \omega_n \sqrt{\zeta^2 - 1}\, t_1 = -1$$

or

$$t_1 = \frac{1}{\omega_n \sqrt{\zeta^2 - 1}} \tanh^{-1}\left(-\frac{x_0 \omega_n \sqrt{\zeta^2 - 1}}{v_0 + \omega_n \zeta x_0} \right).$$

There are again two cases in which finite values of t result:

a) $x_0 > 0$ $v_0 < -x_0 \omega_n(\zeta + \sqrt{\zeta^2 - 1})$,   (b)

b) $x_0 < 0$ $v_0 > -x_0 \omega_n(\zeta + \sqrt{\zeta^2 - 1})$.   (c)

(6.32)

   In these cases the variable crosses $x = 0$ one time and then goes to zero asymptotically as $t \to \infty$. (See Problem 6.99.) In all other cases, $x = 0$ is reached only as $t \to \infty$. Note that if we take $\zeta = 1$ in (6.32a, b) we get the conditions for critical damping. From this point of view, critically damped and overdamped motions are in one category, as opposed to underdamped motion, which is in a distinct category. A plot of an overdamped motion would have characteristics very similar to those in Fig. 6.21 [in which inequalities (a) and (b) above are substituted for those in Fig. 6.21].

Example 12

A particle m is attached to a rigid, weightless bar which is pinned at the wall. The particle is supported by a linear spring k, and the bar has a dashpot c, attached to it a distance a away from the wall. (See Fig. 6.22a.)

Suppose that the values of m, k, c and l are given. Determine the range of the distance a for which the system is (i) underdamped, (ii) critically damped, (iii) overdamped.

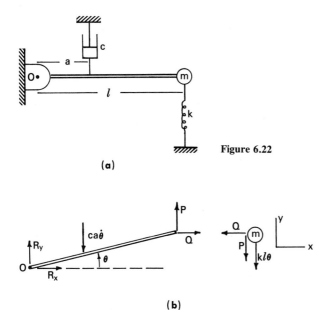

Figure 6.22

(a)

(b)

Solution

Figure 6.22(b) shows free-body diagrams for the particle m and the weightless bar.

We assume that the horizontal position is the equilibrium position of the system. Writing $\mathbf{F} = \dot{\mathbf{p}}$ for the particle, we obtain

$$F_y = ma_y: \qquad -P - kl\theta = ml\ddot{\theta} \tag{a}$$

Writing the equilibrium condition for the weightless bar, we get

$$\mathbf{M}_O = 0: \qquad [Pl - (ca\dot{\theta})a]\mathbf{k} = 0$$

or

$$P = c\left(\frac{a^2}{l}\right)\dot{\theta}. \tag{b}$$

Inserting (b) into (a) gives

$$ml\ddot{\theta} + c \left(\frac{a^2}{l}\right) \dot{\theta} + kl\theta = 0. \tag{c}$$

Let $y = l\theta$ and divide (c) by m:

$$\ddot{y} + \left(\frac{c}{m}\right) \left(\frac{a}{l}\right)^2 \dot{y} + \left(\frac{k}{m}\right) y = 0.$$

We compare this with the standard form (6.17):

$$\ddot{x} + 2\zeta\omega_n\dot{x} + \omega_n^2 x = 0.$$

We have the following correspondence

$$x \rightarrow y$$

$$\omega_n^2 \rightarrow k/m$$

$$\zeta \rightarrow \frac{1}{2} \frac{c}{\sqrt{mk}} \left(\frac{a}{l}\right)^2.$$

For *critical damping*, $\zeta = 1$, and thus

$$a = l \sqrt{\frac{2\sqrt{mk}}{c}}, \qquad \zeta = 1.$$

The system is *underdamped* if

$$a < l \sqrt{\frac{2\sqrt{mk}}{c}}, \qquad \zeta < 1,$$

and *overdamped* if

$$a > l \sqrt{\frac{2\sqrt{mk}}{c}}, \qquad \zeta > 1.$$

. . . .

Example 13

Consider the system in Example 12. With the dashpot removed, the *period* of free oscillation of the mass m is 0.5 sec.

With the dashpot in place, the system is given an initial displacement y_0, and oscillates with decreasing amplitude. The first peak after release has height $0.6y_0$.

   i) Determine ω_n and ζ for the system.

   ii) Given that $a = \frac{1}{2}l$ and $mg = 3.2$ lb, find c and k.

Solution

i) Without the dashpot, the system is undamped. Thus

$$\omega_n = \frac{2\pi}{\tau} \text{ rad/sec.}$$

Thus $\omega_n = 4\pi$ rad/sec or

$$\omega_n = 12.58 \text{ rad/sec.} \tag{a}$$

Since the system oscillates when released, it is underdamped. Thus, from (6.25), we estimate that

$$\zeta \approx \frac{\Delta x/x}{2\pi}, \qquad \text{where } A(0) = x + \Delta x = y_0, \qquad A(\tau) = x = 0.6 y_0.$$

Thus $x = 0.6 y_0$ and $\Delta x = 0.4 y_0$,

$$\zeta \approx \frac{0.4/0.6}{2\pi} = 0.11. \tag{b}$$

ii) From the previous example, we have

$$\omega_n^2 = k/m, \qquad \zeta = \frac{1}{2} \frac{c}{\sqrt{mk}} \left(\frac{a}{l}\right)^2.$$

Thus $k = m\omega_n^2$, $\qquad c = 2\zeta \sqrt{mk} \, (l/a)^2$,

and

$$k = \left(\frac{3.2}{32}\right) (12.58)^2 \text{ lb/ft} = 15.8 \text{ lb/ft.}$$

Thus

$$c = 2(0.111) \sqrt{\frac{(3.2)(15.8)}{32}} \, (2)^2 \qquad \text{or} \qquad c = 1.11 \text{ lb-sec/ft.}$$

7. HARMONICALLY FORCED MOTION OF A DAMPED SYSTEM

We now consider motions described by the differential equation

$$\ddot{x} + 2\zeta\omega_n\dot{x} + \omega_n^2 x = F_0 \sin \omega t. \tag{6.33}$$

We continue our guessing process to find a function x which, when inserted in the left side of (6.33), yields essentially $\sin \omega t$ on the left. If we try $A \sin \omega t$, $\ddot{x}$ and x will be sine functions, but $\dot{x}$ will be cosine. On the other hand, if we try $B \cos \omega t$, $\dot{x}$ will be a sine function, but $\ddot{x}$ and x will be cosine functions. It therefore seems reasonable to try a combination of the two:

$$x = A \sin \omega t + B \cos \omega t. \tag{6.34}$$

Then we have

$$\dot{x} = \omega(A \cos \omega t - B \sin \omega t), \qquad \ddot{x} = -\omega^2(A \sin \omega t + B \cos \omega t).$$

Inserting these into (6.33), we get

$$[(\omega_n^2 - \omega^2)A - 2\zeta\omega_n\omega B] \sin \omega t$$
$$+ [2\zeta\omega_n\omega A + (\omega_n^2 - \omega^2)B] \cos \omega t = F_0 \sin \omega t.$$

Equating the coefficients of $\sin \omega t$ and $\cos \omega t$, we see that if (6.34) is to be a solution, the constants A and B must satisfy:

$$(\omega_n^2 - \omega^2)A - 2\zeta\omega_n\omega B = F_0,$$

$$2\zeta\omega_n\omega A + (\omega_n^2 - \omega^2)B = 0.$$

These are two linear equations in two unknowns, A and B. Such equations have unique solutions if the determinant of the coefficients of A and B is non-zero. The solution can be obtained by Cramer's rule* (or elimination of one unknown). We get

$$A = \frac{\begin{vmatrix} F_0 & -2\zeta\omega_n\omega \\ 0 & (\omega_n^2 - \omega^2) \end{vmatrix}}{\begin{vmatrix} (\omega_n^2 - \omega^2) & -2\zeta\omega_n\omega \\ 2\zeta\omega_n\omega & (\omega_n^2 - \omega^2) \end{vmatrix}}, \qquad B = \frac{\begin{vmatrix} (\omega_n^2 - \omega^2) & F_0 \\ 2\zeta\omega_n\omega & 0 \end{vmatrix}}{\begin{vmatrix} (\omega_n^2 - \omega^2) & -2\zeta\omega_n\omega \\ 2\zeta\omega_n\omega & (\omega_n^2 - \omega^2) \end{vmatrix}},$$

or

$$A = \frac{F_0(\omega_n^2 - \omega^2)}{(\omega_n^2 - \omega^2)^2 + 4(\zeta\omega_n\omega)^2}, \qquad B = \frac{-2F_0\zeta\omega_n\omega}{(\omega_n^2 - \omega^2)^2 + 4(\zeta\omega_n\omega)^2}.$$

Thus

$$x = \frac{F_0}{(\omega_n^2 - \omega^2)^2 + 4(\zeta\omega_n\omega)^2} [(\omega_n^2 - \omega^2) \sin \omega t - 2\zeta\omega_n\omega \cos \omega t]. \qquad (6.35)$$

By means of (6.6), we can re-express (6.35) as

$$x = \frac{F_0/\omega_n^2}{\sqrt{(1 - \omega^2/\omega_n^2)^2 + 4(\zeta\omega/\omega_n)^2}} \sin(\omega t - \phi),$$

$$(6.36)$$

$$\phi = \tan^{-1}\left[\frac{2\zeta(\omega/\omega_n)}{1 - (\omega/\omega_n)^2}\right].$$

From (6.36), we can study the amplitude and phase ϕ of the response x relative to the force $F_0 \sin \omega t$. We note first that the amplitude of x is finite in

_____

* See Appendix 1, Eq. (A1.37).

every case except when $\zeta = 0$ and $\omega = \omega_n$ (the undamped system forced at its natural frequency). If $\zeta > 0$, the amplitude is always finite, no matter what ω is.

Again defining $x_{st} = F_0/\omega_n^2$, the static deflection, we get the ratio of the amplitude of (6.36) to x_{st}, that is, the *magnification factor* μ:

$$\mu = \frac{1}{\sqrt{(1 - \omega^2/\omega_n^2)^2 + 4(\zeta\omega/\omega_n)^2}} . \tag{6.37}$$

We can plot μ as a function of ω/ω_n, as shown in Fig. 6.23.

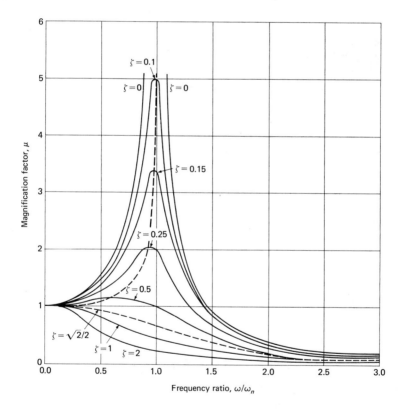

Figure 6.23

We can now ask what the maximum amplitude is for a given value of ζ. We can obtain this by considering the minimum of the function

$$f(\omega) = (\omega_n^2 - \omega^2)^2 + 4(\zeta\omega\omega_n)^2.$$

Setting $f'(\omega) = 0$ gives

$$f'(\omega) = -4(\omega_n^2 - \omega^2)\omega + 8\omega(\zeta\omega_n)^2 = 0$$

or

$$\omega[\omega_n^2(-1 + 2\zeta^2) + \omega^2] = 0.$$

Thus either $\omega = 0$ or $\omega = \omega_n \sqrt{1 - 2\zeta^2}$.

	Damping ratio	Frequency of max. amplitude	Maximum magnification	Resonant magnification
Underdamped	$0 < \zeta < \dfrac{\sqrt{2}}{2}$	$\omega_n \sqrt{1 - 2\zeta^2}$	$\dfrac{1}{2\zeta\sqrt{1 - \zeta^2}}$	$\dfrac{1}{2\zeta}$
	$\dfrac{\sqrt{2}}{2} < \zeta < 1$	$\omega = 0$	1	From $\dfrac{\sqrt{2}}{2}$ to $\frac{1}{2}$
Critically damped	$\zeta = 1$	$\omega = 0$	1	$\frac{1}{2}$
Overdamped	$1 < \zeta$	$\omega = 0$	1	From $\frac{1}{2}$ to 0

The contention that the resonant amplitude (i.e., the amplitude at $\omega = \omega_n$) is a good approximation to the maximum amplitude has to be taken with a grain of salt. This isn't even true throughout the underdamped range. Also the idea that one uses the resonant amplitude because the maximum amplitude is difficult to compute seems suspect, since we have that μ_{max} is either

$$\mu_{max} = \frac{1}{2\zeta\sqrt{1 - \zeta^2}}, \qquad 0 < \zeta < \frac{\sqrt{2}}{2},$$

or

$$\mu_{max} = 1, \qquad \frac{\sqrt{2}}{2} < \zeta.$$

Figure 6.24 is a plot of the phase angle ϕ between the force $F_0 \sin \omega t$ and the response x for various values of the damping ratio ζ.

Repeating (6.36), the phase angle ϕ is given by the expression

$$\phi = \tan^{-1}\left[\frac{2\zeta(\omega/\omega_n)}{1 - (\omega/\omega_n)^2}\right].$$

The phase ϕ indicates how far behind the response is relative to the forcing.

As a final exercise for this chapter, we note that the complete solution for the forced motion of a damped system corresponds to the sum of the forced

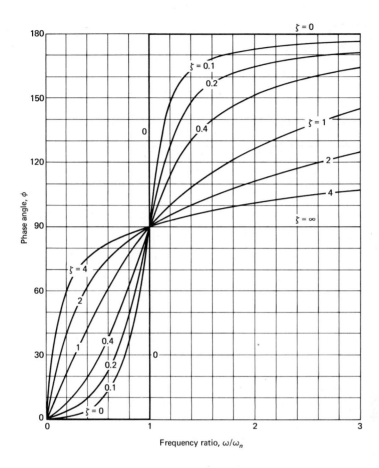

Figure 6.24

solution and the free solution. The underdamped case will be considered here; the overdamped case will appear in the chapter summary.

For $0 < \zeta < 1$, we have

$$x = \frac{F_0/\omega_n^2}{\sqrt{(1 - \omega^2/\omega_n^2)^2 + 4(\zeta\omega/\omega_n)^2}} \sin (\omega t - \phi)$$
$$+ e^{-\zeta\omega_n t}(A \sin \omega_n \sqrt{1 - \zeta^2}\,t + B \cos \omega_n \sqrt{1 - \zeta^2}\,t),$$

where

$$\phi = \tan^{-1}\left[\frac{2\zeta(\omega/\omega_n)}{1 - (\omega/\omega_n)^2}\right]. \tag{6.38}$$

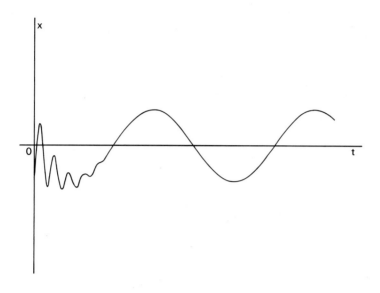

Figure 6.25

If $x(0) = x_0$ and $\dot{x}(0) = v_0$, we determine A and B to get:

$$x = \frac{F_0/\omega_n^2}{\sqrt{(1 - \omega^2/\omega_n^2)^2 + 4(\zeta\omega/\omega_n)^2}} \sin(\omega t - \phi)$$

$$+ \frac{e^{-\zeta\omega_n t}}{\omega_n \sqrt{1 - \zeta^2}} \left[v_0 + \zeta\omega_n x_0 + \frac{F_0/\omega_n^2(\zeta\omega_n \sin\phi - \omega \cos\phi)}{\sqrt{(1 - \omega^2/\omega_n^2)^2 + 4(\zeta\omega/\omega_n)^2}} \right] \cdot \sin\omega_n\sqrt{1 - \zeta^2}\,t$$

$$+ e^{-\zeta\omega_n t} \left[x_0 + \frac{(F_0/\omega_n^2)\sin\phi}{\sqrt{(1 - \omega^2/\omega_n^2)^2 + 4(\zeta\omega/\omega_n)^2}} \right] \cos\omega_n\sqrt{1 - \zeta^2}\,t. \quad (6.39)$$

Considering the size of (6.39), you will be relieved to know that the free part of the solution dies out rapidly, leaving only the first term, the forced solution, for large values of time. Figure 6.25 shows a typical plot.

....

Example 14

Figure 6.26(a) shows a rigid mass sandwiched between two soft rubber pads. The left pad is glued to a wall and to m. The right pad is glued to m and to a rigid metal sheet, which has a specified motion,

$$x_0(t) = A_0 \sin\omega_0 t. \qquad \text{(a)}$$

The rubber pads are identical, and are modeled as a spring k and a dashpot in parallel. Determine the forced motion of m.

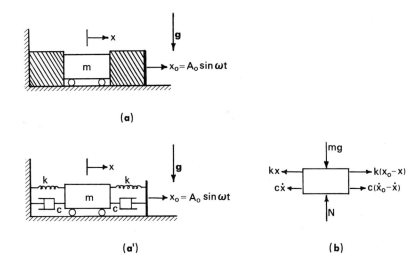

(a)

(a')   (b)

Figure 6.26

Solution

Under the assumptions of the problem, the model is as shown in Fig. 6.26(a').
Figure 6.26(b) shows the free-body diagram for the problem.
Writing $\mathbf{F} = \dot{\mathbf{p}}$, we have

$$F_x = ma_x: \qquad k(x_0 - x) + c(\dot{x}_0 - \dot{x}) - kx - c\dot{x} = m\ddot{x}$$

or

$$m\ddot{x} + 2c\dot{x} + 2kx = kx_0 + c\dot{x}_0.$$

From (a), we get finally

$$\ddot{x} + \left(\frac{2c}{m}\right)\dot{x} + \left(\frac{2k}{m}\right)x = \left(\frac{kA_0}{m}\right)\sin \omega t + \left(\frac{c\omega A_0}{m}\right)\cos \omega t. \qquad (b)$$

Before we get rolling on obtaining the forced solution, let us rewrite the forcing term.
We again use the technique involved in (6.6):

$$\left(\frac{kA_0}{m}\right)\sin \omega t + \left(\frac{c\omega A_0}{m}\right)\cos \omega t = \frac{A_0}{m}\sqrt{k^2 + c^2\omega^2}\,\sin(\omega t - \phi),$$

and further, if we shift time by letting $t_1 = t - \phi/\omega$, our forcing function is

$$F_0(t) = \frac{A_0}{m}\sqrt{k^2 + c^2\omega^2}\,\sin \omega t, \qquad (c)$$

where we have dropped the special notation t_1. (Since we are interested only in the
forced solution, the origin of time is not of any great importance.) Equation (b)

becomes

$$\ddot{x} + \left(\frac{2c}{m}\right)\dot{x} + \left(\frac{2k}{m}\right)x = \frac{A_0}{m}\sqrt{k^2 + c^2\omega^2}\ \sin \omega t$$

which we compare with (6.33):

$$\ddot{x} + 2\zeta\omega_n\dot{x} + \omega_n^2 x = F_0 \sin \omega t.$$

Thus

$$x \rightarrow x$$

$$\omega_n^2 \rightarrow (2k/m)$$

$$\zeta \rightarrow \frac{c}{\sqrt{2mk}}$$

$$F_0 \rightarrow \frac{A_0}{m}\sqrt{k^2 + c^2\omega^2}.$$

From (6.36), we obtain

$$x = \frac{1}{2} A_0 \sqrt{\frac{1 + (c^2\omega^2/k^2)}{(1 - m^2\omega^2/4k^2)^2 + (c^2\omega^2/k^2)}}\ \sin (\omega t - \phi_1), \tag{d}$$

where ϕ_1 is the phase between the force input (c) and the response (d).

....

Example 15

Figure 6.27(a) is a drawing of a motor welded to the end of a weightless, rigid bar, which is pinned to a wall. The motion takes place in the horizontal plane. The motor has mass m, but also an unbalanced mass m_0 at radius R. Suppose that $mg = 16.1$ lb, $m_0 g = 0.5$ lb, $R = 2.0$ in., $\omega = 150$ rpm, and $k = 10$ lb/in. Determine c such that the amplitude of the forced motion is one-half that of the forced motion with the dashpot removed.

Solution

Assume that the position shown in Fig. 6.27(a) is the equilibrium position when m_0 is still. Figure 6.27(b) presents free-body diagrams for the bar, motor, and unbalanced mass. Let y be the vertical motion of the motor and y_0 the vertical motion of the unbalanced mass.

Writing $\mathbf{F} = \dot{\mathbf{p}}$ for the motor and the unbalance gives the following.

Motor $F_y = ma_y$: $-P_1 + P_2 - c\dot{y} = m\ddot{y}.$ \hfill (a)

Unbalance $F_y = ma_y$: $-P_2 = m_0\ddot{y}_0.$ \hfill (b)

Now y and y_0 are related by

$$y_0 = y + R \sin \theta, \qquad \text{where } \theta = \omega t.$$

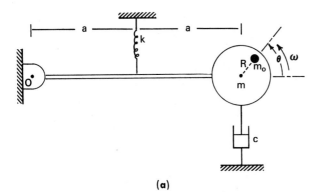

Figure 6.27

(a)

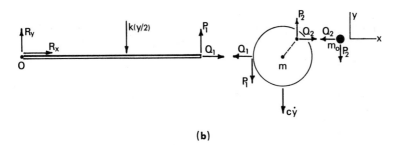

(b)

Thus

$$\ddot{y}_0 = \ddot{y} - \omega^2 R \sin \omega t. \tag{c}$$

And writing the equation of moment equilibrium for the weightless bar (about O), we have

$$\mathbf{M}_O = 0: \qquad \left[-k\left(\frac{y}{2}\right) a + P_1 2a \right] \mathbf{k} = 0$$

or

$$P_1 = \tfrac{1}{4} ky. \tag{d}$$

Adding (a) and (b) and using (c) and (d), we get

$$(m + m_0)\ddot{y} + c\dot{y} + \left(\frac{k}{4}\right) y = \omega^2 m_0 R \sin \omega t. \tag{e}$$

Let $M = m + m_0$. Then we have the equation of motion in standard form,

$$\ddot{y} + \left(\frac{c}{M}\right)\dot{y} + \left(\frac{k}{4M}\right) y = \left(\frac{m_0}{M}\right)\omega^2 R \sin \omega t. \tag{f}$$

Comparing this to (6.33), we get

$$x \rightarrow y$$

$$\omega_n^2 \rightarrow \frac{k}{4M}$$

$$\zeta \rightarrow \frac{c}{\sqrt{Mk}} \tag{g}$$

$$F_0 \rightarrow \left(\frac{m_0}{M}\right) \omega^2 R$$

$$\omega \rightarrow \omega$$

The amplitudes of forced motion for $\zeta = 0$ and $\zeta > 0$ are given in (6.13) and (6.36):

$$A_0 = \left| \frac{F_0/\omega_n^2}{(1 - \omega^2/\omega_n^2)} \right|, \qquad \zeta = 0,$$

$$A_1 = \frac{F_0/\omega_n^2}{\sqrt{(1 - \omega^2/\omega_n^2)^2 + 4(\zeta\omega/\omega_n)^2}}, \qquad \zeta > 0.$$

Our problem is to determine c such that $A_1 = \frac{1}{2}A_0$. Thus (after a bit of manipulation), we get

$$\zeta = \frac{\sqrt{3}}{2} \left| \frac{1 - \omega^2/\omega_n^2}{\omega/\omega_n} \right|,$$

and thus from (g) we get

$$c = \frac{\sqrt{3}}{4\omega} |(k - 4M\omega^2)|. \tag{h}$$

Putting the data into consistent form, we have

$$k = 10 \text{ lb/in.} = 120 \text{ lb/ft},$$

$$M = \frac{16.1 + 0.5}{32} \text{ slug} = 0.515 \text{ slug},$$

$$\omega = 150 \text{ rpm} = 2.5 \text{ rev/sec} = 15.71 \text{ rad/sec}.$$

Thus

$$c = 10.71 \text{ lb-sec/ft}.$$

8. ARBITRARILY FORCED MOTION OF A DAMPED SYSTEM

We now consider the motion of a system described by the differential equation

$$\ddot{x} + 2\zeta\omega_n\dot{x} + \omega_n^2 x = F_0(t), \tag{6.40}$$

where $F_0(t)$ is an arbitrary function of time. As a model for the system described by (6.40), we have the mass–spring–dashpot system in Fig. 6.28.

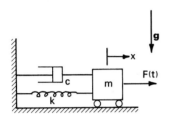

Figure 6.28

We discussed the free solution to (6.40) in Section 6, so here we need to derive only the forced solution. To do this, we begin by defining the *unit impulse response*.

Consider a force which is zero for almost every instant of time except in the immediate vicinity of the time t^*. We also require that the integral of this function (in dynamics terminology, the impulse of the force) is unity. The plot of such a function is shown in Fig. 6.29.

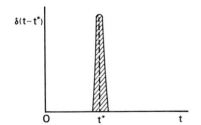

Figure 6.29

We call the function described above a *unit impulse function* or a *Dirac delta function*. It is denoted by

$$\delta(t - t^*).$$

The mathematical description of the delta function is no simple matter. Students with a mathematical bent are referred to the charming book on the subject by M. J. Lighthill.†

---

† M. J. Lighthill, *An Introduction to Fourier Analysis and Generalized Functions*, New York: Cambridge University Press, 1958.

Fortunately, we can deal with the delta function without worrying too much about mathematical details. Consider again the mass–spring–dashpot system subject now to the force $F(t) = \delta(t - t^*)$, as in Fig. 6.29. The equation of motion is

$$m\ddot{x} + c\dot{x} + kx = \delta(t - t^*). \tag{a}$$

We suppose the system to be at rest until time $t = t^*$. Then we apply the unit impulse. Physically, this is exactly the case of hitting the mass with a hammer such that the impulse is unity. From the impulse-momentum theorem, we have then

$$1 = m(\dot{x} - 0).$$

Thus at $t = t^*$, we have

$$x = 0, \qquad \dot{x} = \frac{1}{m}.$$

For $t > t^*$, $F(t) = 0$ and thus the motion is purely free motion. If we consider an underdamped system, we have, by (6.19),

$$
\begin{aligned}
x &= 0 & t &< t^* \\
x &= \frac{1}{m\omega_n \sqrt{1 - \zeta^2}} \, e^{-\omega_n \zeta(t - t^*)} \sin\left[\omega_n \sqrt{1 - \zeta^2}(t - t^*)\right], & t &\geq t^*
\end{aligned}
\tag{b}
$$

The function x of (b) is called the *unit impulse response*, and is denoted by $h(t - t^*)$, the response x to a unit impulse at $t = t^*$. We have then

$$h(t - t^*) = \frac{1}{m\omega_n \sqrt{1 - \zeta^2}} \, e^{-\omega_n \zeta(t - t^*)} \sin\left[\omega_n \sqrt{1 - \zeta^2}(t - t^*)\right], \qquad \zeta < 1. \tag{c}$$

For the undamped case,

$$h(t - t^*) = \frac{1}{m\omega_n} \sin\left[\omega_n(t - t^*)\right]. \tag{d}$$

Now equations (c) and (d) apply to the mass–spring–dashpot system. Let's consider now the equation of damped motion in standard form (in order to relate what we do for an arbitrary system):

$$\ddot{x} + 2\zeta\omega_n\dot{x} + \omega_n^2 x = \delta(t - t^*). \tag{6.41}$$

Equation (6.41) is identical to (a), except that in (6.41) the δ in (a) is replaced by $m\delta$. Thus the response function for $\zeta < 1$ is

$$h(t - t^*) = \frac{1}{\omega_n \sqrt{1 - \zeta^2}} \, e^{-\omega_n \zeta(t - t^*)} \sin\left[\omega_n \sqrt{1 - \zeta^2}(t - t^*)\right], \tag{6.42}$$

and for the undamped case $\zeta = 0$

$$h(t - t^*) = \left(\frac{1}{\omega_n}\right) \sin\left[\omega_n(t - t^*)\right]. \tag{6.43}$$

With a knowledge of (6.42), we now proceed to determine the forced solution to (6.40). First we consider the forcing function $F_0(t)$ (Fig. 6.30).

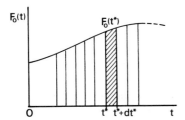

Figure 6.30

We imagine the force $F_0(t)$ to be divided into infinitesimal vertical strips. Let us consider in particular the strip between $t = t^*$ and $t = t^* + dt^*$ (the shaded area). The impulse (area) of this part of the force is

$$F_0(t^*)\, dt^*.$$

Thus the response to this portion of $F_0(t)$ alone would by definition of $h(t - t^*)$ [and by the linearity of (6.40)] be

$$x(t) = \frac{F_0(t^*)\, dt^*}{1} h(t - t^*).$$

We now integrate over all such elements to get

$$x(t) = \int_0^t F_0(t^*) h(t - t^*)\, dt^*. \tag{6.44}$$

The complete solution consists of (6.44) along with the free solution. Equation (6.44) corresponds to a system which is at rest at $t = 0$. Thus the part of the solution which corresponds to (6.44) has $x = 0$, $\dot{x} = 0$ at $t = 0$. If, in fact, $x(0) = x_0$ and $\dot{x}(0) = v_0$, these conditions must be met by the free solution.

Example 16

Consider an undamped mass–spring oscillator forced at its natural frequency. Given that $x(0) = x_0$ and $\dot{x}(0) = v_0$, determine the motion.

Solution

The equation of motion is

$$\ddot{x} + \omega_n^2 x = F_0 \sin \omega_n t.$$

Thus the forced solution can be written using (6.43) and (6.44):

$$x(t) = \frac{F_0}{\omega_n} \int_0^t \sin \omega_n t^* \sin \left[\omega_n(t - t^*)\right] \, dt^*.$$

If we expand

$$\sin \omega_n(t - t^*) = \sin \omega_n t \cos \omega_n t^* - \cos \omega_n t \sin \omega_n t^*,$$

we can determine $x(t)$ as

$$x(t) = \frac{F_0}{2\omega_n} \left(\frac{1}{\omega_n} \sin \omega_n t - t \cos \omega_n t\right). \tag{a}$$

The second portion of (a) is exactly (6.16). The first portion of (a) is part of the free solution. It is that which makes

$$x(0) = \dot{x}(0) = 0.$$

Since $x(0) = x_0$ and $\dot{x}(0) = v_0$, we augment (a) with

$$x(t) = A \sin \omega_n t + B \cos \omega_n t,$$

where A and B are now easily determined:

$$A = \frac{v_0}{\omega_n}, \qquad B = x_0.$$

Thus

$$x(t) = -\frac{F_0 t}{2\omega_n} \cos \omega_n t + \frac{1}{2\omega_n^2} (F_0 + 2\omega_n v_0) \sin \omega_n t + x_0 \cos \omega_n t.$$

....

Example 17

Suppose that $x(0) = \dot{x}(0) = 0$ and $\ddot{x} + \omega_n^2 x = f(t)$, where

$$f(t) = t, \qquad 0 < t < 1,$$

$$= 0, \qquad 1 < t.$$

Find the motion $x(t)$.

Solution

The solution for $0 < t < 1$ is given (entirely) by (6.43) and (6.44):

$$x = \frac{1}{\omega_n} \int_0^t t^* \sin\left[\omega_n(t - t^*)\right] \, dt^*,$$

or integrating

$$x(t) = \frac{1}{\omega_n^3} (\omega_n t - \sin \omega_n t), \qquad 0 \leqslant t \leqslant 1. \tag{a}$$

From $t = 1$ on, the forcing function is zero. Thus from this time forward, the system has free motion. The conditions on x and $\dot{x}$ at $t = 1$ are

$$x(1) = \frac{1}{\omega_n^3} (\omega_n - \sin \omega_n),$$

$$\dot{x}(1) = \frac{1}{\omega_n^2} (1 - \cos \omega_n),$$

and the motion is then

$$x(t) = \frac{1}{\omega_n^3} \left[(1 - \cos \omega_n) \sin \omega_n (t - 1) \right.$$
$$\left. + (\omega_n - \sin \omega_n) \cos \omega_n (t - 1) \right], \qquad 1 < t. \tag{b}$$

....

Those students who have studied differential equations should recognize that the simplest way to evaluate the integral

$$x(t) = \int_0^t F(t^*) h(t - t^*) \, dt^*$$

is by means of Laplace transforms. The integral here has a very special form called *convolution*. This means that the Laplace transform of x is very simple to obtain. The only problem then is to invert the transform of x to get x itself. Tables of transforms are useful in this situation.

9. INTRODUCTION TO VIBRATION ISOLATION

Suppose that we have the system in Fig. 6.31(a). A force $F \sin \omega t$ is applied to a mass–spring system. This force could arise in many ways; e.g., an unbalance in rotating machinery, as in Example 3.

The forced motion of the system in Fig. 6.31(a) is given by (6.11):

$$x_1 = \frac{F/m}{\omega_n^2 - \omega^2} \sin \omega t. \tag{a}$$

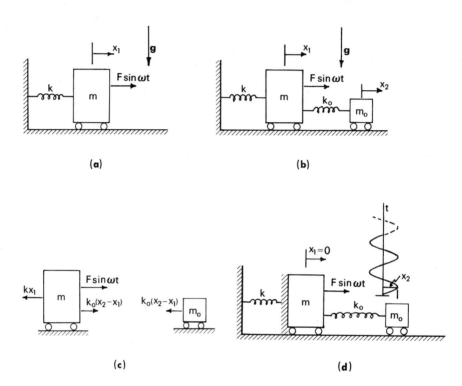

Figure 6.31

In many situations, the motion $x_1(t)$ may be undesirable, especially if ω is close to ω_n.

It is possible to eliminate the forced motion (a) by adding another mass–spring system to the original system, as shown in Fig. 6.31(b). This solution to a vibration problem has found wide application in engineering practice, though, as we shall see, the effectiveness of the solution has one severe restriction. We present this example to illustrate the idea of vibration isolation.

Suppose that we measure the displacements of the two masses m and m_0 by means of x_1 and x_2, their displacements from equilibrium. Let x_1 and x_2 be positive, with $x_2 > x_1$. Figure 6.31(c) gives the free-body diagram for the system (showing horizontal forces only).

Writing $\mathbf{F} = \dot{\mathbf{p}}$ in the horizontal direction for each mass gives

$$F_x = ma_x: \quad -kx_1 + k_0(x_2 - x_1) + F\sin\omega t = m\ddot{x}_1,$$

$$F_x = ma_x: \quad -k_0(x_2 - x_1) = m_0\ddot{x}_2. \tag{b}$$

Rewriting (b) gives

$$m\ddot{x}_1 + (k + k_0)x_1 - k_0x_2 = F \sin \omega t,$$

$$m_0\ddot{x}_2 - k_0x_1 + k_0x_2 = 0. \tag{c}$$

Equations similar to (c) attain a certain amount of transparency when written in matrix* notation. Here we shall express each equation in both standard and matrix form.

$$\begin{bmatrix} m & 0 \\ 0 & m_0 \end{bmatrix} \begin{bmatrix} \ddot{x}_1 \\ \ddot{x}_2 \end{bmatrix} + \begin{bmatrix} k + k_0 & -k_0 \\ -k_0 & k_0 \end{bmatrix} \begin{bmatrix} x_1 \\ x_2 \end{bmatrix} = \begin{bmatrix} F \\ 0 \end{bmatrix} \sin \omega t. \tag{c'}$$

We are interested here in the forced solution to (c). In this kind of problem there are also free solutions (two!). For the present, we seek a pair of functions $[x_1(t), x_2(t)]$ which, when inserted in the left side of (c), yields the right side of (c).

Again we revert to our guessing game. There are several factors to keep in mind. First, there is no damping in this system. The force on the system [not merely on m, but, as (c') clearly shows, on the entire system] is sinusoidal. From our experience with the forced motion of a simple mass–spring oscillator, we are led to guess the forced solution to the present problem in the form

$$x_1(t) = A_1 \sin \omega t, \qquad x_2(t) = A_2 \sin \omega t, \tag{d}$$

or

$$\begin{bmatrix} x_1 \\ x_2 \end{bmatrix} = \begin{bmatrix} A_1 \\ A_2 \end{bmatrix} \sin \omega t, \tag{d'}$$

where A_1 and A_2 are constants to be determined.

Inserting (d) into (c) yields two linear equations for the constants A_1 and A_2:

$$(-m\omega^2 + k + k_0)A_1 - k_0A_2 = F,$$

$$-k_0A_1 + (-m_0\omega^2 + k_0)A_2 = 0, \tag{e}$$

or

$$\begin{bmatrix} (-m\omega^2 + k + k_0) & -k_0 \\ -k_0 & (-m_0\omega^2 + k_0) \end{bmatrix} \begin{bmatrix} A_1 \\ A_2 \end{bmatrix} = \begin{bmatrix} F \\ 0 \end{bmatrix}. \tag{e'}$$

If we can find A_1 and A_2 which satisfy (e), then (d) will be the forced solution to our problem.

---

* See Appendix 1, Section D.

We get the solution to (e) by means of Cramer's rule:*

$$A_1 = \frac{\begin{vmatrix} F & -k_0 \\ 0 & (-m_0\omega^2 + k_0) \end{vmatrix}}{\begin{vmatrix} (-m\omega^2 + k + k_0) & -k_0 \\ -k_0 & (-m_0\omega^2 + k_0) \end{vmatrix}}$$

$$A_2 = \frac{\begin{vmatrix} (-m\omega^2 + k + k_0) & F \\ -k_0 & 0 \end{vmatrix}}{\begin{vmatrix} (-m\omega^2 + k + k_0) & -k_0 \\ -k_0 & (-m_0\omega^2 + k_0) \end{vmatrix}}, \tag{f'}$$

or

$$A_1 = \frac{F(-m_0\omega^2 + k_0)}{mm_0\omega^4 - (mk_0 + m_0k_0 + m_0k)\omega^2 + kk_0},$$

$$A_2 = \frac{Fk_0}{mm_0\omega^4 - (mk_0 + m_0k_0 + m_0k)\omega^2 + kk_0}. \tag{f}$$

With (f) and (d), we have the forced solution to our problem. Now we are left with the design consideration: How can we choose the absorber parameters m_0 and k_0 so that the motion $x_1(t)$ is minimized?

If we set the numerator of A_1 equal to zero, we obtain

$$\frac{k_0}{m_0} = \omega^2. \tag{6.45}$$

In fact, if m_0 and k_0 are chosen to satisfy (6.45), we are assured that $A_1 = 0$ and the forced solution $x_1 = 0$ (see Problem 6.132).

If m_0 and k_0 are chosen to satisfy (g), the motion x_1 is zero. However, the absorber motion, x_2, will not be zero. In fact (see Problem 6.133)

$$x_2 = -(F/k_0) \sin \omega t. \tag{6.46}$$

The forced motion of the system can be visualized as in Fig. 6.31(d).

The main mass m has no motion, while the absorber has motion defined by (6.46). In fact, the total effect of the force on m is to excite m_0. Since m now has no motion, it is as if it were rigidly constrained. Thus the spring could be replaced by a variety of alternative constraints. Suppose, for example, $k = 0$. Our analysis would be identical, with identical results, (g) and (h). On the other hand, we could add elements in parallel to k (see Problem 6.134).

---

* See Appendix 1, Eq. (A1.37).

In theory we could add any element to or in place of the spring k, linear or nonlinear.* In practice, there is a restriction to this statement. Rarely would the excitation to a system be a pure sine wave, $F \sin \omega t$. In a linear system, we can always separate the forced and free solutions. In a linear system, transients which arise from time to time die out (so long as some damping is present). In a nonlinear system, any transient introduced would destroy the argument made above. In nonlinear systems, the free and forced motions cannot be separated.

The most serious restriction of this absorber is that it works at but one frequency, ω. If the input force to m varies in frequency or contains several frequency components, the absorber will be less effective.

Example 18

Suppose we return to Example 15. (See Fig. 6.32.)

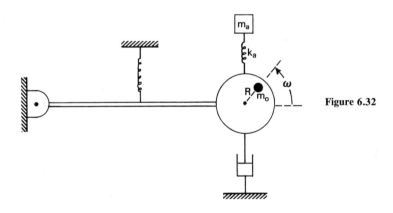

Figure 6.32

Without any of the analysis of Example 15, it should be clear that the frequency of the excitation to the motor due to the unbalanced mass m_0 is ω. Thus we attach the absorber whose parameters are determined by (6.45):

$$\frac{k_a}{m_a} = \omega^2. \tag{a}$$

To determine the amplitude of the absorber motion, we need to know the force amplitude of the excitation. From (e) of Example 15, we see that

$$F = \omega^2 m_0 R. \tag{b}$$

---

* If the effect of $F \sin \omega t$ is entirely spent on m_0, and if at some time t_0, $x_1 = \dot{x}_1 = 0$, Cauchy's theorem in differential equations assures us that $x_1 = 0$ for all time.

Thus from (6.46) we have

$$A_a = \left(\frac{\omega^2 m_0 R}{k_a}\right) \tag{c}$$

for the amplitude of the absorber.

....

In this chapter we have presented the essentials of vibration theory. Of course, in practice, vibration systems consist of rigid bodies, electrical and fluid elements, etc. We shall study the dynamics of rigid bodies in Chapter 8.

The analysis of any dynamic system is one thing. However, once we get a differential equation which can be identified with one of the standard equations of this chapter, we also have the solution. This is true whether we are discussing motion of a particle, motion of a rigid body, the population of a community, the popularity of a president, or a student's savings (which always seem to be in a state of exponential decay).

10. SUMMARY

A. Periodic Functions

τ *period:* the time for the function to repeat itself or go through one cycle

f *frequency:* the number of cycles per second (measured in Hz; i.e., cycles/sec)

ω *circular frequency,* in harmonic motion, radians per second.

For harmonic functions:

$$f = 1/\tau = \omega/2\pi \quad \text{(Hz)}$$

B. Solutions

A few words of explanation of our presentation of the solutions to vibration problems is in order at this point. The solutions are presented in three sections.

1)   *Free motion:* The constants of the free-motion solutions have been determined to meet the following initial conditions:

$$x(0) = x_0, \qquad \dot{x}(0) = v_0.$$

2)   *Forced motion:* Only the forced or particular solution is presented. If there is any damping in the system, the solution given is the actual solution for large times.

3) *Complete solution:* We present here

$$x(t) = \text{(forced motion)} + \text{(free motion)},$$

where the constants of the free-motion solution have been determined in order that the complete solution meet the following initial conditions:

$$x(0) = x_0, \qquad \dot{x}(0) = v_0.$$

---

1. Free-motion solutions

Undamped   Equation of motion: $\ddot{x} + \omega_n^2 x = 0$

$$x = \left(\frac{v_0}{\omega_n}\right) \sin \omega_n t + x_0 \cos \omega_n t$$

Damped   Equation of motion: $\ddot{x} + 2\zeta\omega_n\dot{x} + \omega_n^2 x = 0$

---

a) *Underdamped* $0 < \zeta < 1$

$$x = e^{-\omega_n \zeta t} \left[\frac{v_0 + \omega_n \zeta x_0}{\omega_n \sqrt{1 - \zeta^2}} \sin \omega_n \sqrt{1 - \zeta^2}\, t + x_0 \cos \omega_n \sqrt{1 - \zeta^2}\, t\right]$$

b) *Critically damped* $\zeta = 1$

$$x = [x_0 + (v_0 + \omega_n x_0)t]e^{-\omega_n t}$$

c) *Overdamped* $\zeta > 1$

$$x = e^{-\omega_n \zeta t} \left[\frac{v_0 + \omega_n \zeta x_0}{\omega_n \sqrt{\zeta^2 - 1}} \sinh \omega_n \sqrt{\zeta^2 - 1}\, t + x_0 \cosh \omega_n \sqrt{\zeta^2 - 1}\, t\right]$$

---

2. Forced-motion solutions

Undamped system

Equation of motion	Forced-motion solution
a) $\ddot{x} + \omega_n^2 x = C_0 = \text{const.}$	a) $x = (C_0/\omega_n^2)$
b) $\ddot{x} + \omega_n^2 x = F_0 \sin \omega t, \quad \omega \neq \omega_n$	b) $x = \left[\dfrac{F_0/\omega_n^2}{1 - (\omega^2/\omega_n^2)}\right] \sin \omega t$
c) $\ddot{x} + \omega_n^2 x = F_0 \sin \omega_n t, \quad \omega = \omega_n$	c) $x = -\left(\dfrac{F_0}{2\omega_n}\right) t \cos \omega_n t$

Damped system: $0 < \zeta < \infty$

Equation of motion	Forced-motion solution

---

a) $\ddot{x} + 2\zeta\omega_n\dot{x} + \omega_n^2 x = C_0 = \text{const.}$ a) $x = (C_0/\omega_n^2)$

b) $\ddot{x} + 2\zeta\omega_n\dot{x} + \omega_n^2 x = F_0 \sin \omega t$ b) $x = \dfrac{F_0/\omega_n^2}{\sqrt{(1 - \omega^2/\omega_n^2)^2 + 4(\zeta\omega/\omega_n)^2}} \sin (\omega t - \phi)$

where

$$\phi = \tan^{-1}\left[\frac{2\zeta(\omega/\omega_n)}{1 - (\omega^2/\omega_n^2)}\right]$$

c) $\ddot{x} + 2\zeta\omega_n\dot{x} + \omega_n^2 x = F_0(t)$ c) $x(t) = \displaystyle\int_0^t F_0(t^*)h(t - t^*) \, dt^*$

where $h(t - t^*)$ is the unit impulse response. Note that this solution meets the initial conditions $x(0) = 0$, $\dot{x}(0) = 0$. As such, it will include some part of the free-motion solution. As $t \to \infty$ this portion of the solution will die out if $\zeta > 0$.

---

3. Complete solutions

Undamped systems   Equation of motion: $\ddot{x} + \omega_n^2 x = F_0 \sin \omega t$

---

a) $\omega \neq \omega_n$ $x = \left[\dfrac{F_0/\omega_n^2}{1 - (\omega^2/\omega_n^2)}\right] \sin \omega t$

$$+ \frac{1}{\omega_n}\left[v_0 - \frac{\omega F_0/\omega_n^2}{1 - (\omega^2/\omega_n^2)}\right] \sin \omega_n t + x_0 \cos \omega_n t$$

b) $\omega = \omega_n$ $x = -\dfrac{F_0}{2\omega_n} t \cos \omega_n t + \dfrac{1}{\omega_n}\left[v_0 + \dfrac{F_0}{2\omega_n}\right] \sin \omega_n t + x_0 \cos \omega_n t$

Damped systems: $\zeta \neq 0$ Equation of motion: $\ddot{x} + 2\zeta\omega_n\dot{x} + \omega_n^2 x = F_0 \sin \omega t$

Solution: $x = \dfrac{F_0/\omega_n^2}{\sqrt{(1 - \omega^2/\omega_n^2)^2 + 4(\zeta\omega/\omega_n)^2}} \sin (\omega t - \phi) + x_{\text{free}}$

where $\phi = \tan^{-1}\left[\dfrac{2\zeta(\omega/\omega_n)}{1 - (\omega/\omega_n)^2}\right]$

x_{free}

a) *Underdamped* $0 < \zeta < 1$, all ω

$$x_{\text{free}} = \frac{e^{-\zeta\omega_n t}}{\sqrt{1-\zeta^2}} \left[\frac{v_0}{\omega_n} + \zeta x_0 + \frac{F_0/\omega_n^2 \left(\zeta \sin \phi - \dfrac{\omega}{\omega_n} \cos \phi \right)}{\sqrt{(1 - \omega^2/\omega_n^2)^2 + 4(\zeta\omega/\omega_n)^2}} \right] \sin \omega_n \sqrt{1-\zeta^2}\, t$$

$$+\, e^{-\zeta\omega_n t} \left[x_0 + \frac{(F_0/\omega_n^2)\, \sin \phi}{\sqrt{(1 - \omega^2/\omega_n^2)^2 + 4(\zeta\omega/\omega_n)^2}} \right] \cos \omega_n \sqrt{1-\zeta^2}\, t$$

b) *Critically damped* $\zeta = 1$, all ω

$$x_{\text{free}} = e^{-\omega_n t} \left\{ x_0 + \frac{(F_0/\omega_n^2)\, \sin \phi}{\sqrt{(1 - \omega^2/\omega_n^2)^2 + 4(\zeta\omega/\omega_n)^2}} \right.$$

$$\left. + \left[v_0 + \omega_n x_0 + \frac{F_0/\omega_n^2(\omega_n \sin \phi - \omega \cos \phi)}{\sqrt{(1 - \omega^2/\omega_n^2)^2 + 4(\zeta\omega/\omega_n)^2}} \right] t \right\}$$

c) *Overdamped* $1 < \zeta$, all ω

$$x_{\text{free}} = \frac{e^{-\zeta\omega_n t}}{\sqrt{\zeta^2-1}} \left[\frac{v_0}{\omega_n} + \zeta x_0 + \frac{F_0/\omega_n^2 \left(\zeta \sin \phi - \dfrac{\omega}{\omega_n} \cos \phi \right)}{\sqrt{(1 - \omega^2/\omega_n^2)^2 + 4(\zeta\omega/\omega_n)^2}} \right] \sinh \omega_n \sqrt{\zeta^2-1}\, t$$

$$+\, e^{-\zeta\omega_n t} \left[x_0 + \frac{(F_0/\omega_n^2)\, \sin \phi}{\sqrt{(1 - \omega^2/\omega_n^2)^2 + 4(\zeta\omega/\omega_n)^2}} \right] \cosh \omega_n \sqrt{\zeta^2-1}\, t$$

Arbitrary forcing function   Equation of motion: $\ddot{x} + 2\zeta\omega_n \dot{x} + \omega_n^2 x = F_0(t)$

Solution: $x(t) = \displaystyle\int_0^t F_0(t^*) h(t - t^*)\, dt^* + x_{\text{free}},$

where $h(t - t^*) = $ unit impulse response
$x_{\text{free}} = $ free solution listed in (1)

C. Vibration Isolation

If a system is subjected to a harmonic excitation of amplitude F and frequency ω at some point A, the forced motion due to the excitation can be nullified by the addition at point A of a mass–spring system (absorber) with characteristics m_0 and k_0, where

$$\omega^2 = \frac{k_0}{m_0}.$$

The amplitude of the absorber motion will be

$$A_a = \frac{F}{k_0}.$$

[Note that in the above formula, F must have the dimension of force, not, for example, force/mass.]

PROBLEMS

Section 6.2

For Problems 6.1 through 6.10, write the differential equation of motion for the mass m shown in the accompanying figure. Use the coordinate indicated.

Problem 6.1

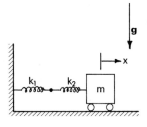

Figure 6.33

Problem 6.2

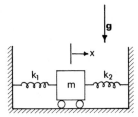

Figure 6.34

Problem 6.3

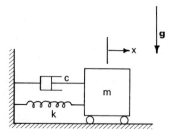

Figure 6.35

Problem 6.4

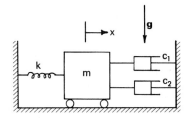

Figure 6.36

Problem 6.5

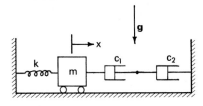

Figure 6.37

Problem 6.6

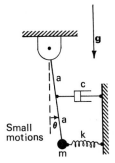

Figure 6.38

Problem 6.7

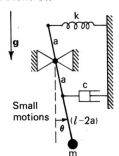

Figure 6.39

Problem 6.8

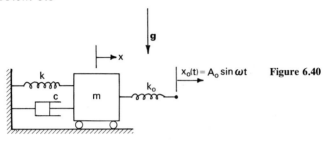

Figure 6.40

$x_0(t) = A_0 \sin \omega t$

Problem 6.9

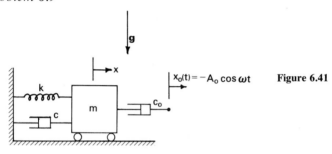

Figure 6.41

$x_0(t) = -A_0 \cos \omega t$

Problem 6.10

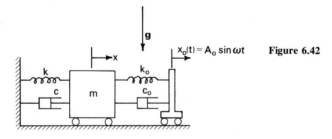

Figure 6.42

$x_0(t) = A_0 \sin \omega t$

Problem 6.11

The cylindrical log shown in Fig. 6.43 has mass m and cross-sectional area A. It floats in a fluid of infinite extent with weight per unit volume γ. If the log is displaced vertically and then released, it will oscillate. Neglecting the motion of the surrounding fluid, write the differential equation of motion of the log.

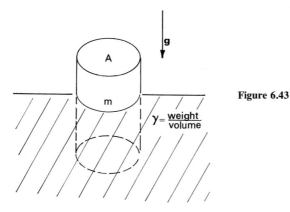

Figure 6.43

Problem 6.12

The piston in the smooth tube shown in Fig. 6.44 has mass m and cross-sectional area A. The gases on either side of the piston are considered to be isothermal: $pV = $ constant, where p is the pressure and V is the volume of the gas. Given that $p = p_0$ and $V = V_0$ on either side of the piston when $x = 0$, show that the differential equation of motion is

$$m\ddot{x} + 2\left(\frac{p_0 V_0 A^2 x}{V_0^2 - A^2 x^2}\right) = 0.$$

Linearize the equation of motion, assuming that x remains small.

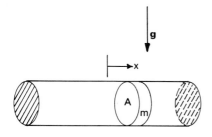

Figure 6.44

★ Problem 6.13

Modify the system in Problem 6.12 in the following way: Let the gases be considered adiabatic, so that $pV^\gamma = $ constant, where $\gamma > 1$. Write the differential equation of motion of the piston. Linearize the differential equation for small motions. [*Hint:* Use the first two terms of a Taylor series expansion of the nonlinear term about $x = 0$. See Appendix A1 Eq. (A1.30a).]

★ A starred problem is one that is more difficult than the average.

Problem 6.14

Consider the differential equations:

a) $\ddot{x} + a(t)x = f(t)$,   b) $\ddot{x} + a(t)x^2 = f(t)$,

where $a(t)$ and $f(t)$ are given functions. Suppose in each case that $x_1(t)$ satisfies either (a) or (b), and suppose that $x_2(t)$ satisfies the homogeneous equation:

a') $\ddot{x} + a(t)x = 0$,   b') $\ddot{x} + a(t)x^2 = 0$.

Show that $x_3 = x_1 + cx_2$ satisfies (a) for all values of c, but that x_3 satisfies (b) only if $c = 0$.

Problem 6.15

Consider the system in the figure. By some internal mechanism, the mass m_0 has the motion $u(t) = A_0 \sin \omega t$ with respect to the main mass m. Find the equation of motion of m.

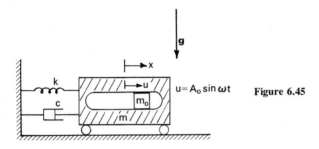

$u = A_0 \sin \omega t$ **Figure 6.45**

Problem 6.16

The cylinder shown in the figure rotates in a vertical plane so that $\dot{\theta} = \omega_0 = $ constant. The mass m moves in a smooth track restrained by a spring with modulus k. Suppose that the force in the spring is zero when a) $r = 0$, b) $r = R$. State the differential equation of motion of m in each case.

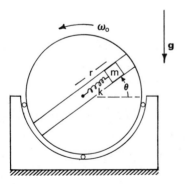

Figure 6.46

Problem 6.17

The system in the figure is similar to that of Problem 6.16, except that the cylinder is mounted in a cart which has a constant horizontal acceleration a_0. Suppose that the force in the spring is zero when $r = R$. Determine the differential equation of motion of m. Take $\theta = 0$ at $t = 0$.

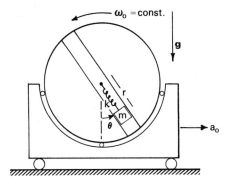

Figure 6.47

Problem 6.18

A pendulum is mounted on a cart which is given the following specified motion:

$$x_0(t) = X_0 \sin \omega t.$$

Assume that the angle θ (see the figure) remains small. Determine the differential equation of motion for m.

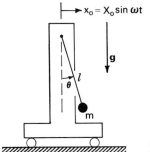

Figure 6.48

Problem 6.19

The mass m in Fig. 6.49 slides with friction on a horizontal surface. The coefficients of static and sliding friction are both μ. At $t = 0$,

$$x = x_0 > \mu \frac{mg}{k}, \qquad \text{and} \qquad \dot{x} = 0.$$

Write the differential equation of motion of m for $t = 0$ until the time at which $\dot{x}$ is again zero.

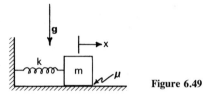

Figure 6.49

Problem 6.20

The particle shown in Fig. 6.50 moves without friction along the x axis restrained by two *constant-force springs*. The force in each spring is F (tension). Assuming that the motion is small so that $(x/h) << 1$, find the differential equation of motion of m.

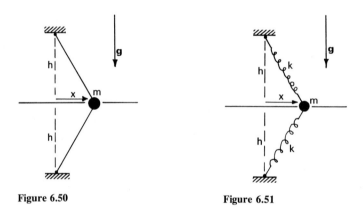

Figure 6.50   Figure 6.51

Problem 6.21

The particle in Fig. 6.51 moves without friction on the x axis, restrained by two linear springs, of modulus k and free length a.

a) Assuming small motions $[(x/h) << 1]$, find the differential equation of motion of m.

b) Is there any circumstance in which the assumption of small motion is unnecessary in order to have a linear differential equation of motion?

Section 6.3

Problem 6.22

Find ω_n for the system of Problem 6.1.

Problem 6.23

Find ω_n for the system of Problem 6.2.

Problem 6.24

Find ω_n for the system of Problem 6.11.

Problem 6.25

Find ω_n for the system of Problem 6.12.

Problem 6.26

Find ω_n for the system of Problem 6.13.

Problem 6.27

Find ω_n for the system of Problem 6.20.

Problem 6.28

Find ω_n for the system of Problem 6.21.

Problem 6.29

In the system shown in Fig. 6.52, an inextensible string passes over two weightless cylinders. Write the differential equation of motion for the mass m, and thus determine the natural frequency ω_n.

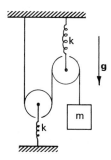

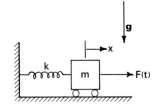

Figure 6.52   **Figure 6.53**

Problem 6.30

At $t = 0$, a force of very short duration, Δt, is applied to the system shown in Fig. 6.53. The velocity of the mass m at $t = 0^+ = \Delta t$ is v_0.

   a) Calculate the impulse of $F(t)$ and the average value of $F(t)$.

   b) Determine the motion of m for $t > 0$.

Problem 6.31

Consider the system shown in the figure. A particle of mass m and speed v_0 strikes and sticks to the mass m of a pendulum.

    a) By considering the momentum of the particles before and after the collision, show that the speed of the combined system just after collision is $\frac{1}{2}v_0$.

    b) Suppose that the motion $\theta(t)$ has a small amplitude. Determine the amplitude. [*Hint:* Use $\cos\theta \approx 1 - \frac{1}{2}\theta^2$.]

    c) Determine the motion $\theta(t)$, given that the collision occurs at $t = 0$.

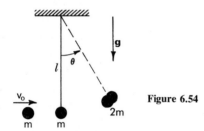

Figure 6.54

Problem 6.32

The horizontal disk shown in the figure rotates at a constant angular speed ω_0. A particle m can slide without friction in a groove cut in the disk, restrained by a linear spring k. The force in the spring is zero when $r = 0$.

    a) What is the differential equation of motion of m?

    b) Determine the natural frequency ω_n, and the values of ω_0 for which m will not oscillate.

    c) At $t = 0$, $r = R$ and $\dot{r} = 0$. Determine $r(t)$, assuming that the system has a natural frequency different from zero.

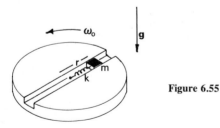

Figure 6.55

Problem 6.33

Suppose that the differential equation of motion for a particle is given by

$$\ddot{x} - a^2 x = 0, \qquad a = \text{constant}, \qquad a > 0.$$

Following the development from (6.3) to (6.5), show that the free solutions have the form

$$x(t) = Ae^{at} + Be^{-at}.$$

Problem 6.34

Suppose that the initial conditions for Problem 6.33 are $x(0) = x_0$, $\dot{x}(0) = v_0$. Show that the motion is

$$x(t) = \tfrac{1}{2}(x_0 + v_0/a)e^{at} + \tfrac{1}{2}(x_0 - v_0/a)e^{-at}.$$

Problem 6.35

The particle shown in the figure can slide without friction in the groove in the horizontal disk. The disk rotates at a constant angular speed ω_0.

a) Determine the differential equation of motion of m.

b) Suppose that at $t = 0$, $r = R$, $\dot{r} = v_0$, where $v_0 = -\omega_0 R$. Determine $r(t)$. [*Hint:* Use the results of Problems 6.33 and 6.34.]

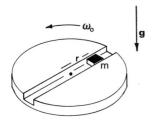

Figure 6.56

In Problems 6.36 through 6.39, write the given function $f(t)$ in the form $A \sin (ct - \phi)$. That is, determine A, c, and ϕ.

Problem 6.36

$f(t) = 3 \sin 6t + 4 \cos 6t$

Problem 6.37

$f(t) = 2 \sin t + \cos t$

Problem 6.38

$f(t) = -2 \sqrt{2} \sin 2t + 2 \sin (2t - \tfrac{1}{4}\pi)$

Problem 6.39

$f(t) = f_1(t) + f_2(t)$, where $f_1(t)$ and $f_2(t)$ are as shown in Fig. 6.57.

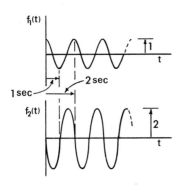

Figure 6.57

For Problems 6.40 through 6.43, determine the natural frequency ω_n for the system in the appropriate figure, using Rayleigh's method.

Problem 6.40

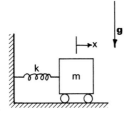

Figure 6.58

Problem 6.41

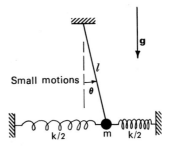

Figure 6.59

Problem 6.42

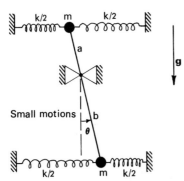

Figure 6.60

Problem 6.43

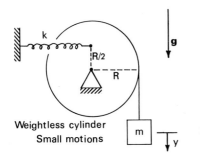

Weightless cylinder
Small motions

Figure 6.61

Section 6.4

Problem 6.44

Determine the frequency f and the period τ for the function shown in the figure.

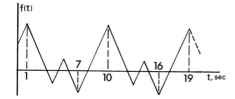

Figure 6.62

Problem 6.45

Determine the frequency f and the period τ for the function shown in the figure.

Figure 6.63

Problem 6.46

Determine the frequency f and the period τ for the function shown in the figure.

Figure 6.64

Problem 6.47

A motion takes place at a frequency of 20 rad/sec. Determine the frequency f and the period τ of the motion.

Problem 6.48

A motion takes place at a frequency of 30 Hz. Find the frequency ω and the period τ, assuming that the motion is harmonic.

Problem 6.49

Consider the function $f(t) = 1000 \sin (35t - 5) + 20 \cos 35t$. Determine the parameters ω, f, and τ for $f(t)$.

Problem 6.50

Determine which of the following functions is periodic, and determine the frequency f and period τ.

a) $f(t) = 10 + 5 \sin 2t$

b) $f(t) = t + 2 \cos 5t$

c) $f(t) = 6 \sin (t^2)$

d) $f(t) = t \sin 2t$

e) $f(t) = (1/t^2) \sin (t^2)$

Problem 6.51

Determine which of the following functions is periodic, and determine the frequency f and the period τ.

a) $f(t) = 2 \sin t + 3 \cos 2t$

b) $f(t) = 5(\sin 6t)(\cos 9t)$

c) $f(t) = 4(\sin 3t)(\sin 7t)$

d) $f(t) = 3 \sin^2 (5t)$

e) $f(t) = 7(\sin 3t)(\cos^2 6t)$

f) $f(t) = (\sin \sqrt{2}t)(\sin 2t)$

Problem 6.52

Determine the parameters ω_n, f, and τ for the free motion of the system in Problem 6.21.

Problem 6.53

You are asked to design a simple pendulum with a free-motion frequency of 20 cycles/min.

a) What are the pertinent design parameters of the pendulum?

b) Given that the mass at the end of the pendulum is 10 slugs, can you determine, from the given data, the maximum force in the string? If not, what additional information is necessary?

c) Determine the natural frequency ω_n of the pendulum.

Problem 6.54

A barrier restricts the motion of the pendulum shown in the figure.

a) Is the restricted motion periodic? If so, what is the period of the motion? (Assume small displacements.)

b) Suppose that $l = 6.44$ ft. Compare the period of the given system to that of an unrestricted pendulum of the same length.

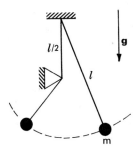

Figure 6.65

Problem 6.55

When the elevator shown in the figure is at rest, the periods of the free motions of the pendulum and the mass–spring system are the same. The elevator is then given a constant acceleration, a_0, upward.

a) Determine m_p and l in terms of m_s, k, and g.

b) Determine the periods of the systems in the moving elevator.

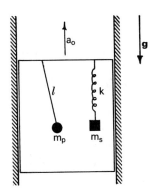

Figure 6.66

Section 6.5

Problem 6.56

Calculate ω_n, F_0, and ω for the system of Problem 6.16.

Problem 6.57

Calculate ω_n, F_0, and ω for the system of Problem 6.17.

Problem 6.58

Calculate ω_n, F_0, and ω for the system of Problem 6.18.

Problem 6.59

For the system of Problem 6.16, suppose that the force in the spring is zero when $r = R$.

a) Determine the forced solution when

$$\omega_0^2 < \frac{k}{m} \quad \text{and} \quad \omega_0^2 \neq \frac{1}{2}\frac{k}{m}.$$

b) Determine the position r about which the mass oscillates.

Problem 6.60

Consider the system of Problem 6.16.

a) Under what circumstance will $\omega = \omega_n$?

b) Let $(k/m) = 256$ (rad/sec)^2, and suppose that the spring force is zero when $r = 0$. Determine ω_0 such that the amplitude of the forced solution is 6 in.

Problem 6.61

Consider the system of Problem 6.17.

a) Determine ω_n, ω, and F_0 for the system.

b) What is the apparent equilibrium position of the mass (about what point r does the mass oscillate)?

c) Which of the parameters a_0, g, m, k, ω_0 affect the natural frequency of the system?

d) Suppose that the cart were accelerating up an inclined plane; which of the parameters ω, ω_n, F_0 would change?

Problem 6.62

Suppose that in Problem 6.18, $mg = 64$ lb, $l = 2$ ft, and the frequency of the cart motion is 0.5 Hz. Calculate X_0 such that the amplitude of the forced solution is $\theta = 7.5°$.

Problem 6.63

Consider the system of Problem 6.19. Suppose that at $t = 0$,

$$x = x_0 > \mu \frac{mg}{k}, \qquad \dot{x} = 0.$$

a) Calculate $x(t)$.

b) From (a), determine the first time and position x at which the speed becomes zero after $t = 0$.

c) Check the value of x in (b), using energy considerations.

Problem 6.64

Consider a mass–spring oscillator subject to a constant applied force:

$$\ddot{x} + \omega_n^2 x = F_0, \qquad F_0 = \text{constant}.$$

The forced solution of this differential equation is called the *static deflection*, x_{st}. Determine x_{st}.

Problem 6.65

The motion of a system is described by the equation

$$2\ddot{x} + 32x = 10 \sin \omega t, \qquad x = \text{feet}.$$

a) Calculate the magnification factor μ, given that (i) $\omega = 2$ rad/sec, (ii) $\omega = 3$ rad/sec, (iii) $\omega = 5$ rad/sec.

b) Calculate the amplitude of the forced solution for the above values of ω.

Problem 6.66

The motion of a system is described by the equation

$$3\ddot{x} + 12x = 6 \sin 3t, \qquad x = \text{feet}.$$

a) Calculate x_{st}.

b) Determine the magnification factor μ.

c) Suppose that the term $6 \sin 3t$ is replaced by $6 \sin t$. Repeat steps (a) and (b).

Problem 6.67

Derive the solution (6.13) from (6.12).

Problem 6.68

Consider the forced motion of the system shown in Fig. 6.67.

a) For what initial conditions will the forced solution be the complete solution (i) if $\omega \neq \omega_n$, (ii) if $\omega = \omega_n$?

b) For the case $\omega = \omega_n$, determine the total energy, $T + V$, as a function of time (consider only the forced solution).

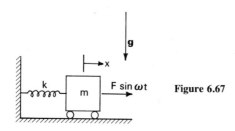

Figure 6.67

Problem 6.69

The motion of a system is described by the equation

$$0.5\ddot{x} + 2x = 4 \sin 3t.$$

At $t = 0$, $x = 0$, $\dot{x} = 0$. Determine $x(t)$.

Problem 6.70

The motion of a system is described by the equation

$$3\ddot{x} + 75x = 12 \sin 3t.$$

At $t = 0$, $x = 1$ ft, $\dot{x} = 3$ ft/sec. Determine $x(t)$.

Problem 6.71

The motion of a system is described by the equation

$$\ddot{x} + \omega_n^2 x = F_0 \sin \omega_n t,$$

(that is, $\omega = \omega_n$). At $t = 0$, $x = x_0$, $\dot{x} = v_0$. Noting that the forced solution is given by (6.16), show that the complete solution is

$$x(t) = -\frac{F_0}{2\omega_n} t \cos \omega_n t + \frac{1}{\omega_n} \left[v_0 + \frac{F_0}{2\omega_n} \right] \sin \omega_n t + x_0 \cos \omega_n t.$$

Problem 6.72

The motion of a system is described by the equation

$$0.25\ddot{x} + 16x = \sin 8t.$$

At $t = 0$, $x = 0$, $\dot{x} = 0$. Determine $x(t)$. [*Hint:* See Problem 6.71 or the summary at the end of this chapter.]

Problem 6.73

The mass shown in the figure is restrained by a spring k, whose free length is y_0.

   a) Write the equation of motion of m, using the variable y.

   b) Write the equation of motion for m, using the displacement from equilibrium λ.

   c) Suppose that at $t = 0$, $y = y_0 + mg/k$ and $\dot{y} = v_0$. Solve the equations of (a) and (b) and show that the solutions are identical.

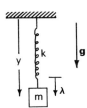   **Figure 6.68**

Section 6.6

Problem 6.74

Find ζ for the system of Problem 6.4.

Problem 6.75

Find ζ for the system of Problem 6.5.

Problem 6.76

Find ζ for the system of Problem 6.6.

Problem 6.77

Find ζ for the system of Problem 6.7.

Problem 6.78

The motion of a system is described by the differential equation

$$3\ddot{x} + \dot{x} + 27x = 0 \text{ lb.}$$

Determine ω_n and ζ for the system.

Problem 6.79

The motion of a system is described by the differential equation

$$0.5\ddot{x} + a\dot{x} + 2x = 0 \text{ lb,}$$

where a is some constant. Find the range of values of the constant a such that the system is (a) underdamped, (b) critically damped, and (c) overdamped.

Problem 6.80

For the system of Problem 6.6, suppose that m, k, and a are given. Determine the value of c such that the system is critically damped.

In Problems 6.81 through 6.87, refer to the formulas

$$\delta = \ln \left[\frac{A(t)}{A(t+\tau)} \right] = \ln \left(\frac{x + \Delta x}{x} \right) \tag{a}$$

$$\delta \approx \left(\frac{\Delta x}{x} \right) \tag{b}$$

for the logarithmic decrement.

Problem 6.81

Suppose that the maximum amplitudes of a system for $n+1$ successive cycles are x_0, $x_1, \cdots, x_n$. Noting that

$$\delta = \ln \left(\frac{x_0}{x_1} \right) = \ln \left(\frac{x_1}{x_2} \right) = \cdots = \ln \left(\frac{x_{n-1}}{x_n} \right),$$

show that

$$\delta = \frac{1}{n} \ln \left(\frac{x_0}{x_n} \right).$$

Problem 6.82

Suppose that δ cannot be assumed to be small. Show that the damping ratio is

$$\zeta = \frac{\delta}{\sqrt{4\pi^2 + \delta^2}}.$$

Problem 6.83

Under what circumstances, heavy or light damping, will formula (b) be a good approximation to formula (a)? Explain.

Problem 6.84

Show that if $(\Delta x/x) < 1$, then formula (b) gives a larger estimate of δ than the actual value from formula (a).

Problem 6.85

For the motion indicated in the figure, determine δ based on formulas (a) and (b).
Using the better value, find ζ and then ω_n.

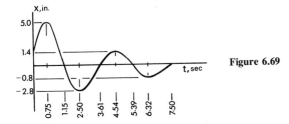

Figure 6.69

Problem 6.86

For the maximum amplitudes shown in the figure, determine δ, ζ, and ω_n for the
system.

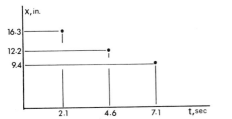

Figure 6.70

Problem 6.87

The maximum amplitudes and the times at which they occurred are tabulated below
for a given system.

	Max. amp., in.	Time, sec
1	6.12	0.15
2	5.88	1.04
3	5.65	1.93

Determine δ, ζ, and ω_n for the system.

Problem 6.88

Characterize the damping ratio ζ for the three plots shown in the figure as either (a) $\zeta = 0$, (b) $0 < \zeta < 1$, or (c) $\zeta \geqslant 1$.

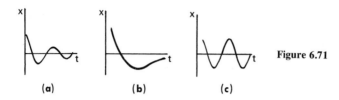

Figure 6.71

(a)   **(b)**   **(c)**

Problem 6.89

Consider the function

$$f(x) = x^a e^{-bx},$$

where a and b are any positive constants. Show that

$$f(x) \to 0 \quad \text{as} \quad x \to \infty.$$

[*Hint:* Recall l'Hôpital's rule from calculus.]

For Problems 6.90 through 6.98, the motion of a system is described by the given differential equation subject to the stated initial conditions. In each case, calculate the free motion of the system.

Problem 6.90

$$0.5\ddot{x} + \dot{x} + 50x = 0$$
$$x(0) = 0, \quad \dot{x}(0) = -5 \text{ ft/sec}$$

Problem 6.91

$$4\ddot{x} + 12\dot{x} + 100x = 0$$
$$x(0) = 3 \text{ in.}, \quad \dot{x}(0) = -4.5 \text{ in./sec}$$

Problem 6.92

$$0.25\ddot{x} + 2\dot{x} + 4x = 0$$
$$x(0) = 0, \quad \dot{x}(0) = 2.5 \text{ ft/sec}$$

Problem 6.93

$$0.143\ddot{x} + 3\dot{x} + 7x = 0$$
$$x(0) = 2 \text{ in.}, \quad \dot{x}(0) = -21 \text{ in./sec}$$

Problem 6.94

$$0.2\ddot{x} + 4\dot{x} + 20x = 0$$
$$x(0) = -1 \text{ in.,} \qquad \dot{x}(0) = 10 \text{ in./sec}$$

Problem 6.95

$$5\ddot{x} + 48\dot{x} + 180x = 0$$
$$x(0) = 0.208 \text{ in.,} \qquad \dot{x}(0) = 0$$

Problem 6.96

$$5\ddot{x} + 60\dot{x} + 45x = 0$$
$$x(0) = 0, \qquad \dot{x}(0) = 3\sqrt{3} \text{ in./sec}$$

Problem 6.97

$$0.02\ddot{x} + 0.6\dot{x} + 4.5x = 0$$
$$x(0) = 0.5 \text{ ft,} \qquad \dot{x}(0) = 0$$

Problem 6.98

$$0.4\ddot{x} + 44\dot{x} + 48.4x = 0$$
$$x(0) = 2.4 \text{ in.,} \qquad \dot{x}(0) = 0$$

Problem 6.99

An overdamped system is given the following initial conditions at $t = 0$: $x = x_0$, $\dot{x} = v_0$.

a) Show that if $x_0 > 0$, then v_0 must be *less than* $[-x_0\omega_n(\zeta + \sqrt{\zeta^2 - 1})]$ if x is to become zero at some finite time.

b) Show that if $x_0 < 0$, then v_0 must be *greater than* $[-x_0\omega_n(\zeta + \sqrt{\zeta^2 - 1})]$ if x is to become zero at some finite time.

Section 6.7

In Problems 6.100 through 6.103, determine expressions for the parameters ω_n, ζ, and F_0 for the system indicated.

Problem 6.100

The system of Problem 6.8.

Problem 6.101

The system of Problem 6.9.

Problem 6.102

The system of Problem 6.10.

Problem 6.103

The system of Problem 6.15.

In Problems 6.104 through 6.107, the motion of a system is described by the given equation. Determine the amplitude and phase of the forced solution.

Problem 6.104

$$5\ddot{x} + 2\dot{x} + 20x = 25 \sin t, \qquad x = \text{inches}$$

Problem 6.105

$$0.33\ddot{x} + 2\dot{x} + 3x = 1.33 \sin 6t, \qquad x = \text{feet}$$

Problem 6.106

$$0.125\ddot{x} + 3\dot{x} + 2x = 0.375 \sin 3t, \qquad x = \text{inches}$$

Problem 6.107

$$0.2\ddot{x} + \dot{x} + 5x = 0.4 \sin 5t, \qquad x = \text{feet}$$

In Problems 6.108 through 6.111, the motion of a system is described by the given equation. Find the amplitude of the forced solution.

Problem 6.108

$$2.5\ddot{x} + \dot{x} + 10x = 7.5 \sin 2t + 10 \cos 2t, \qquad x = \text{inches}$$

Problem 6.109

$$\ddot{x} + 2\dot{x} + 4x = 5 \sin 3t - 12 \cos 3t, \qquad x = \text{feet}$$

Problem 6.110

$$\ddot{x} + 9\dot{x} + 9x = 4 \cos 6t, \qquad x = \text{inches}$$

Problem 6.111

$$\ddot{x} + 6.4\dot{x} + 16x = 2.83 \sin 3t + \cos 3t, \qquad x = \text{feet}$$

In Problems 6.112 through 6.115, determine the following for the equation of motion indicated.

    a) The magnification factor at the given forcing frequency ω

    b) The resonant magnification factor

c) The maximum magnification factor and the forcing frequency ω at which it would occur

Problem 6.112

The equation of motion of Problem 6.104.

Problem 6.113

The equation of motion of Problem 6.106.

Problem 6.114

The equation of motion of Problem 6.109.

Problem 6.115

The equation of motion of Problem 6.111.

Problem 6.116

Determine the complete solution to the system of Problem 6.108, subject to zero initial conditions: $x(0) = \dot{x}(0) = 0$.

Problem 6.117

Determine the complete solution to the system of Problem 6.105, subject to zero initial conditions: $x(0) = \dot{x}(0) = 0$.

Problem 6.118

Determine the complete solution to the system of Problem 6.107, subject to the initial conditions $x(0) = 1$ ft, $\dot{x}(0) = 0$.

Problem 6.119

Determine the complete solution to the system of Problem 6.106, subject to the initial conditions $x(0) = 0$, $\dot{x}(0) = 3$ in./sec.

Problem 6.120

Consider the system of Problem 6.8. Let $k = k_0 = 8$ lb/ft, $mg = 8$ lb, $\omega = 6$ rad/sec.

a) With the dashpot removed, determine A_0 in order that the amplitude of the forced solution is 0.5 ft.

b) With A_0 as determined above, calculate the dashpot constant c in order that the amplitude of the forced solution is reduced to 0.1 ft.

Problem 6.121

For the system of Problem 6.9, let $mg = 16$ lb, $k = 8$ lb/ft, $A_0 = 0.5$ ft, and let $c_0 = c$.

a) Determine c and ω such that: (i) the system is critically damped, and (ii) the system is driven at its resonance frequency.

b) For the system described in (a), determine the complete solution if, at $t = 0$, $x = 0$ and $\dot{x} = -1$ ft/sec.

Problem 6.122

For the system shown in the figure, suppose that $\omega = \omega_n$. Show that the forced solution is

$$x = -\frac{F}{c\omega_n} \cos \omega_n t.$$

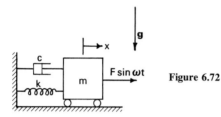

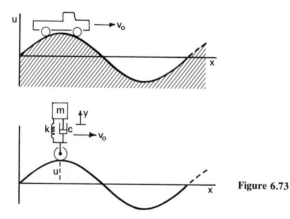

Figure 6.72

Problem 6.123

For the system of Problem 6.122, suppose again that $\omega = \omega_n$. Consider only the forced solution.

a) Compute the work done by the applied force, $F \sin \omega_n t$, over one complete cycle.

b) Show that the work computed in part (a) is equal to the negative of the work of the dashpot over one complete cycle.

Figure 6.73

Problem 6.124

The truck shown in Fig. 6.73 is modeled as a mass–spring–dashpot system. Suppose that the truck has a constant horizontal speed v_0. Let $mg = 3200$ lb, $k = 10^4$ lb/ft, $c = 4 \times 10^2$ lb-sec/ft, and let the shape of the road be $u = 0.5 \sin(1.5x)$, $x = $ ft.

a) Calculate the magnification factor μ as a function of the speed v_0.

b) Calculate the minimum speed v_0 for which the amplitude of the forced motion will be less than 2 in.

Section 6.8

Problem 6.125

Consider the standard system

$$\ddot{x} + 2\zeta\omega_n\dot{x} + \omega_n^2 x = \delta(t),$$

where $\zeta = 1$, critical damping. Show that

$$h(t) = te^{-\omega_n t},$$

and thus obtain

$$h(t - t^*) = (te^{-\omega_n t})e^{\omega_n t^*} - (e^{-\omega_n t})t^* e^{\omega_n t^*}.$$

Problem 6.126

Consider the standard system

$$\ddot{x} + 2\zeta\omega_n\dot{x} + \omega_n^2 x = \delta(t),$$

where $\zeta > 1$, overdamped system. Show that

$$h(t) = \frac{1}{\omega_n \sqrt{\zeta^2 - 1}} e^{-\omega_n \zeta t} \sinh \omega_n \sqrt{\zeta^2 - 1}\, t,$$

and thus that

$$h(t - t^*) = \frac{1}{\omega_n \sqrt{\zeta^2 - 1}} e^{-\omega_n \zeta(t - t^*)} \sinh \omega_n \sqrt{\zeta^2 - 1}\, (t - t^*).$$

Problem 6.127

Consider the differential equation

$$\ddot{x} = \delta(t).$$

a) Calculate $h(t)$ and then $h(t - t^*)$.

b) Find the solution to $\ddot{x} = F_0 \sin \omega t$, subject to zero initial conditions, $x(0) = \dot{x}(0) = 0$, using (6.44).

Problem 6.128

Use (6.44) to find the solution of

$$0.5\ddot{x} + 8x = \cos 3t$$

subject to zero initial conditions: $x(0) = \dot{x}(0) = 0$.

Problem 6.129

Use (6.44) to find the solution of

$$2\ddot{x} + 128x = 10t,$$

subject to zero initial conditions: $x(0) = \dot{x}(0) = 0$.

Problem 6.130

Use (6.44) to find the solution of

$$\ddot{x} + 2\omega_n\dot{x} + \omega_n^2 x = F_0 e^{-\omega_n t}$$

subject to zero initial conditions: $x(0) = \dot{x}(0) = 0$.

Problem 6.131

Consider the differential equation

$$\ddot{x} + \omega_n^2 x = F_0 e^{at}, \qquad \text{where } a \neq 0 \text{ and } x(0) = \dot{x}(0) = 0.$$

a) Guess the form of the forced solution, and then construct the complete solution.

b) Use (6.44) to construct the complete solution.

Section 6.9

Problem 6.132

For the case of a vibration absorber, show that the amplitude of the forced motion of the main mass is zero and not indeterminant:

$$A_1 \neq \frac{0}{0}.$$

[*Hint:* Consider Eq. (f) which precedes (6.45).]

Problem 6.133

Derive the solution for the forced motion of a vibration absorber, Eq. (6.46). [*Hint:* Use Eq. (f) which precedes (6.45).]

Problem 6.134

Consider the problem of the vibration absorber (see the figure). Note that the main system is stationary only if the absorber exerts a force on the system equal and opposite to the excitation force. And thus the force applied to the spring k_0 of the absorber is as shown. Using these facts, give a direct derivation of the results:

$$\omega^2 = \frac{k_0}{m_0} \quad \text{and} \quad x_a = -(F/k_0)\sin\omega t.$$

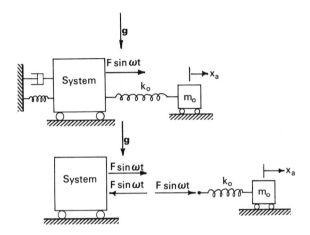

Figure 6.74

Problem 6.135

The clothes dryer in Fig. 6.75 has an unbalanced mass Δm, which causes the dryer to shake in the horizontal plane. A vibration absorber (k_0, m_0) is added to the system. Given that the amplitude of m_0 is designed to be A_0, determine m_0 and k_0.

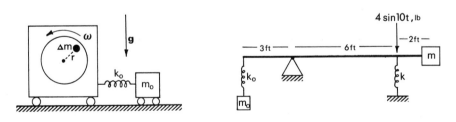

Figure 6.75

Figure 6.76

Problem 6.136

An absorber (m_0, k_0) is added to the system as shown in Fig. 6.76.

a) Calculate k_0, given that m_0 weighs 4 lb.

b) Calculate the amplitude of m_0. [*Hint:* See Problem 6.134.]

Problem 6.137

For the system in the figure, the absorber is added to the bar at a distance a from the fulcrum. If $m_0 g = 16$ lb,

a) find k_0, and

b) find the distance a so that the amplitude of m_0 is 0.2 ft. [*Hint:* See Problem 6.134.]

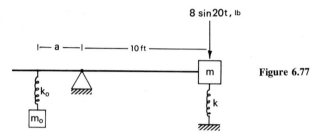

Figure 6.77

7

DYNAMICS OF A
SYSTEM OF
PARTICLES

Stuart spun the wheel over but it was too
late: the Wasp drove her bowsprit straight
into the bag and with a fearful whoosh the
schooner slowed down and came up into
the wind with all sails flapping.

Stuart Little
E. B. White

1. INTRODUCTION*

In Chapters 3 through 6, we dealt with one basic problem: the dynamics of a
single particle. With this chapter, we begin our study of more general collec-
tions of matter. We shall consider the motion of a system of n particles, where
n is a finite number. In this context it is possible to discuss two specialized
classes of problems: (i) collision problems, (ii) so-called *variable-mass*
problems.

Another purpose of this chapter is to give a heuristic introduction to the
dynamics of rigid bodies. In this chapter, we shall consider rigid bodies made
up of a finite number of particles connected by rigid, weightless bars. This will
serve as an introduction to and motivation for the fundamental equations we
shall write for rigid bodies in Chapter 8.†

2. PRELIMINARIES

We can deal with the dynamics of a system of particles in one of two ways: (i)
as a set of individual particles, or (ii) as a system. To begin, let us classify the

---

* This chapter can be skipped without loss of continuity.
† The modern axiomatic view of mechanics regards the concept of a rigid body as
a collection of particles only as motivation for the equations of motion.

forces on the system. The system is taken to be the particles $m_1, m_2, \ldots, m_n$.

Internal Forces. The force which particle j exerts on particle i is denoted by $\mathbf{f}_{ij}$. (Newton's third law then states that $\mathbf{f}_{ij} = -\mathbf{f}_{ji}$ and the two forces lie on the same line.) Again we agree that $\mathbf{f}_{ii} = 0$; a particle cannot exert a force on itself.

External Forces. Any force on a particle of the system which is not an internal force is an external force. The total external force on particle i is denoted by $\mathbf{F}_i$. Thus the total force on particle i is

$$\mathbf{F}_i + \mathbf{f}_{i1} + \mathbf{f}_{i2} + \cdots + \mathbf{f}_{ii-1} + \mathbf{f}_{ii+1} + \cdots + \mathbf{f}_{in}.$$

To illustrate this classification, consider the system of two particles m_1 and m_2 in Fig. 7.1.

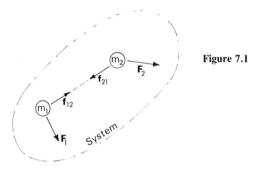

Figure 7.1

The total forces on m_1 and m_2 are as follows.

On m_1: $(\mathbf{F}_1 + \mathbf{f}_{12})$

On m_2: $(\mathbf{F}_2 + \mathbf{f}_{21})$

Note again that $\mathbf{f}_{12}$ and $\mathbf{f}_{21}$ act on the line between the particles m_1 and m_2 and are equal and opposite.

Let O be a fixed point of some inertial reference frame. Let $\mathbf{r}_i$ be the vector from O to m_i. (See Fig. 7.2.)

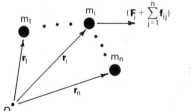

Figure 7.2

The acceleration of m_i is thus

$$\mathbf{a}_i = \ddot{\mathbf{r}}_i.$$

Writing $\mathbf{F} = \dot{\mathbf{p}}$ for m_i, we have

$$\mathbf{F} = \dot{\mathbf{p}}: \quad \mathbf{F}_i + \sum_{j=1}^{n} \mathbf{f}_{ij} = m_i \ddot{\mathbf{r}}_i, \qquad i = 1, 2, \ldots, n. \tag{7.1}$$

To completely determine the motion of our system we could use (7.1). However, the internal forces may be difficult or impossible to evaluate and the number of particles may be so large as to be overwhelming. Thus we look for various ways to write equations for the overall motion of the entire system.

3. MOTION OF THE CENTER OF MASS

Suppose we add together all the equations of (7.1). We would then have

$$\sum_{i=1}^{n} \mathbf{F}_i + \sum_{i=1}^{n} \sum_{j=1}^{n} \mathbf{f}_{ij} = \sum_{i=1}^{n} m_i \ddot{\mathbf{r}}_i. \tag{a}$$

The second term on the left adds to zero because the internal forces occur in equal and opposite pairs (Newton's third law). Thus we define $\mathbf{F}$ to be the resultant of the external forces:

$$\mathbf{F} = \sum_{i=1}^{n} \mathbf{F}_i. \tag{7.2}$$

Now let us examine the right-hand side of (a). First we define the center of mass of our system.

Definition: Let the positions of the particles m_1, m_2, ..., m_n be $\mathbf{r}_1$, $\mathbf{r}_2$, ..., $\mathbf{r}_n$. Then the position of the *center of mass* of the particles is defined to be

$$\mathbf{r}_C = \frac{\displaystyle\sum_{i=1}^{n} m_i \mathbf{r}_i}{\displaystyle\sum_{i=1}^{n} m_i}. \tag{7.3}$$

Employing a notation consistent with (7.2), we denote the total mass of the system by m:

$$m = \sum_{i=1}^{n} m_i = \text{total mass}.$$

Then, by definition,

$$mr_C = \sum_{i=1}^{n} m_i r_i. \tag{b}$$

Noting the form of the right side of (a) leads us to differentiate (b) with respect to time:

$$m\dot{\mathbf{r}}_C = \sum_{i=1}^{n} m_i \dot{\mathbf{r}}_i \tag{c}$$

or

$$\mathbf{p}_C = \sum_{i=1}^{n} \mathbf{p}_i, \tag{7.4}$$

where $\mathbf{p}_C$ is the momentum of the center of mass, $m\mathbf{v}_C$. Differentiating (c) again, and using (a) and (7.2), we obtain

$$\mathbf{F} = m\ddot{\mathbf{r}}_C. \tag{d}$$

> The acceleration of the center of mass of a system of particles is equal to the resultant (external) force on the system divided by the total mass of the system.

With (c) and (7.4), equation (d) can be written:

$$\mathbf{F} = \dot{\mathbf{p}}_C. \tag{7.5}$$

Integrating (7.5), we have

$$\int_{t_1}^{t_2} \mathbf{F} \, dt = \mathbf{p}_C \Big|_{t_1}^{t_2} \tag{7.6}$$

and if $\mathbf{F} = 0$ (no net external force) we have

$$\mathbf{p}_C = m\dot{\mathbf{r}}_C = \text{constant}. \tag{7.7}$$

> The impulse of the resultant external load on a system of particles equals the change in the linear momentum of the center of mass of the system.

Before going any further, we'd better say a few words about the center of mass. The center of mass is the location of the center of the mass distribution of the system. Often this is confused with the concept of the center of gravity, discussed in Chapter 2. The center of gravity of a system of particles is the location of the center of the force distribution (due to gravity) on the system. In many calculations the center of mass coincides with the center of gravity. This need not always be the case, as is shown in Problems 7.6 and 7.7.

Example 1

A boy and his girl friend are separated by a distance L on smooth ice. To get to his girl, the boy walks on a plank, as shown in Fig. 7.3(a). How close to the girl does he get?

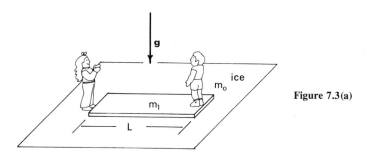

Figure 7.3(a)

Solution

Figure 7.3(b) is a free-body diagram of the forces on the boy–plank system.

The forces $(-m_1g - m_0g + N) = 0$ by the equation of equilibrium in the vertical direction. Thus the force resultant $\mathbf{F}$ is zero. And thus $\ddot{\mathbf{r}}_c = 0$. This means the velocity of the center of mass is constant, and since the boy and plank were not moving initially, the velocity of the center of mass is zero for all times.

Figure 7.3(c) establishes a coordinate system to describe the problem. Although the masses of the system have moved about, the center of mass must remain in the

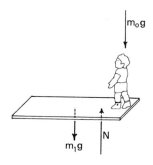

Figure 7.3(b)

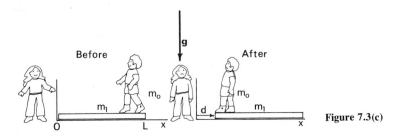

Figure 7.3(c)

same spot. Thus we compute the position of the center of mass before and after the boy walks on the plank:

Before: $x_C = \left(m_0 L + m_1 \dfrac{L}{2}\right) \Big/ (m_0 + m_1)$

After: $x_C = \left[m_0 d + m_1 \left(d + \dfrac{L}{2}\right)\right] \Big/ (m_0 + m_1)$

Equating and solving for d, we get

$$d = \frac{m_0 L}{m_0 + m_1}.$$

Thus $d \neq 0$, and the boy remains separated from his girl. Can you suggest a way for the two to get together?

....

Example 2

Consider a simplified model of the ballistic pendulum. In Fig. 7.4(a) a bullet is shot into a wood block and the maximum angle of the block and bullet is then determined experimentally. From the knowledge of the angle θ_0, we are asked to determine the initial velocity of the bullet.

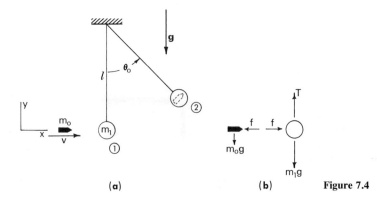

(a)   (b)   **Figure 7.4**

Solution

Suppose we consider an instantaneous collision. Figure 7.4(b) shows the free-body diagram for our system. We see that the force resultant on the system of m_0 and m_1 is

$$\mathbf{F} = -m_0 g \mathbf{j},$$

since $T - m_1 g = 0$ by the equation of vertical equilibrium of the block. Writing (7.5), $\mathbf{F} = \dot{\mathbf{p}}_C$, in the horizontal direction gives

$$F_x = m a_{Cx}: \qquad 0 = m\ddot{x}_C.$$

Thus

$$m\dot{x}_C = \text{constant}.$$

Now

$$m\dot{x}_C = m_0 \dot{x}_0 + m_1 \dot{x}_1.$$

Before the collision, $\dot{x}_0 = v$ and $\dot{x}_1 = 0$. Suppose that just after the collision, the two particles have speed $\dot{x} = v_1$. Then

$$\text{Before collision:} \qquad m\dot{x}_C = m_0 v$$

$$\text{After collision:} \qquad m\dot{x}_C = (m_0 + m_1)v_1$$

Thus

$$v_1 = \left(\frac{m_0}{m_0 + m_1} \right) v. \tag{a}$$

To determine θ_0, we note that after the collision we have a conservative system, since m_0 plus m_1 now appears as a simple pendulum. Writing the equation for energy at points 1 and 2 (Fig. 7.4a), we have

$$\text{At 1:} \qquad T = \tfrac{1}{2}(m_0 + m_1)v_1^2, \qquad V = 0$$

$$\text{At 2:} \qquad T = 0, \qquad V = (m_0 + m_1)gl(1 - \cos \theta_0).$$

And equating $(T + V)$ at 1 and 2 gives

$$\cos \theta_0 = 1 - \frac{v_1^2}{2gl}. \tag{b}$$

And finally, by (a) and (b), we get

$$\cos \theta_0 = 1 - \left(\frac{m_0}{m_0 + m_1} \right)^2 \frac{v^2}{2gl} \tag{c}$$

Now if θ_0 is a small angle (we can arrange for this by making m_1 large compared with m_0), we can approximate

$$\cos \theta_0 \approx 1 - \frac{\theta_0^2}{2}.$$

Thus, by (c),

$$v \approx \sqrt{gl}\left(\frac{m_0 + m_1}{m_0}\right)\theta_0. \tag{d}$$

Was energy conserved during the entire motion, or not? If not (and it isn't), where was the energy dissipated? The answer is that the internal forces in the system have done some work. This is an important point to keep in mind when we consider the work–energy relation for a system of particles.

We shall consider the entire class of collision problems in more detail in Section 7 of this chapter.

....

4. MOTION ABOUT THE CENTER OF MASS, $M = \dot{H}$

We now write $\mathbf{M} = \dot{\mathbf{H}}$ for the system of particles. There are two convenient ways in which this can be done, as we shall see.

In Fig. 7.5, let's suppose that O is the origin of an inertial reference frame. Then the vectors $\mathbf{r}_1, \mathbf{r}_2, \ldots$ from O to $m_1, m_2, \ldots$ give the positions of the particles.

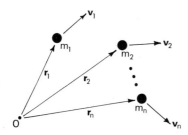

Figure 7.5

We begin by computing the total moment of the forces on the system about O. We denote the total force on particle m_i by $\mathbf{F}_i^t$:

$$\mathbf{F}_i^t = \left(\mathbf{F}_i + \sum_{j=1}^{n} \mathbf{f}_{ij}\right). \tag{a}$$

Writing $\mathbf{F} = \dot{\mathbf{p}}$ for m_i, we have

$$\mathbf{F} = \dot{\mathbf{p}}: \qquad \mathbf{F}_i^t = \dot{\mathbf{p}}_i. \tag{b}$$

We compute the moment resultant about O of all the forces acting on the system:

$$\mathbf{M}_O = \sum_{i=1}^{n} \mathbf{r}_i \times \mathbf{F}_i^t. \tag{c}$$

[In a moment we'll show that (c) is considerably simpler than indicated by (a) and (c).]

We define the *moment of momentum* about O to be

$$\mathbf{H}_O = \sum_{i=1}^{n} \mathbf{r}_i \times \mathbf{p}_i. \tag{d}$$

We wish to show that $\mathbf{M}_O = \dot{\mathbf{H}}_O$. Differentiating (d), we obtain

$$\dot{\mathbf{H}}_O = \sum_{i=1}^{n} \dot{\mathbf{r}}_i \times \mathbf{p}_i + \sum_{i=1}^{n} \mathbf{r}_i \times \dot{\mathbf{p}}_i.$$

The first summation is zero, since each term is

$$\mathbf{v}_i \times m_i \mathbf{v}_i = 0.$$

The second summation can be rewritten using (b) and (c):

$$\dot{\mathbf{H}}_O = \sum_{i=1}^{n} \mathbf{r}_i \times \mathbf{F}_i^t = \mathbf{M}_O,$$

which gives the desired result,

$$\mathbf{M}_O = \dot{\mathbf{H}}_O, \qquad O \text{ a fixed point.} \tag{e}$$

Before leaving this derivation, we show that $\mathbf{M}_O$ is simpler than (c) indicates. From (a), we have

$$\mathbf{M}_O = \sum_{i=1}^{n} \mathbf{r}_i \times \mathbf{F}_i + \sum_{i=1}^{n} \left(\mathbf{r}_i \times \sum_{j=1}^{n} \mathbf{f}_{ij} \right). \tag{f}$$

The final summation of (f) can be shown to be zero. It consists of pairs of terms. For example,

$$(\mathbf{r}_1 \times \mathbf{f}_{12} + \mathbf{r}_2 \times \mathbf{f}_{21}) + \cdots$$

Consider Fig. 7.6. Since $\mathbf{f}_{12} = -\mathbf{f}_{21}$, the first pair can be written

$$(\mathbf{r}_1 \times -\mathbf{f}_{21} + \mathbf{r}_2 \times \mathbf{f}_{21}) = (\mathbf{r}_2 - \mathbf{r}_1) \times \mathbf{f}_{21}. \tag{g}$$

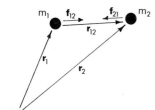

Figure 7.6

From Fig. 7.6 we note that

$$\mathbf{r}_1 + \mathbf{r}_{12} = \mathbf{r}_2 \qquad \text{or} \qquad \mathbf{r}_{12} = \mathbf{r}_2 - \mathbf{r}_1.$$

Thus (g) becomes

$$\mathbf{r}_{12} \times \mathbf{f}_{21} = 0,$$

because $\mathbf{r}_{12}$ and $\mathbf{f}_{21}$ have the same direction. Therefore the total moment of the internal forces acting on the system is zero. And so (c) reduces to

$$\mathbf{M}_O = \sum_{i=1}^{n} \mathbf{r}_i \times \mathbf{F}_i, \tag{7.8}$$

and the moment resultant of all the forces acting on the system equals the moment resultant of the external forces alone.

Then, repeating the definition of $\mathbf{H}_O$:

$$\mathbf{H}_O = \sum_{i=1}^{n} \mathbf{r}_i \times \mathbf{p}_i, \tag{7.9}$$

we have

$$\mathbf{M}_O = \dot{\mathbf{H}}_O. \tag{7.10}$$

> The moment resultant about some fixed point O of the external forces acting on a system equals the rate of change of the moment of momentum about the point O.

The equation $\mathbf{M} = \dot{\mathbf{H}}$ can also be written about the center of mass of the system, even though the center of mass is in motion. This form of the equation is probably more useful than (7.10).

We begin by defining the following vectors:

$\mathbf{r}_i$ vector from O to m_i

$\mathbf{r}_C$ vector from O to the center of mass

$\mathbf{r}_{Ci}$ vector from the center of mass to m_i

When we examine Fig. 7.7, we see that

$$\mathbf{r}_i = \mathbf{r}_C + \mathbf{r}_{Ci}. \tag{h}$$

We define the quantities $\mathbf{M}_C$ and $\mathbf{H}_C$:

$$\mathbf{M}_C = \sum_{i=1}^{n} \mathbf{r}_{Ci} \times \mathbf{F}_i^t \tag{i}$$

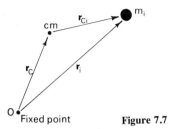

Figure 7.7

and

$$\mathbf{H}_C = \sum_{i=1}^{n} \mathbf{r}_{Ci} \times \mathbf{p}_i. \tag{j}$$

Differentiating (j), we obtain

$$\dot{\mathbf{H}}_C = \sum_{i=1}^{n} \dot{\mathbf{r}}_{Ci} \times \mathbf{p}_i + \sum_{i=1}^{n} \mathbf{r}_{Ci} \times \dot{\mathbf{p}}_i. \tag{k}$$

The first summation can be shown to be zero. Recall that

$$\mathbf{p}_i = m_i \mathbf{v}_i = m_i \dot{\mathbf{r}}_i.$$

From (h) we get

$$\dot{\mathbf{r}}_{Ci} = \dot{\mathbf{r}}_i - \dot{\mathbf{r}}_C.$$

Thus

$$\sum_{i=1}^{n} \dot{\mathbf{r}}_{Ci} \times \mathbf{p}_i = \sum_{i=1}^{n} \dot{\mathbf{r}}_i \times \overset{0}{\overset{\nearrow}{m_i \dot{\mathbf{r}}_i}} - \sum_{i=1}^{n} \mathbf{v}_C \times \mathbf{p}_i.$$

The first summation here is zero because $\dot{\mathbf{r}}_i \times m_i \dot{\mathbf{r}}_i = 0$. The second summation can be written by (7.4):

$$-\mathbf{v}_C \times \sum_{i=1}^{n} \mathbf{p}_i = -\mathbf{v}_C \times \mathbf{p}_C = -\mathbf{v}_C \times m\mathbf{v}_C = 0.$$

Thus, when we use (b) and (i), (k) becomes

$$\dot{\mathbf{H}}_C = \sum_{i=1}^{n} \mathbf{r}_{Ci} \times \dot{\mathbf{p}}_i = \sum_{i=1}^{n} \mathbf{r}_{Ci} \times \mathbf{F}_i^t = \mathbf{M}_C.$$

Once again we can show, in exactly the same way that we derived (7.8), that $\mathbf{M}_C$ is the moment resultant about the center of mass of the external forces acting on the system.

Repeating the final equations for $\mathbf{M}_C$, $\mathbf{H}_C$, we have then

$$\mathbf{M}_C = \sum_{i=1}^{n} \mathbf{r}_{Ci} \times \mathbf{F}_i, \tag{7.11}$$

$$\mathbf{H}_C = \sum_{i=1}^{n} \mathbf{r}_{Ci} \times \mathbf{p}_i, \tag{7.12}$$

and

$$\mathbf{M}_C = \dot{\mathbf{H}}_C. \tag{7.13}$$

> The moment resultant about the center of mass of the external forces acting on a system equals the rate of change of the moment of momentum about the center of mass.

Example 3

Figure 7.8 shows a dumbbell—two particles, $m_1 = m_2$—connected by a weightless bar. The velocities of the particles are indicated at two positions, A and B. Each particle weighs 16.1 lb. Determine $\mathbf{H}_O$ and $\mathbf{H}_C$ in each position.

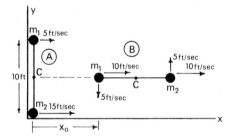

Figure 7.8

Solution

For position A, we compute $\mathbf{H} = \Sigma \mathbf{r} \times m\mathbf{v}$:

$$\mathbf{H}_O = (0 \cdot m_2 \, 15 - 10 \cdot m_1 \, 5)\mathbf{k} \text{ slug-ft}^2/\text{sec},$$

$$\mathbf{H}_C = (5 \cdot m_2 \, 15 - 5 \cdot m_1 \, 5)\mathbf{k} \text{ slug-ft}^2/\text{sec}.$$

And since

$$m_1 = m_2 = \frac{16.1}{32.2} \text{ slug} = 0.5 \text{ slug},$$

A: $\mathbf{H}_O = -25\mathbf{k}$ slug-ft²/sec,

   $\mathbf{H}_C = +25\mathbf{k}$ slug-ft²/sec.

   For position B, we compute

   $\mathbf{H}_O = [-5 \cdot m_1 \ 10 - x_0 \cdot m_1 \ 5 + (x_0 + 10) \cdot m_2 \ 5 - 5 \cdot m_2 \ 10]\mathbf{k}$ slug-ft⁻²/sec,

   $\mathbf{H}_C = [5 \cdot m_1 \ 5 + 5 \cdot m_2 \ 5]\mathbf{k}$ slug-ft²/sec,

and thus

B: $\mathbf{H}_O = -25\mathbf{k}$ slug-ft²/sec,

   $\mathbf{H}_C = +25\mathbf{k}$ slug-ft²/sec.

Example 4

Figure 7.9(a) shows a dumbbell initially at rest in a horizontal, frictionless plane. At $t = 0$, when the dumbbell has position A, a hammer strikes the lower mass, giving it a velocity $\mathbf{v}_0$. The upper mass is at rest at that instant. Show that the position of the center of mass is

   $\mathbf{r}_C = \tfrac{1}{2}v_0 t\mathbf{i} + \tfrac{1}{2}d\mathbf{j}.$

When the dumbbell has the position shown in B, what is the velocity of each particle?

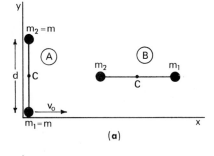

(a)

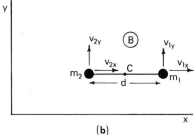

(b)   **Figure 7.9**

Solution

There is no net external force acting on the dumbbell after $t = 0$. Thus $\mathbf{M}_O$ and $\mathbf{F}$ are zero. Writing $\mathbf{F} = \dot{\mathbf{p}}_C$, we get

$$\mathbf{F} = \dot{\mathbf{p}}_C: \qquad 0 = \dot{\mathbf{p}}_C \qquad \text{or} \qquad \mathbf{p}_C = \text{constant}.$$

Initially $\mathbf{p}_C = m_1 v_0 \mathbf{i} + m_2 \cdot 0 = m_1 v_0 \mathbf{i}$. But $\mathbf{p}_C = (m_1 + m_2)\mathbf{v}_C$. Thus

$$\mathbf{v}_C = \frac{m_1 v_0 \mathbf{i}}{m_1 + m_2}.$$

Since $m_1 = m_2$,

$$\mathbf{v}_C = \tfrac{1}{2} v_0 \mathbf{i} \qquad \text{constant}. \tag{a}$$

Thus

$$\frac{d\mathbf{r}_C}{dt} = \tfrac{1}{2} v_0 \mathbf{i}.$$

Integrating and noting that $\mathbf{r}_C(0) = (d/2)\mathbf{j}$, we obtain

$$\mathbf{r}_C = \tfrac{1}{2} v_0 t \mathbf{i} + \frac{d}{2}\mathbf{j}, \qquad \text{as required}.$$

Writing $\mathbf{M}_C = \dot{\mathbf{H}}_C$, we have

$$\mathbf{M}_C = \dot{\mathbf{H}}_C: \qquad 0 = \dot{\mathbf{H}}_C.$$

Thus $\mathbf{H}_C = \text{constant}$. Initially

$$\mathbf{H}_C = \left(\frac{d}{2} m_1 v_0\right) \mathbf{k}. \tag{b}$$

Consider Fig. 7.9(b). The velocity of the center of mass is equal to the value given in (a). Thus

$$\frac{m_1 (v_{1x}\mathbf{i} + v_{1y}\mathbf{j}) + m_2 (v_{2x}\mathbf{i} + v_{2y}\mathbf{j})}{m_1 + m_2} = \tfrac{1}{2} v_0 \mathbf{i}.$$

For $m_1 = m_2$, we obtain

$$\tfrac{1}{2} v_{1x} + \tfrac{1}{2} v_{2x} = \tfrac{1}{2} v_0, \tag{c}$$

$$\tfrac{1}{2} v_{1y} + \tfrac{1}{2} v_{2y} = 0. \tag{d}$$

Since the bar between m_1 and m_2 is rigid, we must have $v_{1x} = v_{2x}$. Thus (c) gives

$$v_{1x} = v_{2x} = v_0. \tag{e}$$

Computing $\mathbf{H}_C$ from Fig. 7.9(b) and equating this to the value in (b):

$$\mathbf{H}_C = \left(\frac{d}{2} m_1 v_{1y} - \frac{d}{2} m_2 v_{2y}\right) \mathbf{k} = \frac{d}{2} m_1 v_0 \mathbf{k}.$$

But from (d), we get (for $m_1 = m_2$)

$$v_{1y} = -v_{2y} = \tfrac{1}{2} v_0.$$

Thus in position B we have

$$\mathbf{v}_1 = \tfrac{1}{2} v_0 (\mathbf{i} + \mathbf{j}), \qquad \mathbf{v}_2 = \tfrac{1}{2} v_0 (\mathbf{i} - \mathbf{j}).$$

....

Example 5

Derive the transformation law for the moment of momentum of a system of particles:

$$\mathbf{H}_A = \mathbf{r}_{AB} \times \mathbf{p}_C + \mathbf{H}_B. \tag{7.14}$$

Solution

Let $\mathbf{r}_A$ and $\mathbf{r}_B$ be vectors from the origin of an inertial reference O to A and B. Let $\mathbf{r}_{Ai}$ and $\mathbf{r}_{Bi}$ be vectors from A and B to the particle m_i (see Fig. 7.10).

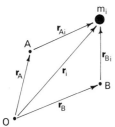

Figure 7.10

Then the position $\mathbf{r}_i$ of m_i can be written either

$$\mathbf{r}_i = \mathbf{r}_A + \mathbf{r}_{Ai} \quad \text{or} \quad \mathbf{r}_i = \mathbf{r}_B + \mathbf{r}_{Bi}. \tag{a}$$

We can now write

$$\mathbf{H}_A = \sum_{i=1}^{n} \mathbf{r}_{Ai} \times \mathbf{p}_i \quad \text{and} \quad \mathbf{H}_B = \sum_{i=1}^{n} \mathbf{r}_{Bi} \times \mathbf{p}_i,$$

or from (a),

$$\mathbf{H}_A = \sum_{i=1}^{n} (\mathbf{r}_i - \mathbf{r}_A) \times \mathbf{p}_i = \sum_{i=1}^{n} \mathbf{r}_i \times \mathbf{p}_i - \mathbf{r}_A \times m\mathbf{v}_C, \tag{b}$$

$$\mathbf{H}_B = \sum_{i=1}^{n} (\mathbf{r}_i - \mathbf{r}_B) \times \mathbf{p}_i = \sum_{i=1}^{n} \mathbf{r}_i \times \mathbf{p}_i - \mathbf{r}_B \times m\mathbf{v}_C, \tag{c}$$

[since, by (7.4), $\Sigma \mathbf{p}_i = \mathbf{p}_C$]. Subtracting (c) from (b) gives

$$\mathbf{H}_A - \mathbf{H}_B = (\mathbf{r}_B - \mathbf{r}_A) \times \mathbf{p}_C = \mathbf{r}_{AB} \times \mathbf{p}_C$$

which shows (7.14).

....

Example 6

Consider a system of particles. Suppose that A is an arbitrary, moving point. Show that

$$\mathbf{M}_A = \dot{\mathbf{H}}_A + \mathbf{v}_A \times \mathbf{p}_C \tag{7.15}$$

Show that Eqs. (7.10) and (7.13) are special cases of (7.15).

Solution

Consider two points: O a fixed point, A an arbitrary point. From (2.13) and (7.14), we have

$$\mathbf{M}_O = \mathbf{r}_{OA} \times \mathbf{F} + \mathbf{M}_A, \tag{a}$$

$$\mathbf{H}_O = \mathbf{r}_{OA} \times \mathbf{p}_C + \mathbf{H}_A. \tag{b}$$

Since O is a fixed point, (7.10) gives

$$\mathbf{M}_O = \dot{\mathbf{H}}_O. \tag{c}$$

From (b), we have

$$\dot{\mathbf{H}}_O = \mathbf{v}_A \times \mathbf{p}_C + \mathbf{r}_{OA} \times \dot{\mathbf{p}}_C + \dot{\mathbf{H}}_A.$$

Thus, by (a) and (c), we get

$$\mathbf{r}_{OA} \times \mathbf{F} + \mathbf{M}_A = \mathbf{v}_A \times \mathbf{p}_C + \mathbf{r}_{OA} \times \dot{\mathbf{p}}_C + \dot{\mathbf{H}}_A$$

or

$$\mathbf{M}_A = \mathbf{v}_A \times \mathbf{p}_C + \dot{\mathbf{H}}_A + \mathbf{r}_{OA} \times (\dot{\mathbf{p}}_C - \mathbf{F}).$$

But by (7.5) the final term is zero, and (7.15) is demonstrated.

If point A is either fixed in space or is the center of mass, the term $\mathbf{v}_A \times \mathbf{p}_C = 0$ and $\mathbf{M}_A = \dot{\mathbf{H}}_A$, as in (7.10) and (7.13), for if A is fixed, then $\mathbf{v}_A = 0$. If A is the center of mass, then $\mathbf{v}_A \times \mathbf{p}_C = \mathbf{v}_C \times m\mathbf{v}_C = 0$.

··

5. THE WORK–ENERGY RELATION

Thus far, in dealing with an arbitrary system of particles, we have derived two equations, $\mathbf{F} = \dot{\mathbf{p}}_C$ and $\mathbf{M} = \dot{\mathbf{H}}$. In neither equation did the internal forces on the system appear. This is one strength of these equations, since, as we have noted, internal forces are often very numerous and difficult to evaluate. On the other hand, by writing $\mathbf{F} = \dot{\mathbf{p}}_C$ or $\mathbf{M} = \dot{\mathbf{H}}$, we may have to settle for something less than a complete formulation of a problem. Fortunately, these two equations give a complete formulation in the case of a rigid body.

In general, the internal forces on a system *do* enter the work–energy relation. That is, internal forces can do work. Consider the system in Fig. 7.11.

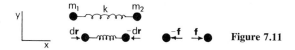

Figure 7.11

Suppose that by some means the two particles are given the displacements $d\mathbf{r}$ and $-d\mathbf{r}$, as shown. The internal forces are $\mathbf{f}$ and $-\mathbf{f}$. Let x be the line of motion of the system. The forces on and displacements of the two particles are

$$m_1:\quad -f\,\mathbf{i},\quad dr\,\mathbf{i},$$
$$m_2:\quad +f\,\mathbf{i},\quad -dr\,\mathbf{i}.$$

Thus the work on each particle is

$$m_1:\quad -f dr,\qquad m_2:\quad -f dr,$$

and the total work is $-2f dr$. Thus the net work is *not* zero.

Once again the rigid body is an important exception to the above situation. In this case, the total work of the internal forces is zero. In the above example, if the system were rigid, we would have to have the displacement of each mass the same (if the particles were restricted to move only along the x axis). The work of the forces would then be $-f(dr) + f(dr) = 0$.

To derive the work–energy relation for a system of particles, we begin by considering Fig. 7.12.

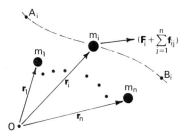

Figure 7.12

The total force on particle m_i is

$$\mathbf{F}_i^t = \left(\mathbf{F}_i + \sum_{j=1}^{n} \mathbf{f}_{ij}\right). \tag{a}$$

Writing $\mathbf{F} = \dot{\mathbf{p}}$ for m_i, we have

$$\mathbf{F}_i^t = \dot{\mathbf{p}}_i. \tag{b}$$

We can now derive the work–energy relation for m_i. We proceed exactly as in Chapter 4 in the derivation of (4.8). We define the kinetic energy of m_i to be

$$T_i = \int_0^{\mathbf{p}_i} \mathbf{v}_i \cdot d\mathbf{p}_i$$

or

$$\frac{dT_i}{dt} = \mathbf{v}_i \cdot \dot{\mathbf{p}}_i. \tag{c}$$

The power of $\mathbf{F}_i^t$ is defined as

$$P_i = \mathbf{F}_i^t \cdot \mathbf{v}_i. \tag{d}$$

Thus, forming the dot product of (b) with $\mathbf{v}_i$, we obtain from (c) and (d):

$$P_i = \frac{dT_i}{dt}. \tag{e}$$

It makes sense to consider the work of all the forces on our system and to relate that to the change of the kinetic energy of the entire system.

The work of all the forces is defined as the sum of the work of the individual forces. And the total kinetic energy is defined as the sum of the kinetic energies of the individual particles. Thus

$$W_{AB} = \sum_{i=1}^{n} \int_{t_A}^{t_B} P_i \, dt = \sum_{i=1}^{n} \int_{A_i}^{B_i} \mathbf{F}_i^t \cdot d\mathbf{r}_i, \tag{f}$$

$$T = \sum_{i=1}^{n} T_i = \sum_{i=1}^{n} \int_0^{\mathbf{p}_i} \mathbf{v}_i \cdot d\mathbf{p}_i, \tag{g}$$

where A and B represent two configurations of the system. (Once again we can re-express the kinetic energy in velocity variables in the context of Newtonian dynamics:

$$T = \sum_{i=1}^{n} \tfrac{1}{2} m_i v_i^2.)$$

The work–energy relation for the system of particles can be obtained by integrating and summing (e):

$$\sum_{i=1}^{n} \int_{t_A}^{t_B} P_i \, dt = \sum_{i=1}^{n} T_i \Big|_A^B$$

or

$$W_{AB} = T(B) - T(A). \tag{7.16}$$

To make any sense of (7.16), we must evaluate the two sides of the equation. From (a) and (f), we get

$$W_{AB} = \sum_{i=1}^{n} \int_{A_i}^{B_i} \left(\mathbf{F}_i + \sum_{j=1}^{n} \mathbf{f}_{ij} \right) \cdot d\mathbf{r}_i,$$

and from (g) we obtain

$$T = \sum_{i=1}^{n} \tfrac{1}{2} m_i v_i^2.$$

Thus

$$\sum_{i=1}^{n} \int_{A_i}^{B_i} \left(\mathbf{F}_i + \sum_{j=1}^{n} \mathbf{f}_{ij} \right) \cdot d\mathbf{r}_i = \left(\sum_{i=1}^{n} \tfrac{1}{2} m v_i^2 \right) \Bigg|_A^B. \tag{7.17}$$

As we have noted in regard to the system shown in Fig. 7.11, the internal forces do work. Thus we can't eliminate the internal forces from (7.17).

There is an alternative way of writing (7.17) which is often useful. Recall (7.5):

$$\mathbf{F} = \dot{\mathbf{p}}_C, \tag{a}$$

where $\mathbf{F}$ is the force resultant acting on the system. Define:

$$T_C = \int_0^{\mathbf{p}_C} \mathbf{v}_C \cdot d\mathbf{p}_C, \tag{b}$$

the *kinetic energy of the center of mass.*

If we dot (a) with $\mathbf{v}_C$, we get, integrating over time,

$$\int_{t_A}^{t_B} \mathbf{F} \cdot \mathbf{v}_C \, dt = \int_0^{\mathbf{p}_C} \mathbf{v}_C \cdot d\mathbf{p}_C$$

or

$$\int_{A_C}^{B_C} \mathbf{F} \cdot d\mathbf{r}_C = \tfrac{1}{2} m v_C^2 \Bigg|_{A_C}^{B_C}. \tag{7.18}$$

If we transform (7.16) and (7.17) into terms involving (a) the motion of the center of mass and (b) motion with respect to the center of mass, we find that (7.18) plays a key role.

Consider Fig. 7.13. We have $\mathbf{r}_i = \mathbf{r}_C + \mathbf{r}_{Ci}$ and thus

$$d\mathbf{r}_i = d\mathbf{r}_C + d\mathbf{r}_{Ci}. \tag{a}$$

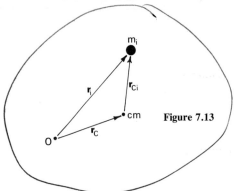

Figure 7.13

First let us transform the work term of (7.17), using (a):

$$W_{AB} = \sum_{i=1}^{n} \int_{A_i}^{B_i} \mathbf{F}_i \cdot (d\mathbf{r}_C + d\mathbf{r}_{Ci}) + \sum_{i=1}^{n} \sum_{j=1}^{n} \int_{A_i}^{B_i} \mathbf{f}_{ij} \cdot (d\mathbf{r}_C + d\mathbf{r}_{Ci}).$$ (b)

Let us evaluate these two sets of integrals separately. For the first, we have

$$\int_{A_i}^{B_i} \left(\sum_{i=1}^{n} \mathbf{F}_i \right) \cdot d\mathbf{r}_C + \sum_{i=1}^{n} \int_{A}^{B_i} \mathbf{F}_i \cdot d\mathbf{r}_{Ci} = \int_{A_C}^{B_C} \mathbf{F} \cdot d\mathbf{r}_C + \sum_{i=1}^{n} \int_{A_i}^{B_i} \mathbf{F}_i \cdot d\mathbf{r}_{Ci}.$$ (c)

Similarly, for the second term, we get

$$\int_{A_i}^{B_i} \left(\sum_{i=1}^{n} \sum_{j=1}^{n} \mathbf{f}_{ij} \right)^{\cancel{0}} \cdot d\mathbf{r}_C + \sum_{i=1}^{n} \int_{A_i}^{B_i} \sum_{j=1}^{n} \mathbf{f}_{ij} \cdot d\mathbf{r}_{Ci}$$ (d)

The first set of terms in (d) is zero, by Newton's third law. The internal forces add to zero. Thus, from (c) and (d), (b) becomes

$$W_{AB} = \int_{A_C}^{B_C} \mathbf{F} \cdot d\mathbf{r}_C + \sum_{i=1}^{n} \int_{A_i}^{B_i} \left(\mathbf{F}_i + \sum_{j=1}^{n} \mathbf{f}_{ij} \right) \cdot d\mathbf{r}_{Ci}.$$ (7.19)

To complete the transformation of (7.17), we rewrite the kinetic energy of the system. From (a), we have

$$\mathbf{v}_i = \mathbf{v}_C + \dot{\mathbf{r}}_{Ci}.$$ (e)

The kinetic energy of m_i is given by

$$T_i = \int_{0}^{\mathbf{p}_i} \mathbf{v}_i \cdot d\mathbf{p}_i = \tfrac{1}{2} m_i v_i^2.$$

From (e) we obtain

$$v_i^2 = \mathbf{v}_i \cdot \mathbf{v}_i = v_C^2 + 2\mathbf{v}_C \cdot \dot{\mathbf{r}}_{Ci} + \dot{r}_{Ci}^2.$$

Thus the total kinetic energy is

$$T = \sum_{i=1}^{n} (\tfrac{1}{2} m_i v_C^2) + \sum_{i=1}^{n} (m_i \mathbf{v}_C \cdot \dot{\mathbf{r}}_{Ci}) + \sum_{i=1}^{n} (\tfrac{1}{2} m_i \dot{r}_{Ci}^2).$$

$$= \tfrac{1}{2} v_C^2 \sum_{i=1}^{n} m_i + \mathbf{v}_C \cdot \sum_{i=1}^{n} m_i \dot{\mathbf{r}}_{Ci} + \sum_{i=1}^{n} (\tfrac{1}{2} m_i \dot{r}_{Ci}^2).$$

The first sum is $\tfrac{1}{2} m v_C^2$. The second sum is zero. To see this, consider

$$\frac{\Sigma m_i \mathbf{r}_{Ci}}{\Sigma m_i} = \mathbf{r}_{CC}$$

where $\mathbf{r}_{CC}$ is the position of the center of mass relative to the center of mass. Thus $\mathbf{r}_{CC} = 0$ and

$$\sum_{i=1}^{n} m_i \mathbf{r}_{Ci} = m \mathbf{r}_{CC} = 0.$$

Differentiating this, we obtain

$$\sum_{i=1}^{n} m_i \dot{\mathbf{r}}_{Ci} = 0.$$

Thus the total kinetic energy is given by

$$T = \tfrac{1}{2} m v_C^2 + \sum_{i=1}^{n} (\tfrac{1}{2} m_i \dot{r}_{Ci}^2). \tag{7.20}$$

The final term of (7.20) is called the *kinetic energy relative to the center of mass*.

With (7.19) and (7.20), (7.17) becomes

$$\int_{A_C}^{B_C} \mathbf{F} \cdot d\mathbf{r}_C + \sum_{i=1}^{n} \int_{A_i}^{B_i} \left(\mathbf{F}_i + \sum_{j=1}^{n} \mathbf{f}_{ij} \right) \cdot d\mathbf{r}_{Ci} = \tfrac{1}{2} m v_C^2 \Big|_{A_C}^{B_C} + \sum_{i=1}^{n} (\tfrac{1}{2} m_i \dot{r}_{Ci}^2) \Big|_{A_i}^{B_i},$$

but from (7.18), we had

$$\int_{A_C}^{B_C} \mathbf{F} \cdot d\mathbf{r}_C = \tfrac{1}{2} m v_C^2 \Big|_{A_C}^{B_C}. \tag{7.21}$$

Thus we get

$$\sum_{i=1}^{n} \int_{A_i}^{B_i} \left(\mathbf{F}_i + \sum_{j=1}^{n} \mathbf{f}_{ij} \right) \cdot d\mathbf{r}_{Ci} = \sum_{i=1}^{n} (\tfrac{1}{2} m_i \dot{r}_{Ci}^2) \Big|_{A_i}^{B_i}. \tag{7.22}$$

Equations (7.21) and (7.22) are often useful, especially in the case of the motion of rigid bodies.

The equations of motion for a system of n particles can be written as follows: When A is a fixed point or the center of mass,

$$\mathbf{F} = \dot{\mathbf{p}}_C, \qquad \mathbf{M}_A = \dot{\mathbf{H}}_A,$$

$$\sum_{i=1}^{n} \int_{A_i}^{B_i} \left(\mathbf{F}_i + \sum_{j=1}^{n} \mathbf{f}_{ij} \right) \cdot d\mathbf{r}_i = \sum_{i=1}^{n} \tfrac{1}{2} m_i v_i^2 \Big|_A^B$$

or

$$\int_{A_C}^{B_C} \mathbf{F} \cdot d\mathbf{r}_C = \tfrac{1}{2} m v_C^2 \Big|_{A_C}^{B_C}$$

$$\sum_{i=1}^{n} \int_{A_i}^{B_i} (\mathbf{F}_i + \sum_{j=1}^{n} \mathbf{f}_{ij}) \cdot d\mathbf{r}_{Ci} = \sum_{i=1}^{n} \tfrac{1}{2} m_i \dot{r}_{Ci}^2 \Big|_{A_i}^{B_i}$$

Example 7

Consider the system in Fig. 7.14, in which m_1 and m_2 move in a gravity field. Determine the work done by the forces on the system.

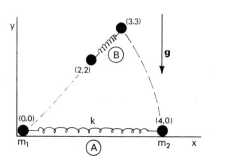

Figure 7.14

Solution

The external forces on the system are the gravity loads; the internal forces are those generated by the spring between the particles. In both cases the forces are conservative, and hence we can evaluate the work they do by taking the difference in potential energy between configuration A and configuration B.

For the external forces, we have

$$V(A) = 0, \qquad V(B) = m_1 g(2) + m_2 g(3).$$

Thus the work of the external forces is

$$W_{\text{ext}} = V(A) - V(B) = -2 m_1 g - 3 m_2 g. \tag{a}$$

Suppose that the undeformed length of the spring is a. Then the potential energy of the spring at A and B is

$$V(A) = \tfrac{1}{2}k(4-a)^2, \qquad V(B) = \tfrac{1}{2}k(\sqrt{2}-a)^2.$$

Thus

$$W_{\text{int}} = V(A) - V(B) = \tfrac{1}{2}k[12 - (8 - 2\sqrt{2})a]. \tag{b}$$

Thus from (a) and (b), the change in the kinetic energy between A and B is

$$T(B) - T(A) = -(2m_1 + 3m_2)g + \tfrac{1}{2}k[12 - (8 - 2\sqrt{2})a].$$

....

At this point, we once again consider the free-body diagram. In Chapter 2, Section 8, we defined the free-body diagram to include all external forces and moments applied to the system under consideration. In the dynamics of many-particle systems, the equations $\mathbf{F} = \dot{\mathbf{p}}_c$ and $\mathbf{M} = \dot{\mathbf{H}}$ involve only external forces and moments. If we write the equation of work–energy, we must also be aware of the internal forces on the system, since, as we have seen, internal forces can do work.

If we are going to write the equation of work–energy for a system, we must first construct a free-body diagram of the system which includes all external forces and moments. We must then evaluate the work done by the internal forces.

Fortunately, in many systems, the internal forces do no net work. This is the case in the motion of a rigid body, as we shall see in the next section, and in Chapter 8.

Example 8

The particles shown in Fig. 7.15 have velocities

$$\mathbf{v}_1 = a\mathbf{i}, \qquad \mathbf{v}_2 = b\mathbf{j}, \qquad \mathbf{v}_3 = c\mathbf{k}.$$

Find the kinetic energy of the center of mass and the kinetic energy relative to the center of mass.

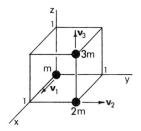

Figure 7.15

Solution

The position of the center of mass is

$$\mathbf{r}_C = \frac{m(0) + 2m(\mathbf{i}+\mathbf{j}) + 3m(\mathbf{i}+\mathbf{j}+\mathbf{k})}{m + 2m + 3m}.$$

Thus

$$\mathbf{r}_C = \tfrac{1}{6}(5\mathbf{i} + 5\mathbf{j} + 3\mathbf{k}).$$ (a)

The velocity of the center of mass is given by

$$\mathbf{v}_C = \frac{ma\mathbf{i} + 2mb\mathbf{j} + 3mc\mathbf{k}}{6m}$$

or

$$\mathbf{v}_C = \tfrac{1}{6}(a\mathbf{i} + 2b\mathbf{j} + 3c\mathbf{k}).$$ (b)

The vectors $\mathbf{r}_1$, $\mathbf{r}_2$, and $\mathbf{r}_3$ are

$$\mathbf{r}_1 = 0, \qquad \mathbf{r}_2 = \mathbf{i}+\mathbf{j}, \qquad \mathbf{r}_3 = \mathbf{i}+\mathbf{j}+\mathbf{k}.$$

Since $\mathbf{r}_{Ci} = \mathbf{r}_i - \mathbf{r}_C$ and $\dot{\mathbf{r}}_{Ci} = \mathbf{v}_i - \mathbf{v}_C$, we have

$$\dot{\mathbf{r}}_{C1} = \tfrac{1}{6}(5a\mathbf{i} - 2b\mathbf{j} - 3c\mathbf{k}), \qquad \dot{r}_{C1}^2 = (25a^2 + 4b^2 + 9c^2)/36,$$

$$\dot{\mathbf{r}}_{C2} = \tfrac{1}{6}(-a\mathbf{i} + 4b\mathbf{j} - 3c\mathbf{k}), \qquad \dot{r}_{C2}^2 = (a^2 + 16b^2 + 9c^2)/36,$$ (c)

$$\dot{\mathbf{r}}_{C3} = \tfrac{1}{6}(-a\mathbf{i} - 2b\mathbf{j} + 3c\mathbf{k}), \qquad \dot{r}_{C3}^2 = (a^2 + 4b^2 + 9c^2)/36.$$

Thus the kinetic energy of the center of mass is

$$\tfrac{1}{2}mv_C^2 = \tfrac{1}{2}(6m)(a^2 + 4b^2 + 9c^2)/36 = m(a^2 + 4b^2 + 9c^2)/12,$$ (d)

and the kinetic energy of motion relative to the center of mass is

$$\sum_{i=1}^{3} \tfrac{1}{2}m_i \dot{r}_{Ci}^2 = \frac{m}{72}[(25a^2 + 4b^2 + 9c^2) + 2(a^2 + 16b^2 + 9c^2) + 3(a^2 + 4b^2 + 9c^2)]$$

$$= m(5a^2 + 8b^2 + 9c^2)/12.$$ (e)

The total kinetic energy should be the sum of (d) and (e):

$$T = \tfrac{1}{2}m(a^2 + 2b^2 + 3c^2).$$ (f)

We can check (f) by computing

$$T = \sum_{i=1}^{3} \tfrac{1}{2}m_i v_i^2 = \tfrac{1}{2}(ma^2 + 2mb^2 + 3mc^2).$$

6. TRANSITION TO THE RIGID BODY

In this section, we consider a particular class of many-particle systems: systems of n particles connected by rigid, weightless members. Consider the system in Fig. 7.16(a).

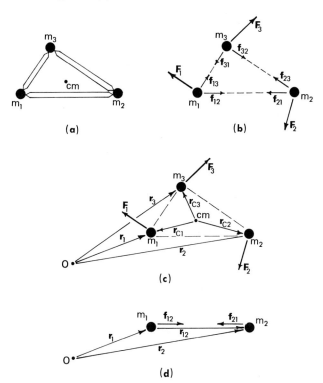

(a)

(b)

(c)

(d)

Figure 7.16

Note that the internal forces $\mathbf{f}_{12}$, $\mathbf{f}_{21}$, . . . $\mathbf{f}_{31}$ act on the particles m_1, m_2, and m_3 to maintain the constraint of rigidity: There is to be no relative motion between the particles.

Let us write the three fundamental equations for this rigid system.

$$\mathbf{F} = \dot{\mathbf{p}}_C: \qquad \mathbf{F}_1 + \mathbf{F}_2 + \mathbf{F}_3 = (m_1 + m_2 + m_3)\ddot{\mathbf{r}}_C. \tag{a}$$

For $\mathbf{M} = \dot{\mathbf{H}}$, consider Figure 7.16(c).

Suppose that m_1 has velocity $\mathbf{v}_1$, m_2 has velocity $\mathbf{v}_2$, and m_3 has velocity $\mathbf{v}_3$. Then

$$\mathbf{M}_O = \mathbf{r}_1 \times \mathbf{F}_1 + \mathbf{r}_2 \times \mathbf{F}_2 + \mathbf{r}_3 \times \mathbf{F}_3,$$

$$\mathbf{M}_C = \mathbf{r}_{C1} \times \mathbf{F}_1 + \mathbf{r}_{C2} \times \mathbf{F}_2 + \mathbf{r}_{C3} \times \mathbf{F}_3,$$

and

$$\mathbf{H}_O = \mathbf{r}_1 \times m_1\mathbf{v}_1 + \mathbf{r}_2 \times m_2\mathbf{v}_2 + \mathbf{r}_3 \times m_3\mathbf{v}_3,$$

$$\mathbf{H}_C = \mathbf{r}_{C1} \times m_1\mathbf{v}_1 + \mathbf{r}_{C2} \times m_2\mathbf{v}_2 + \mathbf{r}_{C3} \times m_3\mathbf{v}_3.$$

Then we can write either

$$\mathbf{M}_O = \dot{\mathbf{H}}_O \tag{b}$$

or

$$\mathbf{M}_C = \dot{\mathbf{H}}_C. \tag{b'}$$

It is not obvious, but Eq. (a) and either (b) or (b′) completely define the motion of the entire system. This will be clear after we have discussed the kinematics of rigid systems in Chapter 8.

We shall now write the work–energy relation for a rigid system. The most important result is that the sum of the work of all the internal forces is zero. For the system in Fig. 7.16(a), consider particles m_1 and m_2 (see Fig. 7.16d).

We wish to consider the work of internal forces $\mathbf{f}_{12}$ and $\mathbf{f}_{21}$ during an incremental displacement of the system. We have

$$dW = \mathbf{f}_{12} \cdot d\mathbf{r}_1 + \mathbf{f}_{21} \cdot d\mathbf{r}_2 \tag{a}$$

But

$$\mathbf{f}_{12} = -\mathbf{f}_{21} \quad \text{and} \quad \mathbf{r}_2 = \mathbf{r}_1 + \mathbf{r}_{12}.$$

Thus

$$d\mathbf{r}_2 = d\mathbf{r}_1 + d\mathbf{r}_{12}. \tag{b}$$

Inserting (b) into (a), we get

$$dW = -\mathbf{f}_{21} \cdot d\mathbf{r}_1 + \mathbf{f}_{21} \cdot d\mathbf{r}_1 + \mathbf{f}_{21} \cdot d\mathbf{r}_{12}$$

or

$$dW = \mathbf{f}_{21} \cdot d\mathbf{r}_{12}. \tag{c}$$

We can show that $d\mathbf{r}_{12}$ is either zero or perpendicular to $\mathbf{r}_{12}$. And in either case (c) will be zero. Consider

$$|\mathbf{r}_{12}|^2 = \mathbf{r}_{12} \cdot \mathbf{r}_{12} = \text{constant}. \tag{d}$$

The magnitude of $\mathbf{r}_{12}$ is the distance between m_1 and m_2, which is constant, because this is a rigid system. We take the differential of (d):

$$d\mathbf{r}_{12} \cdot \mathbf{r}_{12} + \mathbf{r}_{12} \cdot d\mathbf{r}_{12} = 0,$$

or, since $\mathbf{a} \cdot \mathbf{b} = \mathbf{b} \cdot \mathbf{a}$ for any two vectors,

$$2\mathbf{r}_{12} \cdot d\mathbf{r}_{12} = 0. \tag{e}$$

There are only two possibilities for (e)

i) $d\mathbf{r}_{12} = 0,$

ii) $d\mathbf{r}_{12} \perp \mathbf{r}_{12}.$

Returning to (c), we see that in case (i), $dW = 0$. This is also true for case (ii). If $d\mathbf{r}_{12}$ is perpendicular to $\mathbf{r}_{12}$, it is also perpendicular to $\mathbf{f}_{21}$, which lies along $\mathbf{r}_{12}$ (see Fig. 7.16d). Thus

$$dW = 0.$$

Thus it is easy to see that the work of all the internal forces on a rigid system is zero.

> The total work of the internal forces acting on a rigid system of par-
> ticles is zero during any motion of the system.

Example 9

Consider the rigid pendulum system in Fig. 7.17(a). A rigid, weightless bar supports two particles of mass m. Determine the natural frequency of the system.

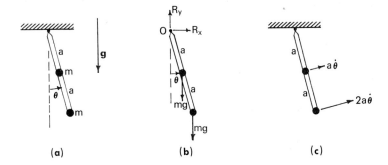

(a)   (b)   (c)   Figure 7.17

Solution

Figure 7.17(b) is the free-body diagram of the system. The moment of the external forces about O is

$$\mathbf{M}_O = (-mga \sin \theta - mg2a \sin \theta)\mathbf{k}. \tag{a}$$

Figure 7.17(c) is a plot of the velocities of the masses. Thus the moment of momentum about O is

$$\mathbf{H}_O = [(ma\dot\theta)a + (m2a\dot\theta)2a]\mathbf{k}. \tag{b}$$

Thus from (a) and (b) we have

$$\mathbf{M}_O = \dot{\mathbf{H}}_O: -3mga \sin\theta = 5ma^2\ddot\theta$$

or

$$\ddot\theta + \left(\frac{3g}{5a}\right)\sin\theta = 0.$$

If θ is small, we approximate $\sin\theta \approx \theta$, and thus

$$\ddot\theta + \left(\frac{3g}{5a}\right)\theta = 0 \quad \text{which gives} \quad \omega_n^2 = \frac{3g}{5a}.$$

....

Example 10

A dumbbell consists of three equal particles, of mass $\frac{1}{3}m$, supported by a weightless bar of length l. Figure 7.18(a) shows the dumbbell supported by strings at A and B. At $t = 0$, the string at B is cut. What is the tension in the string at A for $t < 0$ and just after $t = 0$ ($t = 0^+$)?

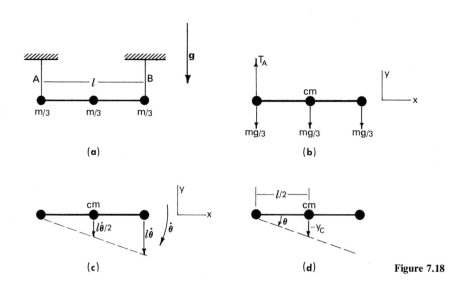

(a)

(b)

(c)

(d)

Figure 7.18

Solution

Before $t = 0$, the tension in each string is half the total weight of the dumbbell. Thus

$$T_A = \tfrac{1}{2}mg, \qquad t < 0. \tag{a}$$

Figure 7.18(b) shows the free-body diagram of the dumbbell just after $t = 0$. The force resultant on the system is

$$\mathbf{F} = (T_A - mg)\mathbf{j}. \tag{b}$$

Thus, writing $\mathbf{F} = \dot{\mathbf{p}}_C$, we get

$$F_x = ma_{Cx}: \qquad 0 = m\ddot{x}_C. \tag{c}$$

$$F_y = ma_{Cy}: \qquad T_A - mg = m\ddot{y}_C, \tag{d}$$

The moment of the external forces on the system about the center of mass is

$$\mathbf{M}_C = \left[(\tfrac{1}{3}mg - T_A)\frac{l}{2} - (\tfrac{1}{3}mg)\frac{l}{2} \right]\mathbf{k}$$

or

$$\mathbf{M}_C = -\tfrac{1}{2}T_A l\,\mathbf{k}. \tag{e}$$

Just after $t = 0$, the velocities of the center of mass and right side of the dumbbell are as shown in Fig. 7.18(c). Thus

$$\mathbf{H}_C = -\left(\frac{m}{3}l\dot{\theta}\right)\frac{l}{2}\mathbf{k}, \qquad t = 0^+,$$

and hence

$$\dot{\mathbf{H}}_C = -\tfrac{1}{6}ml^2\ddot{\theta}\mathbf{k}, \qquad t = 0^+. \tag{f}$$

From (e) and (f), we have

$$\mathbf{M}_C = \dot{\mathbf{H}}_C: \qquad -\tfrac{1}{2}T_A l = -\tfrac{1}{6}ml^2\ddot{\theta}$$

or

$$T_A = \tfrac{1}{3}ml\ddot{\theta}, \qquad t = 0^+. \tag{g}$$

From (d) and (g), we could solve for T_A if we could find the relation between $\ddot{y}$ and $\ddot{\theta}$. From (c), we see that the center of mass moves straight downward at $t = 0^+$. Thus, from Fig. 7.18(d), we have

$$-y_C \approx \frac{l}{2}\theta,$$

and therefore

$$-\ddot{y}_C = \frac{l}{2}\ddot{\theta}, \qquad t = 0^+. \tag{h}$$

Finally (d), (g), and (h) give

$$T_A(0^+) = \tfrac{2}{5} mg.$$

....

7. COLLISION PROBLEMS

We now consider problems in which two particles collide. These problems are characterized by a conservation of linear momentum and, in general, a dissipation of some portion of the initial kinetic energy due to the work of the internal forces acting between the particles during the collision.

Example 11

Let us consider the simple example in Fig. 7.19. Two particles of mass m are sliding without friction along the x axis. A free-body diagram of each particle shows no net external force. If $v_0 > v_1$, the particles collide. During the collision there are strong internal forces acting. Suppose the particles stick together and move as a unit with speed v after collision. Find v, and determine the energy lost in the collision.

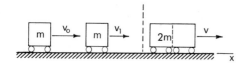

Figure 7.19

Solution

Since the total external force is zero, the momentum of the center of mass, $\mathbf{p}_C$, is a constant. Before the collision,

$$\mathbf{v}_C = \frac{mv_0\mathbf{i} + mv_1\mathbf{i}}{2m} .$$

Thus

$$\mathbf{p}_C = (mv_0 + mv_1)\mathbf{i} = \text{constant.} \tag{a}$$

After the collision, we have

$$\mathbf{p}_C = 2mv\mathbf{i}. \tag{b}$$

Thus, equating (a) and (b), we get

$$v = \tfrac{1}{2}(v_0 + v_1). \tag{c}$$

Consider the kinetic energy of the system before and after the collision. Before the collision, we had

$$T_b = \tfrac{1}{2}mv_0^2 + \tfrac{1}{2}mv_1^2 = \tfrac{1}{2}m(v_0^2 + v_1^2).$$

After the collision, we have, by (c),

$$T_a = \tfrac{1}{2}(2m)\tfrac{1}{4}(v_0^2 + 2v_0v_1 + v_1^2) = \tfrac{1}{2}m\left(\frac{v_0^2}{2} + v_0v_1 + \frac{v_1^2}{2}\right).$$

Thus the work of the internal forces equals

$$W_{\text{int}} = T_a - T_b = \tfrac{1}{2}m\left(-\frac{v_0^2}{2} + v_0v_1 - \frac{v_1^2}{2}\right)$$

$$= -\tfrac{1}{4}m(v_0 - v_1)^2. \tag{d}$$

This quantity is always negative or zero. Thus the internal forces have dissipated some of the initial kinetic energy of the system.

....

In some problems, kinetic energy as well as linear momentum are conserved during the collision. This is the situation in the next example.

Example 12

Figure 7.20 presents a particle m traveling with speed v_0 which collides with a second particle m, initially at rest. There is no dissipation of energy. Calculate the speed of each particle after collision.

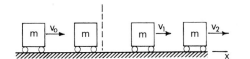

Figure 7.20

Solution

Suppose that, after collision, the particles have velocities $v_1\mathbf{i}$ and $v_2\mathbf{i}$, respectively. Conservation of linear momentum requires that

$$mv_0\mathbf{i} = mv_1\mathbf{i} + mv_2\mathbf{i} \quad \text{or} \quad v_0 = v_1 + v_2. \tag{a}$$

Now if energy is also conserved, we have

$$\tfrac{1}{2}mv_0^2 = \tfrac{1}{2}mv_1^2 + \tfrac{1}{2}mv_2^2 \quad \text{or} \quad v_0^2 = v_1^2 + v_2^2. \tag{b}$$

Equations (a) and (b) are equations for v_1 and v_2. Eliminating v_0, we get

$$v_1^2 + 2v_1v_2 + v_2^2 = v_1^2 + v_2^2$$

or

$$2v_1v_2 = 0. \tag{c}$$

From (c), we see that either v_1 or v_2 must be zero. Thus $v_1 = 0$ and $v_2 = v_0$.

....

Example 13

Figure 7.21 depicts two pendulums which swing and collide. Assuming that there is no energy loss during the motion, determine the period of the combined motion.

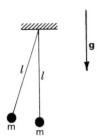

Figure 7.21

Solution

Since the masses are equal and by the result of Example 12, there is an exchange of velocities at collision. Thus the combined motion looks exactly like the motion of a single pendulum of length l. Thus

$$\omega_n^2 = \frac{g}{l}$$

and since $\omega_n \tau = 2\pi$, then

$$\tau = 2\pi \sqrt{\frac{l}{g}}.$$

. . . .

The above example should be compared to Example 15, in which we allow for the possibility that some kinetic energy is dissipated during each collision.

The general problem of the collision of two *finite bodies* is sufficiently complicated to be outside the scope of this book. We shall consider here only that which is termed *the central collision* of two bodies. This will enable us to consider the bodies to be particles, since rotational motions will not be involved.

There are two types of central collision problems: (i) direct collision, (ii) oblique collision. Let us now illustrate both these cases.

Figure 7.22(a) shows a case of a noncentral collision of two bodies. As the two bodies collide, we construct two lines:

l_t, the line* tangent to the colliding surfaces.

l_n, the line* normal to the colliding surfaces through the point of collision.

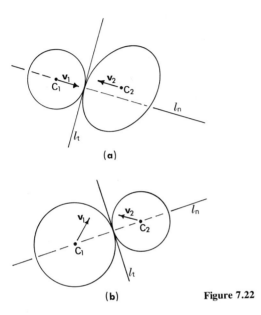

(a)

(b)   Figure 7.22

Central Collision. When the centers of mass of the two colliding bodies lie on l_n, the collision is called *central.* If not, the collision is *noncentral.* Note that any collision of two homogeneous spheres or homogeneous disks is always central. (See Fig. 7.22b.)

Central collisions are called *direct* or *oblique* according to the following definitions.

Direct Central Collision

Let $\mathbf{v}_1$ and $\mathbf{v}_2$ be the velocities of the centers of mass of two bodies just before collision. If $\mathbf{v}_1$ and $\mathbf{v}_2$ lie on the line l_n, the collision is direct.

Oblique Central Collision

If a central collision is not direct, it is oblique. That is, either $\mathbf{v}_1$ or $\mathbf{v}_2$ (or both) do not lie on l_n. The collision shown in Fig. 7.22(b) is oblique.

---

* We are considering collisions in a plane. In general, we would consider the tangent plane and the normal line.

When we are considering the case of finite colliding bodies, we assume that the bodies are smooth, and that they are not in rotation before collision.

The totality of the above assumptions enables us to deal with the problems under consideration as problems of colliding particles.

Consider the two particles m_1 and m_2 in Fig. 7.23. At time $t = 0^-$ the collision begins. At $t = 0^+$ the collision is finished. We assume that the time interval from $t = 0^-$ until $t = 0^+$ is extremely short, and so we call this an instantaneous collision. Let $\mathbf{F}_1$ and $\mathbf{F}_2$ be the external forces on m_1 and m_2 (see Fig. 7.23c). Then $\mathbf{F} = \mathbf{F}_1 + \mathbf{F}_2$ is the resultant external force. By (7.6) we have

$$\int_{0^-}^{0^+} (\mathbf{F}_1 + \mathbf{F}_2) \; dt = \mathbf{p}_C(0^+) - \mathbf{p}_C(0^-). \tag{a}$$

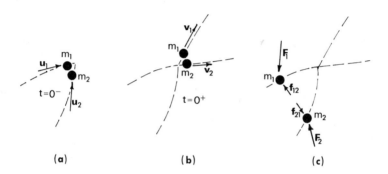

(a)   (b)   (c)   **Figure 7.23**

Let us examine the impulse in (a). Let $(0^+ - 0^-)$ be Δt, a very brief time interval. We can thus approximate the integral by

$$(\mathbf{F}_1 + \mathbf{F}_2) \; \Delta t. \tag{b}$$

If the external forces $\mathbf{F}_1$ and $\mathbf{F}_2$ have finite magnitudes, we see that their impulse from $t = 0^-$ until $t = 0^+$ is zero in the limit as $\Delta t \to 0$. Thus for an instantaneous collision, we have, by (a),

$$\mathbf{p}_C(0^+) = \mathbf{p}_C(0^-). \tag{7.23}$$

In other words, the linear momentum of the system just before the collision equals its linear momentum just after the collision.

> In the case of the instantaneous collision of two particles subject to finite external forces, the linear momentum does not change during the collision.

The above analysis applies to all problems of instantaneous collision, central or noncentral, direct or oblique. However, Eq. (7.23) is not sufficient in itself to determine the velocities of m_1 and m_2 after collision, given the velocities before collision.

When we are considering a collision in the plane, (7.23) is equivalent to two scalar equations. But we wish to determine two velocities $\mathbf{v}_1$ and $\mathbf{v}_2$, or four scalar unknowns v_{1x}, v_{1y}, v_{2x}, v_{2y}.

The additional equations we shall present are based on experimental findings. Let us give some motivation for the form of those equations.

Recall Examples 11 and 12, in which we saw that the kinetic energy of m_1 and m_2 need not be equal before and after the collision.

Suppose we write the kinetic energy of m_1 and m_2 as in (7.20): $T = T_C + T_R$, where

$$T_C = \tfrac{1}{2}(m_1 + m_2)v_C^2, \tag{7.24}$$

$$T_R = \tfrac{1}{2}m_1\dot{r}_{C1}^2 + \tfrac{1}{2}m_2\dot{r}_{C2}^2.$$

Consider the system kinetic energy just before collision, $T(0^-)$, and just after collision, $T(0^+)$. From (7.23), we have

$$(m_1 + m_2)\mathbf{v}_C(0^-) = (m_1 + m_2)\mathbf{v}_C(0^+)$$

or

$$\mathbf{v}_C(0^-) = \mathbf{v}_C(0^+).$$

Thus the kinetic energy of the center of mass during collision is constant,

$$T_C(0^-) = T_C(0^+).$$

And so we can account for any energy dissipated during the collision entirely in terms of a change in the kinetic energy of motion relative to the center of mass:

$$T(0^-) - T(0^+) = T_R(0^-) - T_R(0^+),$$

or, more simply,

$$\Delta T = \Delta T_R. \tag{7.25}$$

Our point here is that it is *not* the *absolute* velocities $\mathbf{u}_1$, $\mathbf{u}_2$ and $\mathbf{v}_1$, $\mathbf{v}_2$ that are of fundamental importance in collision problems; it is the *relative* velocities of the particles.

From the above analysis, we would anticipate using velocities relative to the center of mass. However, it is more common to use another scheme. We'll get back to (7.25) later.

Direct Central Collisions

Consider two particles m_1 and m_2 with velocities $\mathbf{u}_1 = u_1\mathbf{i}$ and $\mathbf{u}_2 = u_2\mathbf{i}$ before collision (see Fig. 7.24).

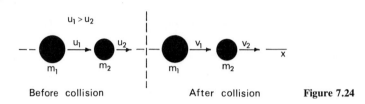

| | Before collision | | After collision | | **Figure 7.24** |

After collision, we assume that m_1 has velocity $\mathbf{v}_1 = v_1\mathbf{i}$ and that m_2 has velocity $\mathbf{v}_2 = v_2\mathbf{i}$. From (7.23) we have

$$m_1 v_1 + m_2 v_2 = m_1 u_1 + m_2 u_2. \tag{7.26}$$

If u_1 and u_2 are given, we need another equation if we are to determine both v_1 and v_2. Suppose we define:

$u_a = u_1 - u_2$ approach speed,

$v_s = v_2 - v_1$ separation speed.

We introduce* the *coefficient of restitution, e* by means of the equation

$$v_s = eu_a$$

or

$$v_2 - v_1 = e(u_1 - u_2). \tag{7.27}$$

The coefficient of restitution is determined experimentally. It depends on the shapes and materials of the colliding bodies. In any event, unless there is a release of energy during collision (as in a bomb), we will show that

$$0 \leqslant e \leqslant 1.$$

If $e = 0$, $v_s = 0$. That is, the particles stick together after collision. This was the case in Example 10. We call this a *plastic collision*.

If $e = 1$, $v_s = u_a$ and no energy is dissipated during the collision, we will show this in Example 14. First let us note that, from (7.26) and (7.27), we have two equations for the two unknowns v_1 and v_2:

$$m_1 v_1 + m_2 v_2 = m_1 u_1 + m_2 u_2, \qquad -v_1 + v_2 = eu_1 - eu_2.$$

---

* The coefficient of restitution constitutes a means of lumping our ignorance of the physics of collision into a single parameter. See Problem 7.67.

We can obtain the solution to these two equations by elimination, or by using Cramer's rule:*

$$v_1 = \frac{1}{m_1 + m_2} [(m_1 - em_2)u_1 + m_2(1 + e)u_2],$$

$$v_2 = \frac{1}{m_1 + m_2} [m_1(1 + e)u_1 + (m_2 - em_1)u_2].$$

(7.28)

Example 14

Two particles m_1 and m_2 collide by direct central collision. Suppose that $e = 1.0$. Show that there is no energy loss.

Solution

The energy before collision is

$$T(0^-) = \tfrac{1}{2}mu_1^2 + \tfrac{1}{2}m_2u_2^2.$$

(a)

After collision, the energy is

$$T(0^+) = \tfrac{1}{2}m_1v_1^2 + \tfrac{1}{2}m_2v_2^2.$$

(b)

If $e = 1$, (7.28) becomes

$$v_1 = \frac{1}{m}[(m_1 - m_2)u_1 + 2m_2u_2],$$

$$v_2 = \frac{1}{m}[2m_1u_1 + (m_2 - m_1)u_2],$$

(c)

where $m = m_1 + m_2$.

Inserting (c) into (b) (which takes some tedious algebra) shows that

$$T(0^-) = T(0^+).$$

....

Thus Example 14 shows that the limiting cases of e give us the following:

$e = 0$ *Plastic collision:* $T_R(0^+) = 0$. All energy relative to the center of mass is dissipated during the collision.

$e = 1$ *Elastic collision:* No energy is dissipated during the collision.

In general, if $0 < e < 1$, some portion of T_R, the kinetic energy relative to the center of mass, is dissipated during collision. We shall prove this after we have given three examples to illustrate the application of (7.28).

---

* See Appendix 1, Eq. (A1.37).

Example 15

Consider the system in Fig. 7.25(a). One of the masses is displaced an initial small angle θ and then released. Given that $e = 0.5$, determine the motion of each mass before the third collision.

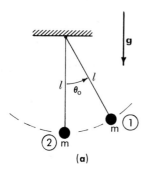

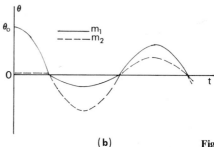

(a)   (b)   **Figure 7.25**

Solution

Assume θ_0 is small enough to take $\sin \theta_0 \approx \theta_0$. We can then write the speed of the first mass just before the first collision as

$$u_1 \approx \sqrt{gl}\ \theta_0.$$

[Write the equation for conservation of energy, noting that $\cos \theta_0 \approx 1 - (\theta_0^2/2)$.] The speed of the second particle, u_2, is zero before collision. Thus, by (7.28), we get

$$v_1 = \tfrac{1}{4}\sqrt{gl}\ \theta_0, \qquad v_2 = \tfrac{3}{4}\sqrt{gl}\ \theta_0. \tag{a}$$

The two particles continue to move to the left, the first particle to an amplitude of α_1,

$$\alpha_1 = \frac{v_1}{\sqrt{gl}} = \tfrac{1}{4}\theta_0, \tag{b}$$

and the second particle to an amplitude α_2,

$$\alpha_2 = \frac{v_2}{\sqrt{gl}} = \tfrac{3}{4}\theta_0. \tag{c}$$

The particles return to the vertical position simultaneously, with speeds having the magnitudes of (a) (though opposite directions). (Can you explain why these statements are true?) For the next collision, we again use (7.28) to get the resulting velocities after collision and the corresponding amplitudes β_1 and β_2:

$$v_1 = -\tfrac{5}{8}\sqrt{gl}\ \theta_0, \qquad \beta_1 = \tfrac{5}{8}\theta_0,$$
$$v_2 = -\tfrac{3}{8}\sqrt{gl}\ \theta_0, \qquad \beta_2 = \tfrac{3}{8}\theta_0.$$

Figure 7.25(b) plots the motions of the two masses.
  See Problem 7.68 for further discussion of this example.

....

Example 16

  i) A particle of mass m and speed u has a direct central collision with another particle m_0 initially at rest. The coefficient of restitution is $e \neq 0$. Determine the speed of m after collision, given that m_0 is very large.
  ii) The particle in Fig. 7.26 is dropped from a height h_0 and rebounds to a height h_1. Calculate e, using the results of part (i).

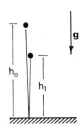

Figure 7.26

Solution

  i) From (7.28), taking $u_1 = u$, $u_2 = 0$, $m_1 = m$, and $m_2 = m_0$, we obtain

$$v_1 = \left(\frac{m - em_0}{m + m_0}\right) u, \qquad v_2 = (1 + e)\frac{m}{m + m_0} u.$$

If $m_0 \to \infty$, we get

$$v_1 \to -eu, \qquad v_2 \to 0. \tag{a}$$

$$\tfrac{1}{2}mu^2 = mgh^0$$

  ii) Given that the particle is released from a height h_0, its speed at collision will be

$$u = \sqrt{2gh_0} \quad \longrightarrow \quad u^2 = 2gh_0 \Longleftarrow \tfrac{1}{2}u^2 = gh_0$$

(write the equation of conservation of energy for the particle). From (a), we know that the speed of the particle after collision will be

$$v = eu = e\sqrt{2gh_0}, \tag{b}$$

and this must be $\sqrt{2gh_1}$. Thus

$$\sqrt{2gh_1} = e\sqrt{2gh_0} \qquad \text{or} \qquad e = \sqrt{\frac{h_1}{h_0}}. \tag{c}$$

....

Example 17

A particle which is thrown upward with speed v_0 collides with the floor with coefficient of restitution $e \neq 0$. Calculate the total distance traveled by the particle during the entire motion.

Solution

Consider Fig. 7.27. The height of the first peak is h_0, the second is h_1, etc. The total distance traveled is

$$d = 2h_0 + 2h_1 + 2h_2 + \cdots \tag{a}$$

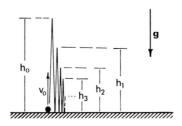

Figure 7.27

From conservation of energy, we can determine that

$$v_0 = \sqrt{2gh_0} \quad \text{or} \quad h_0 = \left(\frac{v_0^2}{2g}\right). \tag{b}$$

The first collision occurs with speed v_0. The speed of rebound is ev_0. Thus

$$ev_0 = \sqrt{2gh_1} \quad \text{or} \quad h_1 = e^2 \left(\frac{v_0^2}{2g}\right) \tag{c}$$

From (b) and (c), we note that subsequent heights are obtained from previous heights by multiplying by e^2:

$$h_n = e^2 h_{n-1}, \quad n = 1, 2, \ldots$$

Thus (a) becomes

$$d = \left(\frac{v_0^2}{g}\right) (1 + e^2 + e^4 + e^6 + \cdots).$$

If we let $r = e^2$,

$$d = \left(\frac{v_0^2}{g}\right) (1 + r + r^2 + r^3 + \cdots).$$

The infinite summation on the right is the *geometric series*. It can be shown* that, if $|r| < 1$,

$$(1 + r + r^2 + \cdots) = \frac{1}{1-r}.$$

Thus if $e < 1$,

$$d = \frac{v_0^2}{g}\left(\frac{1}{1-e^2}\right).$$

. . . .

As a final topic concerned with direct, central collision, let us develop expressions for the loss of energy in a collision. As we have previously remarked, the kinetic energy of the center of mass is constant during a collision. Thus any loss in kinetic energy can be accounted for in the change in kinetic energy relative to the center of mass. Recall (7.25):

$$\Delta T = T_R \Big|_{0^+}^{0^-}, \tag{a}$$

where

$$T_R = \sum_{i=1}^{2} \tfrac{1}{2} m_i \dot{r}_{Ci}^2$$

and $\dot{r}_{Ci}$ is the speed of m_i relative to the center of mass.

Let v_C be the (constant) speed of the center of mass. Then:

Before collision

$$\dot{r}_{C1}(0^-) = u_1 - v_C \qquad \text{and} \qquad \dot{r}_{C2}(0^-) = u_2 - v_C.$$

After collision $\tag{b}$

$$\dot{r}_{C1}(0^+) = v_1 - v_C \qquad \text{and} \qquad \dot{r}_{C2}(0^+) = v_2 - v_C.$$

We begin our development of ΔT by showing that

$$\dot{r}_{C1}(0^+) = -e\dot{r}_{C1}(0^-), \qquad \dot{r}_{C2}(0^+) = -e\dot{r}_{C2}(0^-). \tag{c}$$

First recall (7.27). Noting (b), we write the speed of approach and the speed of separation as

$$u_a = u_1 - u_2 = \dot{r}_{C1}(0^-) - \dot{r}_{C2}(0^-), \qquad v_s = v_2 - v_1 = \dot{r}_{C2}(0^+) - \dot{r}_{C1}(0^+).$$

---

* See G. B. Thomas, *Calculus and Analytic Geometry*, fourth edition, Reading, Mass.: Addison-Wesley, 1968, page 624.

Thus, from (7.27), we get

$$\dot{r}_{C2}(0^+) - \dot{r}_{C1}(0^+) = e[\dot{r}_{C1}(0^-) - \dot{r}_{C2}(0^-)]. \tag{d}$$

Next we note that, before or after collision,

$$m_1 r_{C1} + m_2 r_{C2} = (m_1 + m_2) r_{CC} = 0,$$

since the vectors $\mathbf{r}_{C1}$ and $\mathbf{r}_{C2}$ are measured from the center of mass. Thus

$$m_1 \dot{r}_{C1} + m_2 \dot{r}_{C2} = 0. \tag{e}$$

From (e), we get

$$\dot{r}_{C2}(0^-) = -\left(\frac{m_2}{m_1}\right) \dot{r}_{C1}(0^-)$$

and

$$\dot{r}_{C2}(0^+) = -\left(\frac{m_2}{m_1}\right) \dot{r}_{C1}(0^+).$$

Returning to (d), we have

$$-\left(1 + \frac{m_2}{m_1}\right) \dot{r}_{C1}(0^+) = e\left(1 + \frac{m_2}{m_1}\right) \dot{r}_{C1}(0^-)$$

or

$$\dot{r}_{C1}(0^+) = -e\dot{r}_{C1}(0^-). \tag{f}$$

This is the first of Eqs. (c). Adding (f) to (d) gives the second equation. Finally, using (c) in (a), we get

$$T_R(0^+) = e^2[\tfrac{1}{2} m_1 \dot{r}_{C1}^2(0^-) + \tfrac{1}{2} m_2 \dot{r}_{C2}^2(0^-)],$$

$$T_R(0^-) = [\tfrac{1}{2} m_1 \dot{r}_{C1}^2(0^-) + \tfrac{1}{2} m_2 \dot{r}_{C2}^2(0^-)].$$

Thus

$$T_R(0^+) = e^2 T_R(0^-)$$

and

$$\Delta T = (1 - e^2) T_R(0^-). \tag{7.29}$$

Again we can discuss the limiting cases of collisions:

Plastic Collision

$$e = 0 \qquad \Delta T = T_R(0^-).$$

That is, all the energy relative to the center of mass is lost during the collision.

Elastic Collision

$$e = 1 \qquad \Delta T = 0.$$

That is, there is no energy lost during the collision. We can write ΔT in terms of the initial absolute velocities (see Problem 7.72).

$$\Delta T = \tfrac{1}{2}(1 - e^2)\,\frac{m_1 m_2}{m_1 + m_2}\,(u_1 - u_2)^2. \qquad (7.30)$$

Oblique Central Collisions

All the equations we have developed for direct central collisions can be applied to oblique central collisions, with one important modification. Consider the oblique collision shown in Fig. 7.28(a).

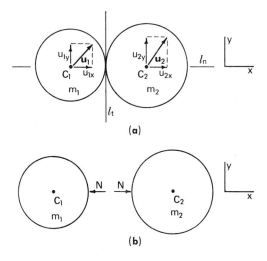

(a)

(b)   Figure 7.28

At least one of the initial velocities $\mathbf{u}_1$, $\mathbf{u}_2$ does not lie on the normal line l_n. Let the x axis be the normal line l_n. Then we write

$$\mathbf{u}_1 = u_{1x}\mathbf{i} + u_{1y}\mathbf{j}, \qquad \mathbf{u}_2 = u_{2x}\mathbf{i} + u_{2y}\mathbf{j}.$$

And after the collision,

$$\mathbf{v}_1 = v_{1x}\mathbf{i} + v_{1y}\mathbf{j}, \qquad \mathbf{v}_2 = v_{2x}\mathbf{i} + v_{2y}\mathbf{j}.$$

(Again we are considering only collisions in a plane. The extension to the three-dimensional case is straightforward.)

Figure 7.28(b) shows a free-body diagram of the two bodies at the moment of collision. Again we needn't show any finite external forces if we are concerned only with the motion from $t = 0^-$ until $t = 0^+$. We assume that the colliding bodies are smooth (frictionless). Thus the only forces on m_1 and m_2 are the normal forces acting at the points of contact.

Writing $\mathbf{F} = \dot{\mathbf{p}}_C$ in the y direction for m_1 and m_2, we have

$$m_1, \quad F_y = ma_{Cy}: \qquad 0 = m_1 a_{C1y}, \qquad 0^- < t < 0^+,$$

$$m_2, \quad F_y = ma_{Cy}: \qquad 0 = m_2 a_{C2y}, \qquad 0^- < t < 0^+.$$

These equations can be integrated to show that

$$v_{1y} = \text{constant}, \qquad 0^- < t < 0^+,$$

$$v_{2y} = \text{constant}, \qquad 0^- < t < 0^+.$$

Thus

$$u_{1y} = v_{1y}, \qquad u_{2y} = v_{2y}. \tag{7.31}$$

The tangential velocities of m_1 and m_2 do not change during the collision.

With respect to the normal velocities u_{1x}, u_{2x} and v_{1x}, v_{2x}, we apply the theory and equations for direct central collision.

Example 18

Two disks m_1 and m_2 collide, as shown in Fig. 7.29(a). Suppose that $\mathbf{u}_1 = 10\mathbf{i}'$ ft/sec, $\mathbf{u}_2 = -5\mathbf{j}'$ ft/sec, $e = 0.5$, $m_1 g = 8$ lb, and $m_2 g = 24$ lb. Calculate $\mathbf{v}_1$ and $\mathbf{v}_2$, the velocities of the bodies after collision.

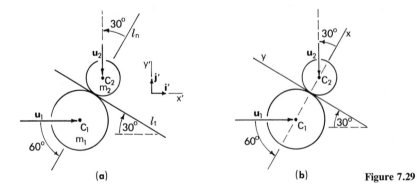

(a)   (b)   **Figure 7.29**

Solution

We align the x axis along the normal line and the y axis along the tangential line. Figure 7.29(b) gives the essential geometry of the problem in more detail.

Let $u_1 = 10$ ft/sec and $u_2 = 5$ ft/sec. Then

$$u_{1x} = u_1 \cos 60°, \qquad u_{2x} = -u_2 \cos 30°,$$

$$u_{1y} = -u_1 \sin 60°, \qquad u_{2y} = -u_2 \sin 30°.$$

Thus

$$u_{1x} = 5 \text{ ft/sec}, \qquad u_{2x} = -4.33 \text{ ft/sec},$$

$$u_{1y} = -8.66 \text{ ft/sec} \qquad u_{2y} = -2.5 \text{ ft/sec}.$$

For the tangential speeds, we have

$$v_{1y} = -8.66 \text{ ft/sec}, \qquad v_{2y} = -2.5 \text{ ft/sec}.$$

And from (7.28), noting that $m_1 + m_2 = 1$ slug, we obtain

$$v_{1x} = [\{0.25 - (0.5)(0.75)\}5 + (0.75)(1.5)(-4.33)] \text{ ft/sec},$$

$$v_{2x} = [(0.25)(1.5)5 + \{0.75 - (0.5)(0.25)(-4.33)\}] \text{ ft/sec}$$

or

$$v_{1x} = -5.50 \text{ ft/sec}, \qquad v_{2x} = -0.83 \text{ ft/sec}.$$

Thus

$$\mathbf{v}_1 = -5.50\mathbf{i} - 8.66\mathbf{j} \text{ ft/sec}, \qquad \mathbf{v}_2 = -0.83\mathbf{i} - 2.50\mathbf{j} \text{ ft/sec}.$$

··

Example 19

In Fig. 7.30, a particle m strikes a smooth, rigid wall at an angle of inclination α from the normal. The coefficient of restitution is e and the initial speed is v_0. Calculate the rebound angle β and the rebound speed v_1.

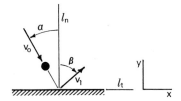

Figure 7.30

Solution

The contact surface is smooth. Thus the speed of the particle in the tangential direction is constant:

$$v_1 \sin \beta = v_0 \sin \alpha. \tag{a}$$

For the speeds in the normal direction, we write

$$v_s = eu_a \qquad \text{or} \qquad v_1 \cos \beta = ev_0 \cos \alpha \tag{b}$$

Dividing (a) by (b), we obtain

$$\tan \beta = \left(\frac{1}{e}\right) \tan \alpha. \tag{c}$$

Equation (c) gives

$$\sin \beta = \frac{\tan \alpha}{\sqrt{e^2 + \tan^2 \alpha}}$$

Thus (a) yields

$$v_1 = v_0 \cos \alpha \sqrt{e^2 + \tan^2 \alpha}$$

. . . .

8. SYSTEMS WHICH ACCUMULATE AND DISBURSE MASS

In some problems concerning systems of particles, we are primarily interested in the motion of a subsystem of the original system of particles. These problems are often (unfortunately) termed "variable-mass" problems. It is only in the sense that some mass is moving into and some other mass is exiting from the subsystem that the mass (of the subsystem) is variable. The mass of individual particles is constant. (We are not talking of relativistic mechanics here!)

We shall consider only systems which exhibit one-dimensional motion. Consider the system in Fig. 7.31. Here we have a container represented by the large block. In the time interval Δt, the container gains mass Δm_i (i standing for input) and loses mass Δm_e (e standing for exit). Δm_i enters the container at speed v_i *relative to the container*. Δm_e leaves the container at speed v_e *relative to the container*. In Fig. 7.31, we show the absolute speeds of Δm_i $(v - v_i)$, and of Δm_e $(v - v_e)$. Here both v_i and v_e are positive quantities.

While all this is going on, we assume that a force F is applied to the container. (This is an external force.) We analyze the motion of this system of particles by writing (7.6) over a short interval of time Δt,

$$F \Delta t = p(t + \Delta t) - p(t), \tag{a}$$

where $p(t)$ is the momentum of the system at time t and $p(t + \Delta t)$ is the momentum at time $t + \Delta t$. In order to write (a), we must talk of exactly the same system of particles.

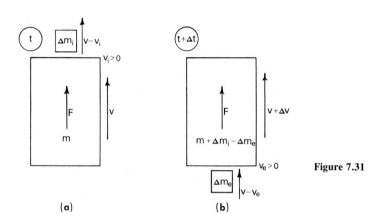

Figure 7.31

At time t, we assume that the container has mass m and speed v. The mass Δm_i has the speed $v - v_i$ (that is, $-v_i$ relative to the container). Thus the total momentum of the system at time t is

$$p(t) = mv + \Delta m_i(v - v_i). \tag{b}$$

At time $t + \Delta t$, Δm_i has entered the container and Δm_e has left it at speed $v - v_e$ (that is, $-v_e$ relative to the container). Assuming that the container has speed $v + \Delta v$ at $t + \Delta t$, we have the total momentum of the system at time $t + \Delta t$ as

$$p(t + \Delta t) = (m + \Delta m_i - \Delta m_e)(v + \Delta v) + \Delta m_e(v - v_e). \tag{c}$$

Writing (a) with (b) and (c), we obtain

$$F = m\frac{\Delta v}{\Delta t} + \left(\frac{\Delta m_i}{\Delta t}\right)v_i - \left(\frac{\Delta m_e}{\Delta t}\right)v_e + (\Delta m_i - \Delta m_e)\frac{\Delta v}{\Delta t}. \tag{d}$$

And if we take the limit in (d) as $\Delta t \to 0$, we obtain

$$F = m\frac{dv}{dt} + \left(\frac{dm_i}{dt}\right)v_i - \left(\frac{dm_e}{dt}\right)v_e, \tag{7.32}$$

where we take pains to repeat that

   F = the (external) force on the container

   m = the instantaneous mass of the container and its contents

   $\dfrac{dm_i}{dt}$ = the rate of mass accumulated by the container

   v_i = the speed, *relative to the container*, at which mass is accumulated by the container

   $\dfrac{dm_e}{dt}$ = the rate at which mass is disbursed from the container

   v_e = the speed, *relative to the container*, at which mass is disbursed from the container

We can formulate many essentially one-dimensional problems by means of Eq. (7.32). Many problems consist of actual containers which gain and lose mass. In other problems, we may have to construct a conceptual container to apply (7.32) properly. In any event, you are advised to keep the derivation of (7.32) in mind rather than merely trying to plug in the terms of the equation.

Example 20

In Fig. 7.32(a), a car filled with water is set in motion in the horizontal direction. A hole in the car allows water to flow out at the rate $c_0 m_0$ (slugs/sec). Calculate the speed of the car as a function of time.

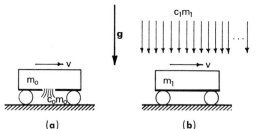

Figure 7.32

(a)   (b)

Solution

We consider the motion of the car in the horizontal direction:

$$F = 0, \qquad \frac{dm_i}{dt} = 0, \qquad \text{and} \qquad \frac{dm_e}{dt} = c_0 m_0.$$

The (absolute) exit speed of the water is the same as that of the car, v. Thus v_e, the horizontal exit speed relative to the car, is $v_e = v - v = 0$. Thus (7.32) becomes

$$0 = m_0(1 - c_0 t) \frac{dv}{dt}. \tag{a}$$

Thus $dv/dt = 0$ and $v = $ const. If the car starts with speed v_0, it will continue to have this speed.

Example 21

A car of mass m_1 (Fig. 7.32b) travels in the horizontal plane in a rainstorm. The constant rate of mass of rain water entering the car is $c_1 m_1$ (slugs/sec). The car starts empty with speed v_0; calculate its speed at time t. Determine $x(t)$.

Solution

We now have

$$F = 0, \qquad \frac{dm_i}{dt} = c_1 m_1, \qquad \text{and} \qquad \frac{dm_e}{dt} = 0.$$

The absolute horizontal speed of the water entering the car is zero. Thus $0 = v - v_i$ or $v_i = v$. Note that v_i and v_e are both positive quantities and should always be such in (7.32). Equation (7.32) becomes

$$0 = m_1(1 + c_1 t) \frac{dv}{dt} + c_1 m_1 v \qquad \text{or} \qquad \int_{v_0}^{v} \frac{dv}{v} = \int_0^t \frac{-c_1 \, dt}{1 + c_1 t}.$$

Integrating both sides, we obtain

$$\ln (v/v_0) = -\ln (1 + c_1 t).$$

And thus

$$v = \frac{v_0}{1 + c_1 t}.$$

Another integration yields

$$x = \frac{v_0}{c_1} \ln (1 + c_1 t).$$

Example 22

A rocket (Fig. 7.33) has an initial mass m_0 and the rate of mass expended is $c_0 m_0$. Determine $v(t)$, the vertical speed of the rocket.

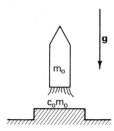

Figure 7.33

Solution

In this problem, we have

$$F = -mg = -m_0(1 - c_0 t)g,$$

$$\frac{dm_i}{dt} = 0, \quad \text{and} \quad \frac{dm_e}{dt} = c_0 m_0.$$

Suppose that the relative exit speed is taken to be v_e. Then (7.32) becomes

$$-m_0(1 - c_0 t)g = m_0(1 - c_0 t)\frac{dv}{dt} - c_0 m_0 v_e$$

or

$$\frac{dv}{dt} = \frac{c_0 v_e}{1 - c_0 t} - g. \tag{a}$$

Integrating (a), noting $v(0) = 0$, gives

$$v = -[v_e \ln (1 - c_0 t) + gt]. \tag{b}$$

There is some concern with (b). The value of $\ln (0) = -\infty$ and the logarithm of negative numbers is undefined. Thus we might worry about the fact that (b) will hold only up to $t = 1/c_0$. This is no problem, however, for t will not reach this value. Note that the mass

$$m = m_0(1 - c_0 t).$$

Thus, unless the rocket disappears, the time values for which (a) is valid are limited:

$$0 \leqslant t < (1/c_0).$$

....

Example 23

Figure 7.34a shows a uniform rope of length l and mass m which coils on a weightless platform. The rope descends to the platform with speed u. Determine the force

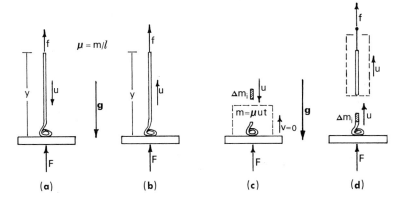

Figure 7.34

required to support the platform F and the force at the top of the rope f. Suppose that at $t = 0$, $y = l$.

Solution

First consider the entire string. The forces on it are

$$F + f - mg. \tag{a}$$

Suppose that the height of the string is y at time t; then the linear momentum is

$$p(t) = -(\mu y)u, \qquad \mu = \text{mass/length}, \tag{b}$$

and, since the height at $t + \Delta t$ is $y - u\,\Delta t$, then

$$p(t + \Delta t) = -\mu(y - u\,\Delta t)u. \tag{c}$$

Writing the equation of impulse–momentum for the entire string gives

$$(F + f - mg)\,\Delta t = \mu u^2\,\Delta t$$

or

$$F + f = mg + \mu u^2. \tag{d}$$

Now we look at Fig. 7.34(c), and consider the (conceptual) container outlined. At t the mass in the container is $m = \mu(ut)$. In time Δt the mass entering the container is $\Delta m_i = \mu u\,\Delta t$. The relative speed of entry is u. Thus (7.32) becomes

$$F - (\mu u t)g = (\mu u)u.$$

Thus

$$F = \mu(u^2 + ugt), \tag{e}$$

and from (d) and $m = \mu l$, we obtain

$$f = \mu g(l - ut). \tag{f}$$

....

Example 24

Now consider the reverse of the previous example. Figure 7.34(b) shows the rope coiled on the weightless platform. It is then lifted off the platform at speed u. Determine the forces F and f in this case.

Solution

Writing equations (a), (b), and (c) of the previous example, we now have

$$F + f - mg \tag{a'}$$

for the resultant force acting on the entire string. The momentum at times t and $t + \Delta t$ are

$$p(t) = (\mu y)u \tag{b'}$$

and

$$p(t + \Delta t) = \mu(y + u\,\Delta t)u. \tag{c'}$$

Thus

$$F + f = mg + \mu u^2. \tag{d'}$$

We note that (d) and (d') are the same. Does this mean that F and f have the same values as in (e) and (f)? The answer is no! The reason is that, as an element of mass $\Delta m_e = \mu u\,\Delta t$ leaves the conceptual container in Fig. 7.34(c), it has a force exerted on it by the rope above the element.

Rather than pursue this logic, refer to Fig. 7.34(d), and consider the mass of the string which has been lifted from the platform. We have for the force resultant

$$f - (\mu u t)g.$$

The mass entering this collection of moving string during a time Δt is $\Delta m_i = \mu u\,\Delta t$. The element enters the motion with relative speed u. Thus (7.32) becomes

$$f - (\mu u t)g = (\mu u)u$$

or

$$f = \mu(u^2 + ugt), \tag{f'}$$

and from (d') and $m = \mu l$, we obtain

$$F = \mu g(l - ut). \tag{e'}$$

The values of F and f are just the reverse of those in Example 23.

....

9. SUMMARY

A. Forces

Internal Forces

Forces between two particles of the system. These are equal in magnitude, opposite in direction, and lie along the line between the particles. The force on m_i due to m_j is denoted by $\mathbf{f}_{ij}$.

External Forces

Any force which is not an internal force is called an external force. The total external force on m_i is denoted by $\mathbf{F}_i$.

B. Motion of the center of mass, $\mathbf{F} = \dot{\mathbf{p}}_C$

The position of the center of mass, $\mathbf{r}_C$, is

$$\mathbf{r}_C = \frac{\sum\limits_{i=1}^{n} m_i \mathbf{r}_i}{\sum\limits_{i=1}^{n} m_i}.$$

Then

$$\mathbf{F} = m\ddot{\mathbf{r}}_C = \dot{\mathbf{p}}_C,$$

where $\mathbf{F}$ is the resultant external force $\sum_{i=1}^{n} \mathbf{F}_i$ and m is the total mass $\sum_{i=1}^{n} m_i$. If $\mathbf{F} = 0$, then $\mathbf{p}_C = $ constant.

N.B. Internal forces do not enter the equation for the motion of the center of mass of the system.

C. Motion about the center of mass, $\mathbf{M} = \dot{\mathbf{H}}$

$$\mathbf{M}_A = \dot{\mathbf{H}}_A$$

where A is either (i) a fixed point or (ii) the center of mass of the system.

$\mathbf{M}_A$ is the resultant moment of the external forces about the point A.

$\mathbf{H}_A$ is the sum of the moment of momentum vectors (about A) for all the particles of the system.

If $\mathbf{M}_A = 0$, then $\mathbf{H}_A = $ constant.

N.B. Internal forces do not enter the equation for the motion of the system about its center of mass.

D. Work–energy relation

The work–energy relation has two forms:

1) $\displaystyle \sum_{i=1}^{n} \int_{A_i}^{B_i} \mathbf{F}_i \cdot d\mathbf{r}_i + \sum_{i=1}^{n} \sum_{j=1}^{n} \int_{A_i}^{B_i} \mathbf{f}_{ij} \cdot d\mathbf{r}_i = T(B) - T(A)$

2) $\displaystyle \int_{A_C}^{B_C} \mathbf{F} \cdot d\mathbf{r}_C = \tfrac{1}{2} m v_C^2 \, \Big|_{A_C}^{B_C}$

$\displaystyle \sum_{i=1}^{n} \int_{A_i}^{B_i} \left(\mathbf{F}_i + \sum_{j=1}^{n} \mathbf{f}_{ij} \right) \cdot d\mathbf{r}_{Ci} = \sum_{i=1}^{n} \tfrac{1}{2} m_i \dot{r}_{Ci}^2 \, \Big|_{A}^{B}$

where $\mathbf{r}_{Ci} = (\mathbf{r}_i - \mathbf{r}_C)$ and the total kinetic energy $T = \tfrac{1}{2} m v_C^2 + \sum_{i=1}^{n} \tfrac{1}{2} m_i \dot{r}_{Ci}^2$, the energy of the center of mass plus the energy relative to the center of mass.

N.B. Internal forces on a system *can* do work. Thus these forces do, in general, enter into the work–energy relation.

E. Collision problems

Central Collision

When two bodies collide, a line normal to the colliding surfaces through the point of contact is defined. If the center of mass of each body lies on the normal line, the collision is called *central*.

Direct Central Collisions

If the *velocity* of the center of mass of each colliding body lies along the normal line, the collision is called *direct*.

$v_s = e u_a,$

$v_s = v_2 - v_1,$ speed of separation,

$u_a = u_1 - u_2,$ speed of approach.

$v_1 = \left(\dfrac{1}{m_1 + m_2} \right) [(m_1 - e m_2) u_1 + m_2 (1 + e) u_2],$

$v_2 = \left(\dfrac{1}{m_1 + m_2} \right) [m_1 (1 + e) u_1 + (m_2 - e m_1) u_2],$

$e = 0$ plastic collision, $e = 1$ elastic collision

Energy Dissipation

$\Delta T = (1 - e^2) T_R (0^-)$

T_R is the kinetic energy relative to the center of mass, or

$$\Delta T = \tfrac{1}{2}(1 - e^2)\, \frac{m_1 m_2}{m_1 + m_2}\, (u_1 - u_2)^2.$$

Oblique Central Collisions

If a central collision is not direct, it is called *oblique*. Let the x axis be the normal line. Given that

$$\mathbf{u}_1 = u_{1x}\mathbf{i} + u_{1y}\mathbf{j} + u_{1z}\mathbf{k},$$

$$\mathbf{u}_2 = u_{2x}\mathbf{i} + u_{2y}\mathbf{j} + u_{2z}\mathbf{k},$$

are the velocities before collision, and

$$\mathbf{v}_1 = v_{1x}\mathbf{i} + v_{1y}\mathbf{j} + v_{1z}\mathbf{k},$$

$$\mathbf{v}_2 = v_{2x}\mathbf{i} + v_{2y}\mathbf{j} + v_{2z}\mathbf{k},$$

are the velocities after collision, then

i) the speeds u_{1x}, u_{2x}, v_{1x}, v_{2x} follow the equations for direct central collision.

ii) If the colliding surfaces are smooth, then

$$u_{1y} = v_{1y}, \qquad u_{1z} = v_{1z},$$

$$u_{2y} = v_{2y}, \qquad u_{2z} = v_{2z}.$$

F. Systems which accumulate and disburse mass

$$F = m\,\frac{dv}{dt} + \left(\frac{dm_i}{dt}\right) v_i - \left(\frac{dm_e}{dt}\right) v_e$$

where

$F =$ the external force on the instantaneous collection of particles

$m =$ the instantaneous mass of the collection of particles

$v =$ the velocity of the collection of particles

$\dfrac{dm_i}{dt} =$ the rate at which mass is added to the collection of particles

$\dfrac{dm_e}{dt} =$ the rate at which mass leaves the collection of particles

$v_i =$ the speed relative to the container of the mass entering the container

$v_e =$ the speed relative to the container of the mass leaving the container

Note: $\dfrac{dm}{dt} = \dfrac{dm_i}{dt} - \dfrac{dm_e}{dt},$ $v_i \geqslant 0,$ $v_e \geqslant 0.$

PROBLEMS

Section 7.3

Problem 7.1

A system consists of three particles: m_1, m_2, and m_3:

	Weight, lb	Position, ft	Velocity, ft/sec
m_1	16	$\mathbf{r}_1 = 0$	$\mathbf{v}_1 = 10\mathbf{i} + 10\mathbf{j}$
m_2	32	$\mathbf{r}_2 = 2\mathbf{i} - 2\mathbf{j}$	$\mathbf{v}_2 = 3\mathbf{k}$
m_3	48	$\mathbf{r}_3 = 5\mathbf{k}$	$\mathbf{v}_3 = -2\mathbf{i} - 3\mathbf{j}$

a) Calculate the position of the center of mass of the system.

b) Calculate the velocity and the momentum of the center of mass, $\mathbf{v}_C$ and $\mathbf{p}_C$.

Problem 7.2

For the system in the figure, let $m_1 = m_2 = m_3 = m_4 = m$, where $mg = 32$ lb. At the instant shown, the velocities of the particles are

m_1: $\mathbf{v}_1 = 4\mathbf{i} - 2\mathbf{j}$ ft/sec, m_3: $\mathbf{v}_3 = 4\mathbf{i} + 2\mathbf{j}$ ft/sec,

m_2: $\mathbf{v}_2 = 6\mathbf{i}$ ft/sec, m_4: $\mathbf{v}_4 = 2\mathbf{i}$ ft/sec.

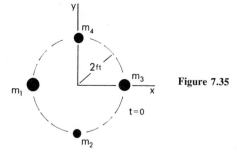

Figure 7.35

a) What is the position of the center of mass at the instant shown?

b) What is the velocity of the center of mass at the instant shown?

Problem 7.3

For the system of Problem 7.2, assume that there are no external forces acting. Calculate the position of the center of mass as a function of time. Let $t = 0$ be the instant described in Problem 7.2.

Problem 7.4

For the system of Problem 7.2, suppose that gravity is the only external force acting. Let **k** be the direction of the local vertical. Calculate the position of the center of mass as a function of time. Let $t = 0$ be the instant described in Problem 7.2.

Problem 7.5

Suppose that the velocities of the particles in Problem 7.2 are

$$\mathbf{v}_1 = 3\mathbf{i} - 2\mathbf{j} \text{ ft/sec}, \qquad \mathbf{v}_3 = -\mathbf{i} + 4\mathbf{j} \text{ ft/sec},$$
$$\mathbf{v}_2 = 4\mathbf{i} + 3\mathbf{j} \text{ ft/sec}, \qquad \mathbf{v}_4 = -2\mathbf{i} - \mathbf{j} \text{ ft/sec}.$$

a) Calculate the velocity of the center of mass at the instant of the above data.

b) Suppose that gravity acts on each particle and that $t = 0$ is the instant indicated in the figure; calculate the position of the center of mass of the system as a function of time if **k** is the direction of the local vertical.

Problem 7.6

Determine the position of (a) the center of mass, (b) the center of gravity of the dumbbell shown in the figure. The dumbbell is in the gravity field defined by Newton's law of universal gravitation:

$$\mathbf{F} = \frac{-\gamma m_0 m}{\rho^2} \mathbf{e}_\rho.$$

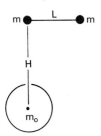

Figure 7.36

Problem 7.7

Determine the position of (a) the center of mass, (b) the center of gravity of the uniform, thin bar of length L and mass m in the position shown in Fig. 7.37. Use

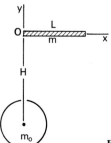

Figure 7.37

Newton's law of universal gravitation to define the forces on the bar:

$$dF = \frac{-\gamma m_0 dm}{\rho^2} \, e_\rho.$$

[*Hint:* Compute dF_{Ry}, F_{Ry}, $\mathbf{M}_0$, and then determine x_{cg} from

$$x_{cg} F_{Ry} = M_0,$$

where $\mathbf{F}_R$ is the force resultant of the forces on the bar, etc.]

Problem 7.8

At $t = 0^-$, the bomb shown in the figure has the velocity $v_0 \mathbf{i}$. At $t = 0$, the bomb separates into two parts. One part has velocity $10 v_0 \mathbf{j}$. What is the velocity of the other half?

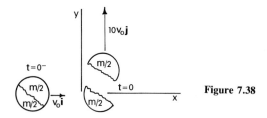

Figure 7.38

Problem 7.9

At $t = 0$, the mass M in Fig. 7.39 is at rest on smooth ice. The mass m is then moved relative to M by some internal means. Suppose that $u(t) = A \sin \omega t$ and at $t = 0$, $x = 0$. Compute $x(t)$.

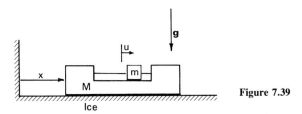

Figure 7.39

Problem 7.10

A bullet of mass m_0 is fired into a block of mass m which is initially at rest (see the figure).

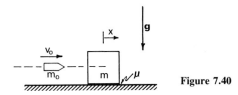

Figure 7.40

a) The coefficient of sliding friction between the block and the horizontal plane is μ, and the block comes to rest in a distance d. Calculate the initial speed v_0.

b) Suppose that oil has been spread on the horizontal plane so that the total friction force is $f = -c\dot{x}$. Determine the motion $x(t)$ of the block after collision, assuming that v_0 is given.

Problem 7.11

A particle of mass $\frac{1}{2}m$ falls from rest a distance h, and then sticks to a second particle of mass $\frac{1}{2}m$ which is supported by a spring k. (See the figure.) The instant of collision is $t = 0$.

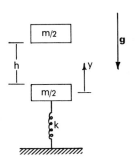

Figure 7.41

a) What is the new equilibrium position of the system?

b) Calculate the motion $y(t)$ of the combined particle after collision.

Section 7.4

For Problems 7.12 through 7.14, determine the moment of momentum **H** of the systems indicated about the point $\mathbf{r} = 0$.

Problem 7.12

The system of Problem 7.1.

Problem 7.13

The system of Problem 7.2.

Problem 7.14

The system of Problem 7.5.

In Problems 7.15 through 7.17, it is useful to recall the moment transformation law, (2.13):

$$\mathbf{M}_O = \mathbf{r}_{OA} \times \mathbf{F} + \mathbf{M}_A.$$

Problem 7.15

The system in the figure consists of three equal particles. For the position shown, determine:

a) the force resultant **F** acting on the system,

b) the moment resultant about (i) $\mathbf{r} = 0$, (ii) $\mathbf{r} = \mathbf{r}_{\text{cm}}$.

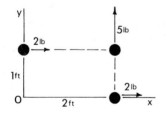

Figure 7.42

Problem 7.16

For the system of four particles of mass $2m$, $2m$, m, and m shown in the figure, find:

a) the force resultant **F** acting on the system,

b) the moment resultant about (i) $\mathbf{r} = 0$, (ii) $\mathbf{r} = \mathbf{r}_{\text{cm}}$.

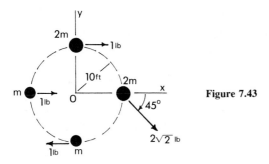

Figure 7.43

Problem 7.17

The free length of the spring in the figure is 10 in. For the system consisting of the two particles, find

a) the force resultant **F** acting on the system,

b) the moment resultant about the point (i) $\mathbf{r} = 0$, (ii) $\mathbf{r} = \mathbf{r}_{cm}$.

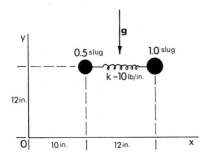

Figure 7.44

Problem 7.18

The system in the figure consists of two equal particles connected by a spring. There are no external forces acting on the system, and at the instant shown, the velocity of the center of mass is zero. Particle 1 has velocity components v_n and v_t, as shown.

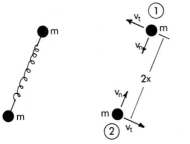

Figure 7.45

Prove that particle 2 will have the velocity components shown for it, and show that v_t is given by

$$v_t = \frac{H_C}{2mx},$$

where H_C is a constant and $2x$ is the distance between the particles.

Problem 7.19

The two particles in Fig. 7.46, which have masses m and $\frac{1}{2}m$, are connected by a rigid, weightless rod of length $2R$. This rod is at rest until a second particle of mass $\frac{1}{2}m$ collides with and sticks to the first mass, $\frac{1}{2}m$.

   a) What is the velocity of the center of mass after the collision?

   b) What is the angular rate of the rod after the collision?

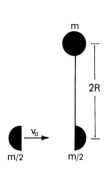

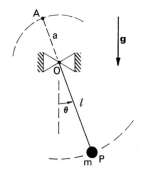

Figure 7.46   Figure 7.47

Problem 7.20

Suppose that A is a moving point whose velocity is always parallel to the velocity of the center of mass of a system. Show that $\mathbf{M}_A = \dot{\mathbf{H}}_A$.

Problem 7.21

Consider the pendulum shown in Fig. 7.47. Let A be a moving point a distance a from O on the line OP.

   a) Show that for this system $\mathbf{M}_A = \dot{\mathbf{H}}_A$.

   b) Determine the differential equation of motion for the pendulum, using the equation of part (a).

Section 7.5

In Problems 7.22 through 7.24, determine:

a) the kinetic energy of the center of mass

b) the kinetic energy of motion about the center of mass for the system indicated.

Problem 7.22

The system of Problem 7.1.

Problem 7.23

The system of Problem 7.2.

Problem 7.24

The system of Problem 7.5.

Problem 7.25

Suppose that the internal forces of a given system are conservative and are described by a potential energy function V_0. Show that the work–energy relation can be written

$$W_{AB} = \sum_{i=1}^{n} \int_{A_i}^{B_i} \mathbf{F}_i \cdot d\mathbf{r}_i + V_0(A) - V_0(B).$$

Problem 7.26

The internal forces acting on a given system are conservative and are described by the potential energy function V_0. The external forces are also conservative and are described by the potential energy function V_1. Show that

$$W_{AB} = [V_0(A) + V_1(A)] - [V_0(B) + V_1(B)].$$

Show, however, that the following equations are *not valid*:

$$V_1(A_C) - V_1(B_C) = \tfrac{1}{2} m v_C^2 \Big|_{A_C}^{B_C},$$

$$V_0(A) - V_0(B) = \sum_{i=1}^{n} \tfrac{1}{2} m_i \dot{r}_{Ci}^2 \Big|_{A_i}^{B_i}$$

Problem 7.27

For the bomb in Problem 7.8, determine the kinetic energy before and after the explosion.

Problem 7.28

For the system of Problem 7.19:

    a) Determine the total kinetic energy before and after collision.

    b) After collision, what is the kinetic energy of the center of mass and what is the kinetic energy of motion relative to the center of mass?

Problem 7.29

For the system of Problem 7.18, suppose that the free length of the spring is $2a$. Let E be the (constant) total energy, $T + V$, and let H_C be the constant moment of momentum. Show that

$$v_t = \frac{H_C}{2mx} \quad \text{and} \quad v_n^2 = \frac{1}{m}\left[E - 2k(x-a)^2 - \frac{H_C^2}{4mx^2}\right].$$

Problem 7.30

The two blocks of mass m_1 and m_2 are supported by a pulley which is both weightless and frictionless. Let the system be the two blocks shown in Fig. 7.48.

    a) What is the force resultant acting on the system?

    b) Write the differential equation, $\mathbf{F} = \dot{\mathbf{p}}_C$, for the system.

    c) Show that the internal forces do no work.

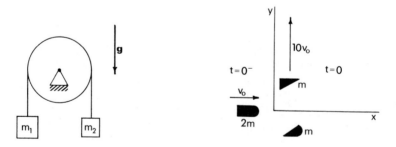

Figure 7.48   Figure 7.49

Problem 7.31

At $t = 0^-$, the bomb in Fig. 7.49 has velocity $v_0\mathbf{i}$. At $t = 0$, the bomb explodes into two equal parts. One part has velocity $10v_0\mathbf{j}$.

    a) What is the velocity of the other part at $t = 0$?

    b) What energy was released in the explosion?

Problem 7.32

A bomb initially at rest explodes into two parts, m_1 and m_2, with the velocities shown in the figure. Suppose that the internal energy E is dissipated in the explosion. Determine m_1 and m_2 in terms of E, v_1, and v_2.

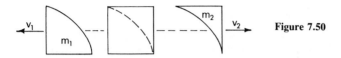

Figure 7.50

Problem 7.33

Consider the system $m_1 = m_2 = m$ shown in the figure. The bead m_1 slides on the smooth horizontal curve $y = a + b \sin x$. The bead m_2 slides on the smooth *horizontal* curve $y = -b \sin x$ in such a way that m_1 and m_2 always have the same displacement x. The free length of the spring k is a. When $x = 0$, $\dot{x} = v_0$.

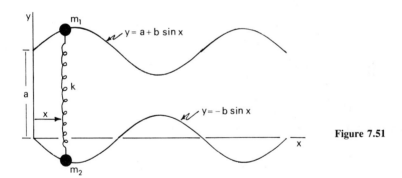

Figure 7.51

a) Show that the external forces acting on the system do no work.

b) Determine $\dot{x}$ as a function of x. [*Hint:* Use energy.] What forces are doing work in this problem?

c) Determine the minimum value of v_0 in order that the system will move over the entire curve.

d) Suppose that the y axis were vertical and gravity were considered. Would this alter your answers to parts (a), (b), and (c)?

Problem 7.34

The two particles shown in the figure are connected by a spring with free length of 3 ft. The particles move under the influence of gravity from configuration A to configuration B. Let $m = 1$ slug and $k = 50$ lb/ft.

a) What is the force resultant acting on the system?

b) Calculate the work of the external forces.

c) Calculate the work of the internal forces.

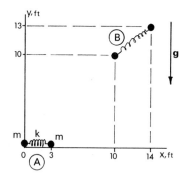

Figure 7.52

Section 7.6

Problem 7.35

Consider the *rigid system* shown in the figure. Let $\mathbf{r}_{AB}$ be the vector from A to B and let $\mathbf{v}_A$ and $\mathbf{v}_B$ be the velocities of the points A and B.

a) Show that $\mathbf{v}_A \cdot \mathbf{r}_{AB} = \mathbf{v}_B \cdot \mathbf{r}_{AB}$.

b) Show that $\mathbf{v}_A \times \mathbf{r}_{AB} \neq \mathbf{v}_B \times \mathbf{r}_{AB}$. (Give a counterexample.)

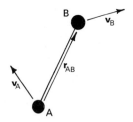

Figure 7.53

Problem 7.36

Suppose that a plane, rigid system moves in the plane so that point A has the velocity $\mathbf{v}_A$, and the system has the angular speed $\dot{\theta}$. (See Fig. 7.54.) Let $\mathbf{r}_{AB}$ be the vector from point A to point B.

a) Show that

$$\frac{d\mathbf{r}_{AB}}{dt} = \mathbf{v}_B - \mathbf{v}_A.$$

b) Noting that the length of $\mathbf{r}_{AB}$ is constant, show that $\dot{\mathbf{r}}_{AB}$ is perpendicular to $\mathbf{r}_{AB}$ and has magnitude $r_{AB}\dot{\theta}$. Thus $\mathbf{v}_B$ has the geometrical construction indicated in the figure.

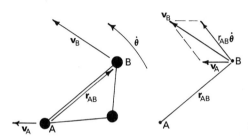

Figure 7.54

Problem 7.37

Consider the rigid system shown in Fig. 7.55. The velocities of particles (1) and (2) are

$$\mathbf{v}_1 = 10\mathbf{i}\ \text{ft/sec}, \qquad \mathbf{v}_2 = 20\mathbf{i}\ \text{ft/sec}.$$

a) Determine the angular speed $\dot{\theta}$ at the given instant.

b) Determine the velocity of particle (3). [*Hint:* Use Problem 7.36.]

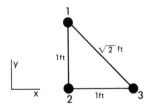

Figure 7.55   Figure 7.56

Problem 7.38

Consider small motions of the system shown in Fig. 7.56.

a) Determine the differential equation of motion of the system.

b) Determine the natural frequency of the system ω_n.

Problem 7.39

Consider small motions of the system shown in the figure.

a) Determine the differential equation of motion of the system.

b) Determine the natural frequency of the system ω_n. Compare the result to that obtained in Problem 7.38, part (b). Show that, as $a \to 0$, the two results agree.

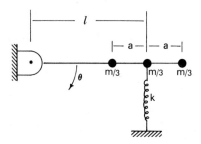

Figure 7.57

Problem 7.40

For the system shown in the figure, the string is cut at $t = 0$. Calculate the pin reaction:

a) for $t < 0$,

b) just after $t = 0$,

c) when the bar passes through the vertical position.

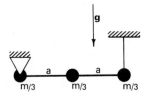

Figure 7.58

Problem 7.41

Consider the system in Fig. 7.59. Given that the bar moves straight up and down during the motion:

a) Calculate the differential equation of motion of the system;

b) Calculate the natural frequency of the system ω_n.

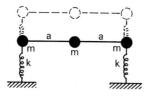

Figure 7.59

Problem 7.42

Consider the system shown in the figure. Its motion consists of a pure rotation about the center of mass.

a) Determine the differential equation of motion of the system.

b) Calculate the natural frequency of the system ω_n. Compare the result with that of Problem 7.41, part (b).

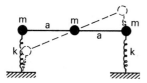

Figure 7.60

Problem 7.43

Suppose the dumbbell system in the figure consists of two equal particles of mass m. Suppose that (1) slides on the smooth vertical wall and (2) slides on the smooth horizontal surface. The system is released from rest when $\theta = 0$.

a) Calculate $\dot{\theta}$ when the bar has the angle θ.

b) Calculate the speed of (1) when it hits the horizontal surface.

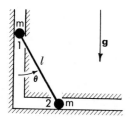

Figure 7.61

Problem 7.44

For the system shown in Fig. 7.62, suppose that the angular speed $\dot{\theta}$ is given.

    a) Calculate $\mathbf{H}_O$ for the system.

    b) Calculate the kinetic energy T for the system.

    c) Write $\mathbf{M}_O = \dot{\mathbf{H}}_O$ to obtain the differential equation of motion for the system.

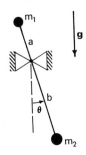

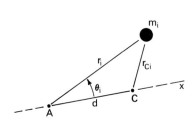

    Figure 7.62   **Figure 7.63**

Problem 7.45

Suppose that one point O of a plane rigid system is fixed in space. Let the distance of m_i from O be r_i. The system rotates in the plane with angular speed $\dot{\theta} = \omega$. Show that

    a) $\mathbf{H}_O = I^0 \omega \mathbf{k}$

    b) $T = \frac{1}{2} I^0 \omega^2$

where $I^0 = \sum\limits_{i=1}^{n} m_i r_i^2$ and $\mathbf{k}$ is the unit vector perpendicular to the plane of motion.

Problem 7.46

Suppose that a plane, rigid system moves in its plane so that v_C is the speed of its center of mass and ω is its angular speed. Show that the kinetic energy is

$$T = \frac{1}{2} m v_C^2 + \frac{1}{2} I^C \omega^2,$$

where $I^C = \Sigma m_i r_{Ci}^2$ and r_{Ci} is the distance of m_i from the center of mass. The term m is the total mass.

Problem 7.47

Consider a plane, rigid system of particles, $m_1, m_2, \ldots, m_n$ with center of mass, C. Let A be another point. Let r_i and r_{Ci} be the distances of m_i from A and C, respectively, as shown in Fig. 7.63. Given that the *moments of inertia* about A and C are defined as

$$I^C = \sum_{i=1}^{n} (m_i r_{Ci}^2) \qquad \text{and} \qquad I^A = \sum_{i=1}^{n} (m_i r_i^2),$$

show that $I^A = I^C + md^2$, where m is the total mass and d is the distance from C to A.

Section 7.7

For Problems 7.48 through 7.50, two particles slide along the smooth surface and then collide. The particles are shown before and after collision. In each case,

a) determine the missing speed,

b) take $m = 10$ slugs and evaluate the change in energy during the collision.

Problem 7.48

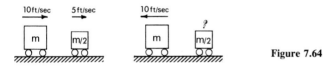

Figure 7.64

Problem 7.49

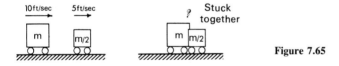

Figure 7.65

Problem 7.50

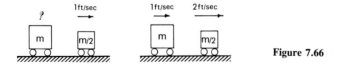

Figure 7.66

Problem 7.51

Consider the instantaneous collision between two particles (Fig. 7.23). Show that the moment of momentum about a fixed point O is constant during the collision $\mathbf{H}_O = \text{const}$.

Problem 7.52

Suppose two equal particles collide as shown. Given that one-half of the initial energy is lost during the collision, find the speeds of the particles after collision.

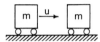

Figure 7.67

Problem 7.53

Consider the system of Problem 7.52. Suppose that one-fourth of the initial energy is lost during the collision. Calculate the speeds of the particles after the collision.

Problem 7.54

For the system shown in the figure, the two pendulums are released from rest in the position indicated. Suppose that one-third of the energy before each collision is lost during the collision. Assuming small amplitudes

$$\cos \theta_0 \approx \left(1 - \frac{\theta_0^2}{2}\right),$$

a) calculate the amplitude of each pendulum after the first collision,

b) calculate the amplitude of each pendulum after the second collision and after the nth collision.

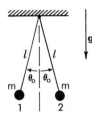

Figure 7.68

Problem 7.55

In the system shown in Fig. 7.69, the two equal particles are released from rest at angles θ_1 and θ_2. The path is smooth and the particles stick together after collision. Determine the amplitude of the motion which occurs after the collision. Assume that the angles are small enough so that $\cos \theta_1 \approx 1 - \frac{1}{2}\theta_1^2$ and $\cos \theta_2 \approx 1 - \frac{1}{2}\theta_2^2$.

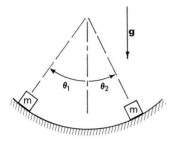

Figure 7.69

Problem 7.56

Consider the system of Problem 7.55, and assume that there is no energy loss during the collision. Calculate the maximum amplitude of each particle after the first collision.

Problem 7.57

For the system shown in the figure, the free lengths of the springs are such that the blocks just touch in the equilibrium position. Initially, each cart has speed u_0, as shown. Suppose that one-tenth of the energy prior to each collision is lost during the collision. Determine the maximum spring deflections:

a) on the first compression,

b) after the first collision,

c) after the second collision and after the nth collision.

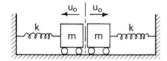

Figure 7.70

Problem 7.58

Consider the system in the figure. Suppose that $m_1 = m_2$, and that the collision is elastic. Show that the final speeds are $v_1 = u_2$, $v_2 = u_1$.

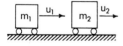

Figure 7.71

Problem 7.59

Consider the system of Problem 7.58. Suppose that $e = \frac{1}{2}$, $u_1 = u_0$, and $u_2 = 0$. Given that $m_1 = m$, determine m_2 such that $v_1 = 0$. Calculate v_2.

Problem 7.60

What is the coefficient of restitution in Problem 7.48?

Problem 7.61

What is the coefficient of restitution in Problem 7.49?

Problem 7.62

What is the coefficient of restitution in Problem 7.50?

Problem 7.63

What is the coefficient of restitution in Problem 7.52?

Problem 7.64

What is the coefficient of restitution in Problem 7.53?

Problem 7.65

What is the coefficient of restitution in Problem 7.54?

Problem 7.66

What is the coefficient of restitution in Problem 7.57?

★ *Problem 7.67*

To appreciate the concept of the coefficient of restitution, consider the motion of the system shown in the figure. A mass m is attached to a weightless platform by means of a spring and a dashpot. The system collides with a wall with speed v_0. When the platform leaves contact with the wall, the speed of m is v_1. Assume that the system is underdamped, $\zeta < 1$.

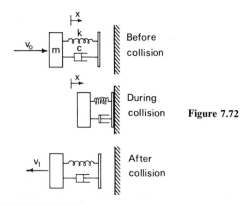

Figure 7.72

★ Starred problems are those which are more difficult than average.

a) Solve for the position of the mass, $x(t)$, after collision.

b) Determine the force between the platform and the wall as a function of time. [*Hint:* Write the force

$$f = kx + c\dot{x} = m\omega_n^2 x + 2m\zeta\omega_n\dot{x}.]$$

c) Determine v_1 in terms of v_0, given that $\zeta = 0.5$.

d) What is the coefficient of restitution for the system in part (c)? [*Hint:* Recall Example 7.16.]

Problem 7.68

Both pendulum systems shown in the figure are released from rest in the configuration shown. In system (i), the coefficient of restitution is $e = 0.01$; in system (ii), $e = 0.99$. Neglect air drag, and *do not* assume small angles.

a) For each system, determine the amplitude of each mass as $t \to \infty$ as a function of θ_0.

b) Calculate the difference between the initial $(t = 0)$ and final $(t \to \infty)$ total energy, $T + V$, for each system.

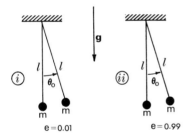

Figure 7.73

Problem 7.69

Consider the system of Problem 7.54. Given that the pendulums collide indefinitely, calculate the total distance traveled by each pendulum.

Problem 7.70

Consider the system of Problem 7.57. Assuming that the two carts continue to collide indefinitely, calculate the total distance traveled by each cart.

Problem 7.71

When the tennis ball in Fig. 7.74 is dropped from a height h, it rebounds to a height $h/3$. The ball is then thrown at a speed $v_0 = 21.1$ ft/sec at an angle of $60°$ at a smooth vertical wall 6 ft away. Determine the distance from the wall at which the ball will land.

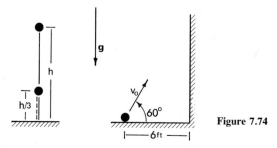

Figure 7.74

Problem 7.72

Derive Eq. (7.30).

Problem 7.73

Disk (1) collides with disk (2), initially at rest, as shown in the figure. Given that $m_1 = 1$ slug, $m_2 = 2$ slugs, and $e = 0.75$, determine the velocities of the two disks after the collision.

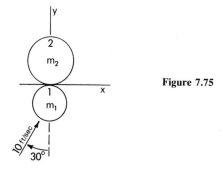

Figure 7.75

Problem 7.74

Two circular disks, each of radius R, collide as shown in the figure. Calculate the velocities of the disks after collision, given that $e = 1$.

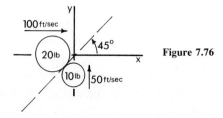

Figure 7.76

Problem 7.75

Consider the system of Problem 7.74. Repeat the problem with $e = \frac{1}{2}$.

Section 7.8

Problem 7.76

Consider the system of Example 21. Suppose that $c_1 m_1 = 0.1$ slug/sec and $m_1 = 2$ slugs.

a) Given that the initial speed is v_0, compute the time required to have the speeds: (i) $\frac{1}{2} v_0$, (ii) $\frac{1}{4} v_0$.

b) Determine the position x when the speed is (i) $\frac{1}{2} v_0$, (ii) $\frac{1}{4} v_0$.

Problem 7.77

In Example 22, suppose that $m_0 = 3$ slugs and $v_e = 1000$ ft/sec.

a) Determine the minimum value of c_0 in order that there will be lift-off at $t = 0$.

b) Suppose that c_0 is larger than the value determined in part (a). Given that $y(0) = 0$, calculate $y(t)$.

Problem 7.78

The chain shown in the figure has a density of μ lb/ft. It is coiled on the table top with a small amount of overhang.

a) State the differential equation which will define the motion of the end of the chain.

b) Is the total energy, $T + V$, constant in this problem?

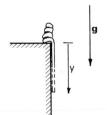

Figure 7.77

Problem 7.79

A chain of density μ lb/ft is stretched out on a smooth table top, as shown in Fig. 7.78. The length of the chain is l.

a) State the differential equation which will define the motion of the end of the chain. Compare the result with that of Problem 7.78, part (a).

b) Given that $y(0) = y_0$, and the chain is released from rest, calculate $y(t)$.

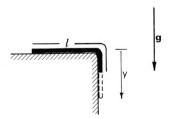

Figure 7.78

Problem 7.80

Consider the system of Problem 7.79. Suppose now that the table top is no longer smooth. Let μ_s be the coefficient of both static and sliding friction, and let the initial overhang be y_0. The system is released from rest at this position.

a) For what values of y_0 will the chain remain at rest?

b) Suppose that the chain slides when released. Determine the work done by the friction forces when the end moves from y_0 to y.

c) Determine the speed $\dot{y}$ when the chain end has position y.

Problem 7.81

The railroad car shown in the figure has an unloaded mass m_0 and length l, and rolls without friction. Above the car is a hopper from which sand pours at the rate c slugs/sec. The initial speed of the car is v_0.

a) Determine the speed of the car after it has cleared the hopper.

b) What is the total mass of sand in the car after it has cleared the hopper? Will the mass be evenly distributed in the car?

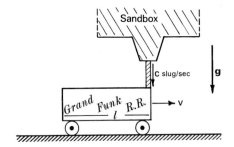

Figure 7.79

Problem 7.82

Consider again the system of Problem 7.81. A horizontal force F is applied to the car, starting when it first comes beneath the hopper. The idea is to have the speed of the car remain constant, v_0, so that the sand will be evenly distributed.

a) Calculate the force F.

b) What is the total mass of sand loaded in the car?

c) What is the work done by F?

d) How much energy is dissipated by the horizontal collisions of the sand with the car?

8

RIGID BODIES IN PLANE MOTION

1. INTRODUCTION

Consider the body in Fig. 8.1. Given that A and B are *any* two points of the body and the distance d_{AB} does not change, the body is called *rigid*.

In this chapter, we shall deal with various aspects of rigid-body motion (kinematics) and rigid-body dynamics (kinetics). We shall restrict ourselves to the important case of plane motion.

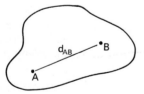

Figure 8.1

The kinematics and kinetics of rigid bodies in plane motion are straightforward when compared with the case of general, three-dimensional motion. Fortunately many important problems can be dealt with in the context of plane motion.

2. KINEMATICAL PRELIMINARIES: RELATIVE MOTION

To begin our study of the motion of rigid bodies, we introduce the concept of relative motion, which will enable us to decompose velocities and accelerations into a number of simple terms.

Consider the point P in Fig. 8.2. The vector from O' to A is denoted by $\mathbf{r}_{A/O'}$; the vector from O' to P by $\mathbf{r}_{P/O'}$, and the vector from A to P by $\mathbf{r}_{P/A}$. We give them the names:

$\mathbf{r}_{A/O'}$ the position of A with respect to O'

$\mathbf{r}_{P/O'}$ the position of P with respect to O'

$\mathbf{r}_{P/A}$ the position of P with respect to A

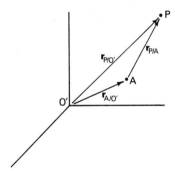

Figure 8.2

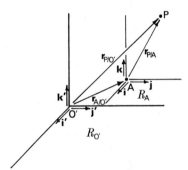

Figure 8.3

It is clear from Fig. 8.2 that these vectors are related, in that

$$\mathbf{r}_{P/O'} = \mathbf{r}_{A/O'} + \mathbf{r}_{P/A}. \tag{8.1}$$

Because (8.1) gives the decomposition of the position vector, we can ask for the decomposition of the velocity and acceleration of P. This requires a certain amount of preliminary thinking.

We label the reference frame which passes through O' by the symbol $R_{O'}$. Another reference frame R_A passes through the point A. Let the axes of R_A be parallel to the axes of $R_{O'}$. (See Fig. 8.3.)

Let us differentiate (8.1) with respect to time. We get

$$\frac{d}{dt_{O'}} \mathbf{r}_{P/O'} = \frac{d}{dt_{O'}} \mathbf{r}_{A/O'} + \frac{d}{dt_{O'}} \mathbf{r}_{P/A}. \tag{8.2}$$

The notation $d/dt_{O'}$ requires explanation. When we differentiate an equation, we must differentiate each side with respect to the same parameter; in this case, time.

Now since R_A is in translation with respect to $R_{O'}$, we can relate the unit vectors in the two systems:

$$\mathbf{i'} = \mathbf{i}, \qquad \mathbf{j'} = \mathbf{j}, \qquad \mathbf{k'} = \mathbf{k}.$$

And this means that the derivative of a vector with respect to $R_{O'}$ is the same as the derivative of the vector with respect to R_A. Thus we write (8.2) as

$$\mathbf{v}_{P/O'} = \mathbf{v}_{A/O'} + \mathbf{v}_{P/A}, \tag{8.3}$$

where we now realize that the term $\mathbf{v}_{P/A}$ can be computed by considering the change of $\mathbf{r}_{P/A}$ with respect to a reference frame passing through A and always parallel to $R_{O'}$.

Differentiating (8.3), we obtain the equation for the acceleration of P:

$$\mathbf{a}_{P/O'} = \mathbf{a}_{A/O'} + \mathbf{a}_{P/A}. \tag{8.4}$$

In most cases of concern in this book, the point O' will be the origin of an inertial reference frame. Then $\mathbf{v}_{P/O'}$, $\mathbf{a}_{P/O'}$ and $\mathbf{v}_{A/O'}$, $\mathbf{a}_{A/O'}$ are the absolute velocity and acceleration of the points P and A. Thus in the present context we shall write (8.1), (8.3), and (8.4):

$$\mathbf{r}_P = \mathbf{r}_A + \mathbf{r}_{P/A}, \tag{a}$$

$$\mathbf{v}_P = \mathbf{v}_A + \mathbf{v}_{P/A}, \tag{b} \tag{8.5}$$

$$\mathbf{a}_P = \mathbf{a}_A + \mathbf{a}_{P/A}. \tag{c}$$

Example 1

The pendulum in Fig. 8.4(a) is mounted on a cart which has a constant horizontal acceleration. The cart is at rest at $t = 0$. At time t, the inclination of the pendulum is θ, and the angular motion is characterized by $\dot{\theta}$ and $\ddot{\theta}$.

Calculate the velocity and acceleration of P at time t. Write the expressions in terms of components along the pendulum and perpendicular to the pendulum.

Solution

In Fig. 8.4(b) we introduce the two coordinate frames R_0 and $R_{0'}$. The velocity and acceleration of the point O are

$$\mathbf{v}_O = a_0 t\, \mathbf{i'}, \qquad \mathbf{a}_O = a_0\, \mathbf{i'}$$

To compute $\mathbf{v}_{P/O}$ and $\mathbf{a}_{P/O}$, we note that these are the velocity and acceleration of P with respect to R_0, the translating frame. But these are the velocity and acceleration

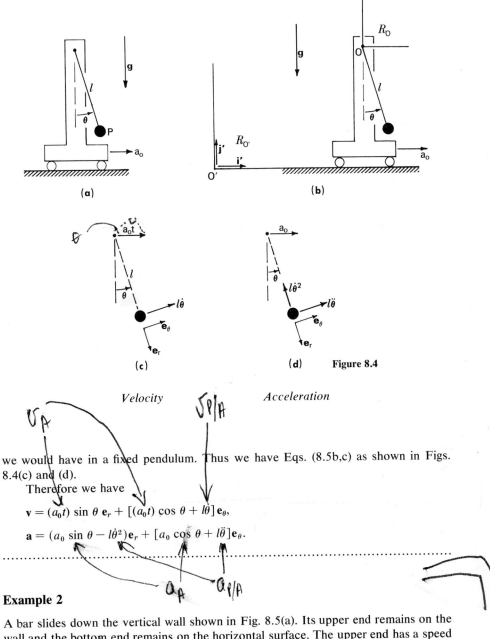

(a)

(b)

(c)

(d)   **Figure 8.4**

Velocity   *Acceleration*

we would have in a fixed pendulum. Thus we have Eqs. (8.5b,c) as shown in Figs. 8.4(c) and (d).

Therefore we have

$$\mathbf{v} = (a_0 t) \sin \theta \, \mathbf{e}_r + [(a_0 t) \cos \theta + l\dot{\theta}] \, \mathbf{e}_\theta,$$

$$\mathbf{a} = (a_0 \sin \theta - l\dot{\theta}^2) \mathbf{e}_r + [a_0 \cos \theta + l\ddot{\theta}] \mathbf{e}_\theta.$$

Example 2

A bar slides down the vertical wall shown in Fig. 8.5(a). Its upper end remains on the wall and the bottom end remains on the horizontal surface. The upper end has a speed of $v_A = 5$ ft/sec and the bar is $l = 10$ ft long. What is the angular rate $\dot{\theta}$ of the bar, given that the inclination to the vertical is $\theta = 30°$?

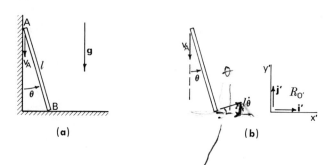

(a)   (b)   Figure 8.5

Solution

We write the velocity of B as

$$\mathbf{v}_B = \mathbf{v}_A + \mathbf{v}_{B/A}. \tag{a}$$

We can evaluate (a) graphically as in Fig. 8.5(b). Thus

$$\mathbf{v}_B = (l\dot{\theta} \cos \theta)\mathbf{i}' + (l\dot{\theta} \sin \theta - v_A)\mathbf{j}'.$$

But if B is to remain on the horizontal surface, $v_{By} = 0$. Thus

$$l\dot{\theta} \sin \theta - v_A = 0$$

or

$$\dot{\theta} = \frac{v_A}{l \sin \theta}. \tag{b}$$

Or, with the numerical values of the problem,

$$\dot{\theta} = \frac{5 \text{ ft/sec}}{10 \text{ ft } (\frac{1}{2})} = 1.0 \text{ rad/sec.}$$

....

Thus far we referred motion to two reference frames $R_{O'}$ and R_A, where the two frames were always parallel to each other. Let us now consider the plane motion of a rigid body (Fig. 8.6a).

Writing (8.5b),

$$\mathbf{v}_P = \mathbf{v}_A + \mathbf{v}_{P/A},$$

we note that if A is *a point of the body,** the motion of P about A is circular. Given that the body rotates with angular rate $\dot{\theta}$, the relative velocity $\mathbf{v}_{P/A}$ is as shown in Fig. 8.6(b).

---

* The point A need not be literally a "point of the body." If the x,y plane is imbedded in the body, A can be any fixed point of that plane.

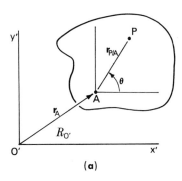

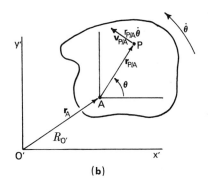

(a)

(b)

Figure 8.6

Note that $\mathbf{v}_{P/A}$ is perpendicular to $\mathbf{r}_{P/A}$ and has magnitude $(r_{P/A}\dot{\theta})$. In order to simplify our future equations, we introduce the notion of the *angular velocity* of a rigid body:

$$\boldsymbol{\omega} = \dot{\theta}\mathbf{k}', \tag{8.6}$$

where $\dot{\theta}$ is the angular rate of the rigid body relative to the frame R_0 and $\mathbf{k}'$ is the unit vector perpendicular to the plane of motion. The quantity $\dot{\theta}$ is taken to be positive if it is counterclockwise.

Since $\boldsymbol{\omega}$ is perpendicular to $\mathbf{r}_{P/A}$, we note that

$$|\boldsymbol{\omega} \times \mathbf{r}_{P/A}| = |\boldsymbol{\omega}|\,|\mathbf{r}_{P/A}|\sin(90°) = \dot{\theta}\,r_{P/A}.$$

This is the magnitude of $\mathbf{v}_{P/A}$ (see Fig. 8.6b). But $\boldsymbol{\omega} \times \mathbf{r}_{P/A}$ has the same direction as $\mathbf{v}_{P/A}$. Thus*

$$\mathbf{v}_{P/A} = \boldsymbol{\omega} \times \mathbf{r}_{P/A}, \tag{8.7}$$

and now (8.5b) reads

$$\mathbf{v}_P = \mathbf{v}_A + \boldsymbol{\omega} \times \mathbf{r}_{P/A}. \tag{8.8}$$

Suppose that we write (8.8) employing a reference frame R_1 which *moves with the rigid body.* (See Figs. 8.7a, b.)

Figures 8.7(a) and (b) represent two views of the same situation. The reference frame R_1 (that is, x, y, z) is embedded in the rigid body. We say that the body is in plane motion if the x, y plane always remains in the x', y' plane.

---

* We have established (8.7) for the very special case of plane motion. In fact, (8.7) can be shown to be valid for the general motion of a rigid body. In that case, however, the definition of $\boldsymbol{\omega}$ is more difficult than the definition given here.

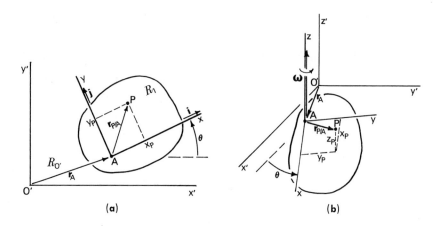

(a)   **(b)**

Figure 8.7

The advantage of the description given by Fig. 8.7 is that the vector $\mathbf{r}_{P/A}$ can be written

$$\mathbf{r}_{P/A} = x_P\mathbf{i} + y_P\mathbf{j} + z_P\mathbf{k}, \tag{8.9}$$

where x_P, y_P, and z_P are constants for a given point P.

If we are interested in determining the velocity of point P, it is probably advisable to apply (8.5b) or (8.8) directly. However, if we wish the velocity for the purpose of writing $\mathbf{M} = \dot{\mathbf{H}}$ or the kinetic energy, it is useful to employ (8.9). From (8.9) and (8.8) we get

$$\mathbf{v}_P = \mathbf{v}_A + \begin{vmatrix} \mathbf{i} & \mathbf{j} & \mathbf{k} \\ 0 & 0 & \dot{\theta} \\ x_P & y_P & z_P \end{vmatrix}$$

or

$$\mathbf{v}_P = \mathbf{v}_A + \dot{\theta}(-y_P\mathbf{i} + x_P\mathbf{j}). \tag{8.10}$$

3. THE INSTANT CENTER

Suppose we could find a point IC of the rigid body whose velocity $\mathbf{v}_{IC} = 0$. Then we could write (8.8) as

$$\mathbf{v}_P = \boldsymbol{\omega} \times \mathbf{r}_{P/IC}. \tag{8.11}$$

In other words, the velocity field of the body instantaneously appears to be that of pure circular motion about point IC. We shall elaborate on this statement below. First let us prove that, at each instant, a point IC, called the *instant center,* does exist.

Suppose we write $\mathbf{v}_A$ in terms of its components in the moving x,y,z coordinate system. Then, by (8.10),

$$\mathbf{v}_P = (v_{Ax} - \dot{\theta}y_P)\mathbf{i} + (v_{Ay} + \dot{\theta}x_P)\mathbf{j}.$$

But $\mathbf{v}_{IC} = 0$. Thus

$$v_{Ax} - \dot{\theta}y_{IC} = 0, \qquad v_{Ay} + \dot{\theta}x_{IC} = 0,$$

and if $\dot{\theta} \neq 0$, we can solve for

$$x_{IC} = -\frac{v_{Ay}}{\dot{\theta}}, \qquad y_{IC} = +\frac{v_{Ax}}{\dot{\theta}}. \tag{8.12}$$

Equation (8.12) gives the location of the instant center IC in terms of its position relative to the moving x,y plane.*

In many instances, there is a more direct method for determining the position of the instant center than the application of equations (8.12). Recall (8.11):

$$\mathbf{v}_P = \boldsymbol{\omega} \times \mathbf{r}_{P/IC}.$$

This equation states that the velocity of any point P of the body

i)   is perpendicular to $\mathbf{r}_{P/IC}$

ii) has a magnitude directly proportional to the distance from IC to P.

We now use the above properties to determine the position of the instant center. Suppose that at some instant we know the *directions* of the velocities of two points of the body (see Fig. 8.8a). We know that the point IC lies along a line perpendicular to each velocity vector. We construct the lines for both points, and if these lines intersect in a unique point, that point must be the instant center IC.

It can happen that the two lines just constructed are the same line l (see Fig. 8.8b). In this case, there is not a unique intersection of the lines. However, if we happen to know the magnitude and direction of the velocities of the two points, we can determine the position of IC.

We know that the point IC lies along the line perpendicular to the velocities. We also know that the magnitude of the velocity of any point is proportional to its distance from the point IC. This means that if we take a line through IC, the velocities of the various points on the line will appear as

---

* In fact, z_{IC} has not been determined by (8.12). This means that all points on a vertical line through x_{IC}, y_{IC} have $\mathbf{v} = 0$ at the given instant.

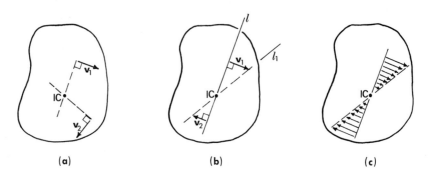

(a)   (b)   (c)

Figure 8.8

in Fig. 8.8(c). Thus the velocities along a line through *IC* have a linear dis-tribution.

Returning to Fig. 8.8(b), we now draw a line l_1 through the endpoints of the velocities $\mathbf{v}_1$ and $\mathbf{v}_2$. The line l_1 will intersect the line *l* in the instant center.

We should take some pains to point out that, at a given instant, the in-stant center need not be a point of the actual rigid body. It is a point of the *x,y,z* system embedded in the body. Examples 4 and 5 below illustrate this point.

It is possible that there is not an instant center for the motion. By (8.12), this occurs when $\dot{\theta} = 0$. Since $\boldsymbol{\omega} = \dot{\theta}\mathbf{k}$, we would then have $\boldsymbol{\omega} = 0$. Thus, by (8.8), we would have

$$\mathbf{v}_P = \mathbf{v}_A \quad \text{if} \quad \dot{\theta} = 0. \tag{8.13}$$

That is, every point of the rigid body has the same velocity instantaneously. This is called a *pure translation* (i.e. no rotation).

> At every instant, either a body in plane motion is in a pure translation or an instant center for the velocity field exists. The velocity of any point appears as the result of a pure rotation about the instant center.

Example 3

Let us return to Example 9 of Chapter 3, which concerned the motion of a cylinder restricted to roll without slip (see Fig. 8.9). We saw that the point of the body which had, instantaneously, zero velocity was the point of contact of the cylinder and the

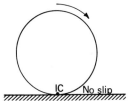

Figure 8.9

plane. The point of contact is thus the instant center. We also noted that the acceleration of the point of contact is not zero.

....

Example 4

The cylinder in Fig. 8.10(a) rolls without slip on a cart which has a speed v_0. Given
that $v_0 = 10$ ft/sec, $R = 1$ ft, and $\dot{\theta} = 60$ rpm, determine the position of the instant
center and the velocity of the center of the cylinder.

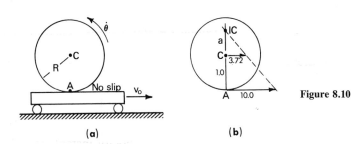

Figure 8.10

(a)   (b)

Solution

If the cart were not moving, the point of contact A in Fig. 8.10(a) would be the point
IC. Now this point has speed $v_A = v_0$, since the point of contact shares the velocity of
the cart. To get the velocity of the center C, we write

$$\mathbf{v}_C = \mathbf{v}_A + \boldsymbol{\omega} \times \mathbf{r}_{C/A}$$

or

$$\mathbf{v}_C = 10\mathbf{i} - \left(\frac{60 \cdot 2\pi}{60}\right) 1\mathbf{i} = 3.72 \text{ ft/sec.}$$

Thus we can construct the velocities in Fig. 8.10(b). We have, by similar triangles,

$$\frac{1+a}{10} = \frac{a}{3.72} \qquad \text{or} \qquad 10a = 3.72 + 3.72a$$

from which we obtain

$$a = 0.592 \text{ ft.}$$

Thus the instant center is 1.59 ft above the point of contact.

....

Example 5

We return to consideration of the problem in Example 2 of this chapter. The point A has the speed v_A when the inclination of the bar is θ. (See Fig. 8.11.) Find the position of the point IC and find v_B assuming that v_A, l, and θ are given.

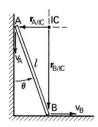

Figure 8.11

Solution

We assume that the ends of the bar remain on the vertical wall and horizontal surface. This means that the velocities of A and B must have the directions shown. Constructing lines perpendicular to these directions determines the point IC.

The speed of A can be written using the instant center:

$$v_A = r_{A/IC}\,\dot\theta = (l \sin \theta)\dot\theta. \tag{a}$$

But

$$v_B = r_{B/IC}\dot\theta = (l \cos \theta)\dot\theta. \tag{b}$$

Therefore from (a) and (b), we get

$$\dot\theta = \frac{v_A}{l \sin \theta} \qquad \text{and thus} \qquad v_B = v_A \cot \theta.$$

....

Example 6

The system in Fig. 8.12 consists of two rigid bodies: a cylinder of radius R, which rolls without slip, and a bar which is pinned at the edge of the cylinder and is restricted to slide over a projection a distance l from the point of contact of the cylinder. The top of the cylinder is known to have a velocity v_0, and we are asked to determine the values of $\dot\theta_1$ and $\dot\theta_2$ for the cylinder and bar when the bar is horizontal.

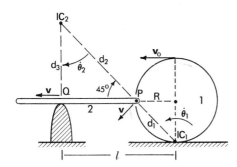

Figure 8.12

Solution

The instant center IC_1 for the cylinder is the point of contact. Thus

$$v_0 = (2R)\dot\theta_1 \quad \text{or} \quad \dot\theta_1 = \tfrac{1}{2}(v_0/R). \tag{a}$$

We turn now to the bar: Consider first the pin joint. The pin is a point of both the cylinder and the bar. Using the instant center for the cylinder, IC_1, we know that the velocity of the pin is perpendicular to the line IC_1P. In turn, the instant center for the bar, IC_2, must lie along the line perpendicular to the velocity of P, that is, the line IC_1P.

Now consider the point of the bar Q that is in contact with the projection. Suppose we write the velocity of this point in two components, one along the bar and one perpendicular to it. Since the bar is restricted to slide on the projection, this latter component must be zero. Thus the velocity of Q has the direction shown in Fig. 8.12. And IC_2 must then lie along a line through Q perpendicular to the bar. The intersection of the two lines just described gives IC_2.

The distance from IC_2 to P is easily computed to be

$$d_2 = \sqrt{2}(l - R).$$

Thus the velocity of P has the magnitude

$$v_P = d_2\dot\theta_2 = \sqrt{2}(l - R)\dot\theta_2. \tag{b}$$

The distance from IC_1 to P is $d_1 = \sqrt{2}R$, and thus, by (a), we can compute

$$v_P = d_1\dot{\theta}_1 = \sqrt{2}R\dot{\theta}_1 = \sqrt{2}(\tfrac{1}{2}v_0). \tag{c}$$

From (b) and (c), we then obtain

$$\dot{\theta}_2 = \frac{v_0}{2(l - R)}. \tag{d}$$

For completeness, we now consider the acceleration of an arbitrary point P of a rigid body. One might well wish to have this information in order to write $\mathbf{F} = \dot{\mathbf{p}}_C$. (The other two fundamental equations, $\mathbf{M} = \dot{\mathbf{H}}$ and the work–energy equation, can be written in terms of velocities alone.)

4. ACCELERATION OF A POINT OF A RIGID BODY

From (8.5b) and (8.8), we had

$$\mathbf{v}_P = \mathbf{v}_A + \mathbf{v}_{P/A}, \tag{a}$$

$$\mathbf{v}_P = \mathbf{v}_A + \boldsymbol{\omega} \times \mathbf{r}_{P/A}, \tag{b}$$

where we again take A to be a point of the rigid body. To obtain the equation for the acceleration of P, we differentiate (b):

$$\frac{d}{dt}\mathbf{v}_P = \frac{d}{dt}\mathbf{v}_A + \frac{d\boldsymbol{\omega}}{dt} \times \mathbf{r}_{P/A} + \boldsymbol{\omega} \times \frac{d\mathbf{r}_{P/A}}{dt}. \tag{c}$$

Given that $\mathbf{v}_P$ and $\mathbf{v}_A$ are the absolute velocities of P and A, we write the first two terms of (c) as $\mathbf{a}_P$ and $\mathbf{a}_A$, the absolute accelerations of P and A.

Now consider the final term of (c). From (8.5) and (8.7), we note that

$$\frac{d\mathbf{r}_{P/A}}{dt} = \mathbf{v}_{P/A} = \boldsymbol{\omega} \times \mathbf{r}_{P/A}.$$

Thus (c) becomes

$$\mathbf{a}_P = \mathbf{a}_A + [\dot{\boldsymbol{\omega}} \times \mathbf{r}_{P/A} + \boldsymbol{\omega} \times (\boldsymbol{\omega} \times \mathbf{r}_{P/A})]. \tag{8.14}$$

We note that, by comparison with (8.5c), the term in brackets is $\mathbf{a}_{P/A}$.

Equation (8.14) applies to the general motion of a rigid body. However, for plane motion, we can evaluate the terms easily. First note that

$$\boldsymbol{\omega} = \dot{\theta}\mathbf{k} \quad \text{and thus} \quad \dot{\boldsymbol{\omega}} = \ddot{\theta}\mathbf{k}$$

Thus, in the case of plane motion, the vectors $\boldsymbol{\omega}$ and $\dot{\boldsymbol{\omega}}$ are both perpendicular to the plane of motion. Figure 8.13 illustrates the components of the acceleration of P.

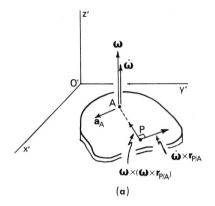

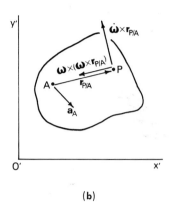

(a)   (b)

Figure 8.13

Given that $r_{P/A}$ is the magnitude of the vector $\mathbf{r}_{P/A}$ from A to P, we can evaluate the magnitudes of the components:

$$|\dot{\boldsymbol{\omega}} \times \mathbf{r}_{P/A}| = r_{P/A}\ddot{\theta} \quad \text{and} \quad |\boldsymbol{\omega} \times (\boldsymbol{\omega} \times \mathbf{r}_{P/A})| = r_{P/A}\dot{\theta}^2.$$

Referring again to Fig. 8.7(b), we can write

$$\mathbf{r}_{P/A} = x_P\mathbf{i} + y_P\mathbf{j} + z_P\mathbf{k} \quad \text{and} \quad \mathbf{a}_A = a_{Ax}\mathbf{i} + a_{Ay}\mathbf{j}.$$

Then (8.14) will read [see Problem 8.29]

$$a_x = a_{Ax} - \ddot{\theta}\, y_P - \dot{\theta}^2 x_P, \qquad a_y = a_{Ay} + \ddot{\theta}\, x_P - \dot{\theta}^2 y_P. \tag{8.15}$$

We could repeat the analysis which preceded the definition of the instant center to define an *instant center for accelerations.* Such a point can be shown to exist [see Problems 8.30 and 8.31], but it is not nearly so useful as the instant (velocity) center, Keep in mind that in general the instant velocity center and the instant acceleration center do not coincide.

Example 7

Consider the cylinder in Fig. 8.14(a), which rolls without slipping at the constant rate $\dot{\theta} = \text{constant}$. Describe the acceleration of various points of the body.

Solution

Since $\dot{\theta}$ is a constant, we have $\ddot{\theta} = 0$. And it is clear that the center of the cylinder moves along the line indicated with a constant speed $v = R\dot{\theta}$. Thus $\mathbf{a}_C = 0$, and C is the

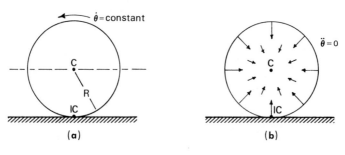

Figure 8.14

instant acceleration center. The point of contact IC is, of course, the instant (velocity) center.

Writing the accelerations $\mathbf{a}_P$ about C by (8.14), we have

$$\mathbf{a}_P = \dot{\boldsymbol{\omega}} \times \mathbf{r}_{P/C} + \boldsymbol{\omega} \times (\boldsymbol{\omega} \times \mathbf{r}_{P/C}).$$

In the present case, $\dot{\boldsymbol{\omega}} = 0$ since $\ddot{\theta} = 0$. Thus,

$$\mathbf{a} = \boldsymbol{\omega} \times (\boldsymbol{\omega} \times \mathbf{r}_{P/C}),$$

and we have the acceleration field shown in Fig. 8.14(b). For a point a distance r from C, $|\mathbf{a}| = r\dot{\theta}^2$.

....

Example 8

The cylinder in Example 7 rolls without slip; now suppose that

$$\frac{\ddot{\theta}}{\dot{\theta}^2} = k, \tag{a}$$

where k is a (nondimensional) constant. Determine the position of the instant acceleration center.

Solution

In Fig. 8.15(a) we set up a coordinate system to study the problem. The point C has the acceleration, $-R\ddot{\theta}\mathbf{i}$. Writing (8.15) and noting (a), we obtain for the acceleration of a point (x,y),

$$a_x = \ddot{\theta}\left(-R - y - \frac{x}{k}\right), \qquad a_y = \ddot{\theta}\left(x - \frac{y}{k}\right). \tag{b}$$

If we wish to find the instant acceleration center, we set $a_x = a_y = 0$ and solve for x

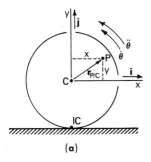

(a)

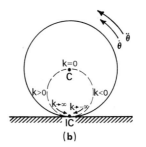

(b)   Figure 8.15

and y:

$$x = \frac{-Rk}{1 + k^2}, \qquad y = \frac{-Rk^2}{1 + k^2}. \tag{c}$$

Figure 8.15(b) is a plot of the locus of the points given by (c) for various values of k. (Can you show that the locus consists of the circle shown?)

Notice that the instant (velocity) center IC and the instant acceleration center coincide only if

$$\frac{\ddot{\theta}}{\dot{\theta}^2} \rightarrow \pm\infty.$$

This example is noticeably academic.* Yet there is at least one important lesson to be learned here. One cannot use the instant (velocity) center to determine accelerations. That is, the acceleration field is not that of circular motion about IC. Also, in general, $\mathbf{a}_{IC} \neq 0$.

....

5. EQUATIONS FOR THE DYNAMICS OF
A RIGID BODY IN PLANE MOTION

In the modern view of dynamics, there are two axioms for the motion of a body:

$$\mathbf{F} = \dot{\mathbf{p}}_C, \tag{a}$$

$$\mathbf{M}_O = \dot{\mathbf{H}}_O, \tag{b}$$

_____

* An interesting feature of this problem is that the point set determined for various values of k can be shown to be the locus of all points of the cylinder whose trajectories are at inflection points. This is shown by using the Euler–Savary equation in advanced kinematics.

where $\mathbf{F}$ is the force resultant acting on the body, $\mathbf{M}_O$ is the moment resultant acting on the body about a fixed point O of some inertial reference frame, and $\mathbf{H}_O$ is the moment of momentum of the body* about O:

$$\mathbf{H}_O = \int_V \mathbf{r} \times \mathbf{v} \, dm.$$

This is clearly a generalization of the definition of $\mathbf{H}_O$ encountered in Chapters 4 and 7:

$$\mathbf{H}_O = \sum_{i=1}^{n} \mathbf{r}_i \times \mathbf{p}_i.$$

We can now see that the material in Chapter 7, particularly the material concerned with the motion of rigid systems of particles (Section 7.6), is motivation (if not proof) of Eqs. (a) and (b). Let us evaluate (a) and (b) in detail.

$$\mathbf{F} = \dot{\mathbf{p}}_C$$

Here $\mathbf{F}$ is the *force resultant* acting on the body and $\mathbf{p}_C = m\mathbf{v}_C$, where m is the *total mass* of the body and $\mathbf{v}_C$ is the *absolute velocity* of the *center of mass*. Thus

$$\mathbf{F} = \dot{\mathbf{p}}_C: \qquad \mathbf{F} = m\mathbf{a}_C. \tag{8.16}$$

$$\mathbf{M}_O = \dot{\mathbf{H}}_O$$

Here $\mathbf{M}_O$ is the *moment resultant* acting on the body about a fixed point O of some inertial reference frame. Unless the point O happens to be a point of the body (one point of the body is fixed in space), an explicit expression for $\mathbf{H}_O$ is not useful.

We shall consider two cases: (i) the body has a fixed point O, and (ii) the body has no fixed point. In the later case, we shall write $\mathbf{M} = \dot{\mathbf{H}}$ about the center of mass.

---

* This is a volume integral extending over the volume V of the body. In fact, $\mathbf{H}_O$ is stated here in terms of a Stieltjes integral. If we have a continuous distribution of matter in the body, so that ρ is the mass density, we can write $dm = \rho \, dV$ and $\mathbf{H}_O$ can be expressed as the ordinary Riemann integral,

$$\mathbf{H}_O = \int_V \mathbf{r} \times \mathbf{v} \rho \, dV.$$

i) *One Point Fixed*

Let O be the fixed point of the body (Fig. 8.16). The velocity of dm is, by (8.8) and (8.10),

$$\mathbf{v} = \boldsymbol{\omega} \times \mathbf{r} = \dot{\theta}(-y\mathbf{i} + x\mathbf{j}).$$

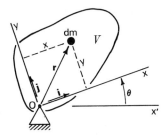

Figure 8.16

Thus with $\mathbf{r} = x\mathbf{i} + y\mathbf{j} + z\mathbf{k}$,

$$\mathbf{r} \times \mathbf{v} = \begin{vmatrix} \mathbf{i} & \mathbf{j} & \mathbf{k} \\ x & y & z \\ -\dot{\theta}y & +\dot{\theta}x & 0 \end{vmatrix}$$

$$= -\mathbf{i}\dot{\theta}xz - \mathbf{j}\dot{\theta}yz + \mathbf{k}\dot{\theta}(x^2 + y^2).$$

Thus $\mathbf{H}_O$ can be written

$$\mathbf{H}_O = -\mathbf{i}\dot{\theta}\int_V xz \, dm - \mathbf{j}\dot{\theta}\int_V yz \, dm + \mathbf{k}\dot{\theta}\int_V (x^2 + y^2) \, dm.$$

Since the axis system x, y, z is attached to the rigid body, the integrals

$$I^O_{xz} = \int_V xz \, dm, \qquad I^O_{yz} = \int_V yz \, dm, \qquad I^O_{zz} = \int_V (x^2 + y^2) \, dm \qquad (8.17)$$

are constants. These quantities are called the *products of inertia* (I^O_{xz} and I^O_{yz}) and the *moment of inertia* (I^O_{zz}). We shall see that only I^O_{zz} is crucial for determining the motion in problems in this chapter. In Section 6, we shall discuss theoretical and experimental determination of the moments and products of inertia of the body.

With (8.17), we have

$$\mathbf{H}_O = -(I^O_{xz} \dot{\theta})\mathbf{i} - (I^O_{yz} \dot{\theta})\mathbf{j} + (I^O_{zz} \dot{\theta})\mathbf{k}. \qquad (8.18)$$

We wish now to write $\mathbf{M}_O = \dot{\mathbf{H}}_O$. We note again that if x, y, z are axes fixed in the body, I^O_{xz}, I^O_{yz} and I^O_{zz} are three constants.

The derivatives of $\mathbf{i}$, $\mathbf{j}$, and $\mathbf{k}$ are

$$\left.\begin{array}{l} \dot{\mathbf{i}} = \boldsymbol{\omega} \times \mathbf{i} = \dot{\theta}\mathbf{j} \\ \dot{\mathbf{j}} = \boldsymbol{\omega} \times \mathbf{j} = -\dot{\theta}\mathbf{i} \\ \dot{\mathbf{k}} = \boldsymbol{\omega} \times \mathbf{k} = 0 \end{array}\right\} \quad \text{where } \boldsymbol{\omega} = \dot{\theta}\mathbf{k}.$$

Thus, from (8.18), we get

$$\dot{\mathbf{H}}_O = (-I^O_{xz}\,\ddot{\theta} + I^O_{yz}\,\dot{\theta}^2)\,\mathbf{i} + (-I^O_{yz}\,\ddot{\theta} - I^O_{xz}\,\dot{\theta}^2)\,\mathbf{j} + (I^O_{zz}\,\ddot{\theta})\,\mathbf{k}. \tag{8.19}$$

Now suppose that we write the components of the moment resultant $\mathbf{M}_O$ in the moving x, y, z frame:

$$\mathbf{M}_O = M_{Ox}\mathbf{i} + M_{Oy}\mathbf{j} + M_{Oz}\mathbf{k}. \tag{8.20}$$

Then $\mathbf{M}_O = \dot{\mathbf{H}}_O$ becomes

$$\begin{aligned} M_{Ox} &= (-I^O_{xz}\,\ddot{\theta} + I^O_{yz}\,\dot{\theta}^2), \\ M_{Oy} &= (-I^O_{yz}\,\ddot{\theta} - I^O_{xz}\,\dot{\theta}^2), \\ M_{Oz} &= I^O_{zz}\,\ddot{\theta}. \end{aligned} \tag{8.21}$$

The first two equations of (8.21) are relevant only in determining the moments M_{Ox} and M_{Oy} which are necessary to maintain an unsymmetrical body in plane motion. In order to determine the (angular) motion of the body, the only equation which is necessary is

$$M_{Oz} = I^O_{zz}\,\ddot{\theta}. \tag{8.22}$$

Example 9

A rigid body which weighs 20 lb is pinned at point O, a distance $d = 1$ ft from its center of mass (Fig. 8.17a). When the body is given a small displacement from equilibrium, it oscillates with period $\tau = 1.0$ sec. Determine I^O_{zz} about O.

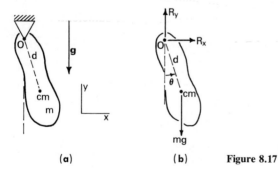

(a)   (b)   **Figure 8.17**

Solution

Figure 8.17(b) is the free-body diagram for the body. We have $M_{Oz} = -mg\, d \sin \theta$ and by (8.22) $\dot{H}_{Oz} = I^0_{zz}\, \ddot{\theta}$. Thus

$$\mathbf{M}_O = \dot{\mathbf{H}}_O: \quad -mg\, d \sin \theta = I^0_{zz}\, \ddot{\theta}.$$

If θ is small, we approximate $\sin \theta \approx \theta$, and then we get

$$\ddot{\theta} + \left(\frac{mgd}{I^0_{zz}}\right)\theta = 0.$$

The natural frequency ω_n is

$$\omega_n = \sqrt{\frac{mgd}{I^0_{zz}}}$$

and since $\omega_n \tau = 2\pi$,

$$\tau = 2\pi \sqrt{\frac{I^0_{zz}}{mgd}}.$$

Thus

$$I^0_{zz} = \frac{mgd\tau^2}{4\pi^2}.$$

With the numerical data, we have

$$I^0_{zz} = \frac{(20 \text{ lb})(1 \text{ ft})(1)^2 \sec^2}{4(9.88)} \quad \text{or} \quad I^0_{zz} = 0.507 \text{ lb-ft-sec}^2.$$

Note that $1 \text{ lb} = 1 \text{ slug-ft/sec}^2$. Thus

$$I^0_{zz} = 0.507 \text{ slug-ft}^2.$$

....

The above example clearly shows that if a rigid body has one point fixed, the entire motion can be determined from Eq. (8.22). If we wish to determine the forces which cause the motion, we may wish to write $\mathbf{F} = \dot{\mathbf{p}}_C$. But if we are interested only in the motion of the body, (8.22) is sufficient.

ii) *General Motion — No Fixed Point*

The second axiom for the motion of a rigid body is

$$\mathbf{M}_O = \dot{\mathbf{H}}_O, \tag{8.23}$$

where

$$\mathbf{H}_O = \int_V \mathbf{r} \times \mathbf{v}\, dm,$$

and O is a fixed point of an inertial reference frame. We wish to show that this equation is equivalent to

$$\mathbf{M}_C = \dot{\mathbf{H}}_C, \tag{8.24}$$

where C is the center of mass of the body.

The proof consists of two parts.

i)   Transformation equation for the moment resultant:

$$\mathbf{M}_O = \mathbf{r}_C \times \mathbf{F} + \mathbf{M}_C. \tag{8.25}$$

ii) Transformation equation for the moment of momentum and the derivative of the moment of momentum:

$$\mathbf{H}_O = \mathbf{r}_C \times \mathbf{p}_C + \mathbf{H}_C \tag{a}$$

and   (8.26)

$$\dot{\mathbf{H}}_O = \mathbf{r}_C \times \dot{\mathbf{p}}_C + \dot{\mathbf{H}}_C. \tag{b}$$

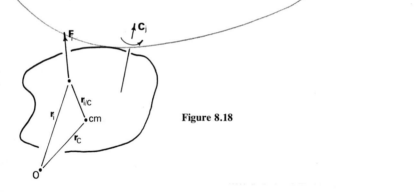

Figure 8.18

To prove (8.25), consider Fig. 8.18. Suppose that the following forces and couples act on the body: $\mathbf{F}_1, \mathbf{F}_2, \ldots \mathbf{F}_n; \mathbf{C}_1, \mathbf{C}_2, \ldots \mathbf{C}_{n'}$. Writing the resultant moment about O and C, the center of mass, we get

$$\mathbf{M}_O = \sum_{i=1}^{n} \mathbf{r}_i \times \mathbf{F}_i + \sum_{j=1}^{n'} \mathbf{C}_j,$$

$$\mathbf{M}_C = \sum_{i=1}^{n} \mathbf{r}_{i/C} \times \mathbf{F}_i + \sum_{j=1}^{n'} \mathbf{C}_j.$$

Subtracting gives

$$\mathbf{M}_O - \mathbf{M}_C = \sum_{i=1}^{n} (\mathbf{r}_i - \mathbf{r}_{i/C}) \times \mathbf{F}_i.$$

But $\mathbf{r}_i = \mathbf{r}_C + \mathbf{r}_{i/C}$; thus $(\mathbf{r}_i - \mathbf{r}_{i/C}) = \mathbf{r}_C$ and

$$\mathbf{M}_O - \mathbf{M}_C = \mathbf{r}_C \times \sum_{i=1}^{n} \mathbf{F}_i.$$

And finally, noting that $\Sigma \mathbf{F}_i = \mathbf{F}$, the force resultant, we get

$$\mathbf{M}_O = \mathbf{r}_C \times \mathbf{F} + \mathbf{M}_C,$$

which is (8.25). Note that (8.25) is a special case of (2.13).

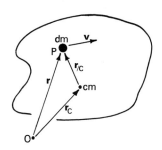

Figure 8.19

To establish Eqs. (8.26), we begin by defining $\mathbf{H}_O$ and $\mathbf{H}_C$. Refer to Fig. 8.19, and suppose that point P of the body has velocity $\mathbf{v}$. Then we define

$$\mathbf{H}_O = \int_V \mathbf{r} \times \mathbf{v} \, dm, \qquad \mathbf{H}_C = \int_V \mathbf{r}_{/C} \times \mathbf{v} \, dm.$$

Subtracting, we obtain

$$\mathbf{H}_O - \mathbf{H}_C = \int_V (\mathbf{r} - \mathbf{r}_{/C}) \times \mathbf{v} \, dm.$$

But, from Fig. 8.19, we see that

$$\mathbf{r} = \mathbf{r}_C + \mathbf{r}_{/C} \qquad \text{or} \qquad (\mathbf{r} - \mathbf{r}_{/C}) = \mathbf{r}_C,$$

the position of the center of mass. Thus

$$\mathbf{H}_O - \mathbf{H}_C = \int_V \mathbf{r}_C \times \mathbf{v} \, dm = \mathbf{r}_C \times \int_V \mathbf{v} \, dm.$$

The position of the center of mass $\mathbf{r}_C$ is given by

$$\mathbf{r}_C = \frac{\displaystyle\int_V \mathbf{r} \, dm}{\displaystyle\int_V dm} = \frac{1}{m} \int_V \mathbf{r} \, dm,$$

where m is the total mass of the body. Thus

$$m\mathbf{r}_C = \int_V \mathbf{r} \, dm.$$

Differentiating, we obtain

$$m\mathbf{v}_C = \int_V \dot{\mathbf{r}} \, dm = \int_V \mathbf{v} \, dm.$$

But $m\mathbf{v}_C = \mathbf{p}_C$, the momentum of the center of mass. Thus we have

$$\mathbf{H}_O - \mathbf{H}_C = \mathbf{r}_C \times \mathbf{p}_C \qquad \text{or} \qquad \mathbf{H}_O = \mathbf{H}_C + \mathbf{r}_C \times \mathbf{p}_C,$$

which proves Eq. (8.26a). Differentiating to get $\dot{\mathbf{H}}_O$, we have

$$\dot{\mathbf{H}}_O = \dot{\mathbf{H}}_C + \mathbf{v}_C \times \mathbf{p}_C + \mathbf{r}_C \times \dot{\mathbf{p}}_C.$$

The middle term on the right is zero, for $\mathbf{v}_C \times \mathbf{p}_C = \mathbf{v}_C \times m\mathbf{v}_C = 0$. Thus

$$\dot{\mathbf{H}}_O = \mathbf{r}_C \times \dot{\mathbf{p}}_C + \dot{\mathbf{H}}_C,$$

which is (8.26b).

Finally, to prove (8.24), we write (8.23), using (8.25) and (8.26b):

$$\mathbf{r}_C \times \mathbf{F} + \mathbf{M}_C = \mathbf{r}_C \times \dot{\mathbf{p}}_C + \dot{\mathbf{H}}_C. \tag{a}$$

But from (8.16), $\mathbf{F} = \dot{\mathbf{p}}_C$. Thus

$$\mathbf{r}_C \times \mathbf{F} = \mathbf{r}_C \times \dot{\mathbf{p}}_C. \tag{b}$$

Subtracting (b) from (a) gives $\mathbf{M}_C = \dot{\mathbf{H}}_C$, which proves (8.24).

Let us now evaluate $\mathbf{H}_C$ and $\dot{\mathbf{H}}_C$ for the case of plane motion. Consider Fig. 8.20. Writing (8.10), with the point A identified with the center of mass,

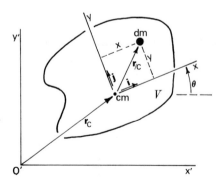

Figure 8.20

we obtain

$$\mathbf{v} = \mathbf{v}_C + \dot{\theta}(-y\mathbf{i} + x\mathbf{j}).$$

Writing $\mathbf{r}_{/C} = x\mathbf{i} + y\mathbf{j} + z\mathbf{k}$, we get

$$\mathbf{r}_{/C} \times \mathbf{v} = \mathbf{r}_{/C} \times \mathbf{v}_C - \mathbf{i}\dot{\theta}xz - \mathbf{j}\dot{\theta}yz + \mathbf{k}\dot{\theta}(x^2 + y^2).$$

Integrating to get $\mathbf{H}_C$:

$$\mathbf{H}_C = \int_V \mathbf{r}_{/C} \times \mathbf{v} \ dm$$

$$= \int_V \mathbf{r}_{/C} \times \mathbf{v}_C \ dm - \mathbf{i}\dot{\theta} \int_V xz \ dm - \mathbf{j}\dot{\theta} \int_V yz \ dm + \mathbf{k}\dot{\theta} \int_V (x^2 + y^2) \ dm.$$

The integral

$$\int_V \mathbf{r}_{/C} \times \mathbf{v}_C \ dm = \left(\int_V \mathbf{r}_{/C} \ dm \right) \times \mathbf{v}_C.$$

But $\int_V \mathbf{r}_{/C} \ dm = 0$, since this gives the total mass times the position of the center of mass *referred to the center of mass;* that is,

$$\mathbf{r}_{C/C} = 0.$$

Note that the coordinate system x, y, z is fixed to the rigid body through the center of mass. Thus we write

$$I_{xz}^C = \int_V xz \ dm, \qquad I_{yz}^C = \int_V yz \ dm, \qquad I_{zz}^C = \int_V (x^2 + y^2) \ dm,$$

and we again note that these quantities are constant. Thus $\mathbf{H}_C$ becomes

$$\mathbf{H}_C = -(I_{xz}^C \ \dot{\theta})\mathbf{i} - (I_{yz}^C \ \dot{\theta})\mathbf{j} + (I_{zz}^C \ \dot{\theta})\mathbf{k}.$$

Differentiating to get $\dot{\mathbf{H}}_C$ and noting that

$$\dot{\mathbf{i}} = \dot{\theta}\mathbf{j}, \qquad \dot{\mathbf{j}} = -\dot{\theta}\mathbf{i}, \qquad \dot{\mathbf{k}} = 0,$$

we get

$$\dot{\mathbf{H}}_C = (-I_{xz}^C \ \ddot{\theta} + I_{yz}^C \ \dot{\theta}^2)\mathbf{i} + (-I_{yz}^C \ \ddot{\theta} - I_{xz}^C \ \dot{\theta}^2)\mathbf{j} + (I_{zz}^C \ \ddot{\theta})\mathbf{k}.$$

Writing $\mathbf{M}_C = M_{Cx}\mathbf{i} + M_{Cy}\mathbf{j} + M_{Cz}\mathbf{k}$, we get

$$M_{Cx} = (-I_{xz}^C \ \ddot{\theta} + I_{yz}^C \ \dot{\theta}^2), \qquad \text{(a)}$$

$$M_{Cy} = (-I_{yz}^C \ \ddot{\theta} - I_{xz}^C \ \dot{\theta}^2), \qquad \text{(b) (8.27)}$$

$$M_{Cz} = I_{zz}^C \ \ddot{\theta}. \qquad \text{(c)}$$

The quantities I_{xz}^C, I_{yz}^C and I_{zz}^C are called the *products and moment of inertia of the body about the center of mass.* Again note that only Eq. (8.27c) need be used if we are interested only in the angular motion of the body.

Example 10

It is given that the moment of inertia of a homogeneous cylinder of radius R and mass m is $I_{zz}^C = \frac{1}{2}mR^2$. The cylinder in Fig. 8.21(a) rolls without slipping on the horizontal surface. Show that the cylinder oscillates with period $\tau = 2\pi \sqrt{3m/2k}$ sec.

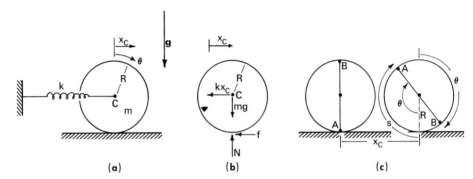

(a)   (b)   (c)

Figure 8.21

Solution

Suppose that x_C is a measure of the horizontal displacement of the center of mass of the cylinder and θ is a measure of the rotation of the body. Figure 8.21(b) is a free-body diagram for the cylinder in a displaced position.

Writing $\mathbf{F} = \dot{\mathbf{p}}_C$ and $\mathbf{M}_C = \dot{\mathbf{H}}_C$, we have

$$F_x = ma_{Cx}: \qquad -kx_C - f = m\ddot{x}_C, \tag{a}$$

$$M_{Cz} = I_{zz}^C \, \ddot{\theta}: \qquad fR = (\tfrac{1}{2}mR^2)\ddot{\theta}. \tag{b}$$

Two points are worth making at this juncture: (1) The force of friction f is static friction. Thus, until we have solved the problem, we don't know the direction of f. (2) The moment equation (b) has been written consistent with the definition of a positive rotation θ as defined by Fig. 8.21(a).

Now we need another equation in order to solve for the three unknowns, f, x_C, and θ, in (a) and (b). We haven't used the fact that the cylinder rolls without slip. Consider Fig. 8.21(c). When the cylinder rolls through an angle θ, the center of the cylinder moves a distance x_C. If there is no slipping, the arc length $s = R\theta$ must equal x_C. Thus $x_C = R\theta$. From (a), (b), and (c), we get

$$\tfrac{3}{2}m\ddot{x}_C + kx_C = 0 \qquad \text{or} \qquad \ddot{x}_C + \left(\frac{2}{3}\frac{k}{m}\right)x_C = 0.$$

Thus

$$\omega_n = \sqrt{\frac{2}{3}\frac{k}{m}},$$

and since $\omega_n \tau = 2\pi$ radians, we obtain

$$\tau = 2\pi \sqrt{\frac{3m}{2k}}.$$

....

Example 11

Consider the system in Fig. 8.22(a). The homogeneous cylinder has moment of inertia $I_{zz}^C = \frac{1}{2}mR^2$, and it rolls without slipping on the lower plane. On top of the cylinder is a rigid bar of mass m_1, pinned at O. The coefficient of sliding friction between the cylinder and the bar is μ.

When $x_C = 0$, the speed of the center of the cylinder is v_0. Calculate its speed after the cylinder has rolled a distance x_C from the original position.

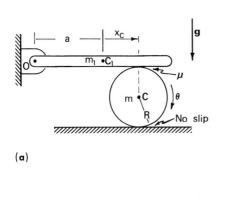

(a)

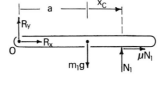

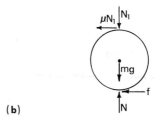

(b)

Figure 8.22

Solution

Figure 8.22(b) shows free-body diagrams for the cylinder and the bar. Writing $\mathbf{F} = \dot{\mathbf{p}}_C$ and $\mathbf{M}_C = \dot{\mathbf{H}}_C$ for the cylinder, we have

$$F_x = ma_{Cx}: \qquad -f - \mu N_1 = m\ddot{x}_C, \tag{a}$$

$$M_{Cz} = I_{zz}^C \ddot{\theta}: \qquad fR - \mu N_1 R = (\tfrac{1}{2}mR^2)\ddot{\theta}. \tag{b}$$

Again, the no-slip condition gives

$$x_C = R\theta. \tag{c}$$

Equations (a), (b), (c) contain four unknowns, f, N_1, x_C, and θ. We can get an expression for N_1 by writing the moment equation for the bar (about O):

$$\mathbf{M}_O = 0: \ [-m_1 ga + N_1(a + x_C)]\mathbf{k} = 0,$$

or

$$N_1 = \frac{m_1 ga}{a + x_C} \tag{d}$$

From Eqs. (a), (b), (c) and (d), we obtain finally:

$$\tfrac{3}{2} m\ddot{x}_C + \frac{2\mu m_1 ga}{a + x_C} = 0$$

or

$$\ddot{x}_C + \left(\frac{4}{3}\frac{\mu m_1 ga}{m}\right)\frac{1}{a + x_C} = 0. \tag{e}$$

To determine the velocity from this equation, let $\dot{x}_C = v$. Then

$$\ddot{x}_C = \dot{v} = \frac{dv}{dx_C}\frac{dx_C}{dt} = \frac{dv}{dx_C} v.$$

Thus (e) becomes

$$v\, dv = -\left(\frac{4}{3}\frac{\mu m_1 ga}{m}\right)\frac{dx_C}{a + x_C}.$$

Integrating and noting that $v = v_0$ at $x_C = 0$, we get

$$\int_{v_0}^{v} v\, dv = -\left(\frac{4}{3}\frac{\mu m_1 ga}{m}\right)\int_{0}^{x_C}\frac{dx_C}{a + x_C}$$

or

$$\tfrac{1}{2}(v^2 - v_0^2) = -\left(\frac{4}{3}\frac{\mu m_1 ga}{m}\right)\ln\left(1 + \frac{x_C}{a}\right)$$

or

$$v = \sqrt{v_0^2 - \left[\frac{8}{3}\frac{\mu m_1 ga}{m}\ln\left(1 + \frac{x_C}{a}\right)\right]}.$$

....

In summary, we can write the equations for the determination of the plane motion of a rigid body as follows.

A rigid body is in plane motion.

i) O, a fixed point of the body

$$M_{Oz} = I_{zz}^O \, \ddot{\theta}$$

ii) C, center of mass of the body

$$F_x = ma_{Cx}, \qquad F_y = ma_{Cy}, \qquad M_{Cz} = I_{zz}^C \, \ddot{\theta}.$$

If the body has no fixed point, we use equations (ii). If the body has a fixed point, either set of equations is applicable, but set (i) will be simpler.

6. MOMENT OF INERTIA

In the case of the plane motion of a rigid body, we can describe the rotary motion of the body by (8.22) or (8.27):

$$M_{Oz} = I_{zz}^O \, \ddot{\theta} \quad \text{(a)} \qquad \text{or} \qquad M_{Cz} = I_{zz}^C \, \ddot{\theta}, \quad \text{(b)}$$

where I_{zz} is the moment of inertia of the body about the z axis through either O (if this is a fixed point) or the center of mass C.

If A is a point of the body, the moment of inertia I_{zz}^A is defined as

$$I_{zz}^A = \int_V (x^2 + y^2) \, dm, \qquad (8.28)$$

where x and y are referred to axes passing through A (see Fig. 8.23).

There are three ways in which I_{zz}^A is determined for a body.

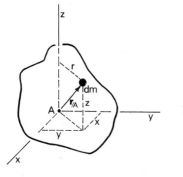

Figure 8.23

i) *Exact Integration*

If the body has a simple configuration, (8.28) may be evaluated directly. The results for several simple bodies of constant density are included in a Table of Inertias in Appendix 2 at the end of the book.

ii) *Approximate Numerical Integration*

The integration in (8.28) can be performed approximately by some numerical technique. This is discussed in Appendix 2.

iii) *Experimental Determination*

A simple experiment for determining I_{zz}^C is also given in Appendix 2. If we want to find the moment of inertia about some other point, we can use the parallel-axis theorem, which will be developed in this section.

Of the three methods, there can be little debate that the experimental determination of I_{zz}^C is the most important from the point of view of industrial applications. Thus it is realistic in our examples and problems to assume that the value of I_{zz}^C is given.

To give a feeling for the moment of inertia, let's compute this quantity for a number of simple configurations. We refer to Fig. 8.23, and let $r = \sqrt{x^2 + y^2}$ be the distance of dm from the z axis. Then (8.28) becomes

$$I_{zz}^A = \int_V r^2 \, dm. \tag{8.29}$$

Example 12

In Fig. 8.24, a particle m has a position a distance R from the z axis. By (8.29), we have

$$I_{zz}^0 = mR^2. \tag{8.30}$$

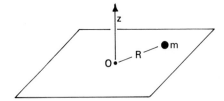

Figure 8.24

Example 13

Figure 8.25 shows a thin ring of radius R and mass m. The z axis passes through the center of the ring O; we see that every element of mass dm lies a distance R from the z

axis. Thus (8.29) becomes

$$I_{zz}^0 = \int_V R^2 \, dm = R^2 \int_V dm.$$

Thus

$$I_{zz}^0 = mR^2. \tag{8.31}$$

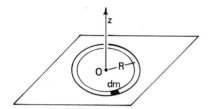

Figure 8.25

Example 14

Figure 8.26(a) shows a thin, homogeneous bar of mass m and length L, and we consider z to be an axis through one end of the bar.

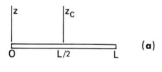

(a)

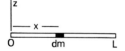

(b)   Figure 8.26

In Fig. 8.26(b) we let the rod lie along the x axis. Then

$$I_{zz}^0 = \int_0^L x^2 \, dm.$$

Let μ be the (constant) *mass per unit length* of the rod. Then $dm = \mu \, dx$ and

$$I_{zz}^0 = \mu \int_0^L x^2 \, dx = \tfrac{1}{3}\mu L^3.$$

We usually find it convenient to express the moment of inertia in terms of the total mass of the body, m. This is $m = \mu L$. Thus we have

$$I_{zz}^0 = \tfrac{1}{3}mL^2 \text{ (about end).} \tag{8.32}$$

Suppose we now compute I_{zz}^C about z_C, the axis through the center of mass. The computation now becomes

$$I_{zz}^C = \mu \int_{-L/2}^{+L/2} x^2 \, dx = \tfrac{1}{12} \mu L^3.$$

Thus

$$I_{zz}^C = \tfrac{1}{12} m L^2 \text{ (center of mass).} \tag{8.32'}$$

We shall see that we could obtain either (8.32) or (8.32') once we know the other by a simple method called the *parallel-axis theorem*.

....

Example 15

Figure 8.27(a) shows a homogeneous disk of radius R and mass m. The z axis passes through the center of mass perpendicular to the plane of the disk.

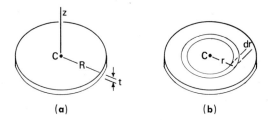

(a) (b) **Figure 8.27**

In Figure 8.27(b) we consider the disk to consist of a series of concentric rings of radius r and width dr; we take the thickness to be t. Then if μ is the mass per unit volume, the mass of the ring is

$$dm = \mu t 2\pi r \, dr.$$

And thus, by (8.31), we have the contribution of the ring at r to be

$$dI_{zz}^C = 2\pi \mu t r^3 \, dr.$$

We now integrate over all such rings:

$$I_{zz}^C = 2\pi \mu t \int_0^R r^3 \, dr = \frac{\mu \pi R^4 t}{2}.$$

The total mass of the ring is

$$m = \mu \pi R^2 t.$$

Thus

$$I^C_{zz} = \tfrac{1}{2} mR^2. \tag{8.33}$$

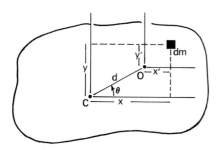

Figure 8.28

Let us now prove the parallel-axis theorem. Two parallel axes z and z' are separated by a distance d (see Fig. 8.28). Suppose that z passes through the center of mass of the body. Then we have

$$I^C_{zz} = I_{zz} = \int_V (x^2 + y^2)\, dm, \tag{a}$$

$$I^0_{zz} = I_{z'z'} = \int_V (x'^2 + y'^2)\, dm, \tag{b}$$

where

$$x' = x - d \cos \theta, \qquad y' = y - d \sin \theta,$$

and θ is the angle defined in Fig. 8.28. Thus

$$x'^2 + y'^2 = x^2 - 2xd \cos \theta + d^2 \cos^2 \theta + y^2 - 2yd \sin \theta + d^2 \sin^2 \theta$$

or

$$x'^2 + y'^2 = (x^2 + y^2) + d^2 - 2d(x \cos \theta + y \sin \theta).$$

Then, by (a) and (b), we have

$$I_{z'z'} = \int_V (x^2 + y^2)\, dm + d^2 \int_V dm - 2d \cos \theta \int_V x\, dm - 2d \sin \theta \int_V y\, dm.$$

The first sum is I_{zz}. The second is md^2. The final terms are zero, since

$$\int_V x\, dm = mx_C, \qquad \int_V y\, dm = my_C.$$

But the x,y position of the center of mass is $(0,0)$. Thus we get

$$I_{z'z'} = I_{zz} + md^2$$

or

$$I_{zz}^O = I_{zz}^C + md^2. \tag{8.34}$$

> If z passes through the center of mass of a body and z' is an axis par-
> allel to z and separated from it by a distance d, the moment of inertia
> about z' equals the moment of inertia about the parallel axis through
> the center of mass z, plus md^2, where m is the total mass of the body.

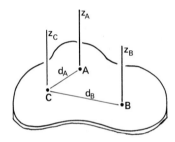

Figure 8.29

Suppose now that we know I_{zz}^A, and wish to know I_{zz}^B, where neither A
nor B is the center of mass axis (see Fig. 8.29). Then by (8.34), we have

$$I_{zz}^A = I_{zz}^C + md_A^2, \qquad I_{zz}^B = I_{zz}^C + md_B^2.$$

Subtracting, we obtain

$$I_{zz}^B = I_{zz}^A + m(d_B^2 - d_A^2). \tag{8.35}$$

Example 16

The body in Fig. 8.30(a) consists of two parts: a section bounded by a parabola and a
circular section. The thickness of the body is t ft and the constant density is μ slug/ft³.
Calculate I_{zz}^O.

Solution

We begin by computing the moment of inertia of the parabolic section. In Fig. 8.30(b)
we isolate an element of mass with dimensions $(t)(dy)(dx)$. Thus $dm = \mu t \, dy \, dx$. The
distance of this element from the origin is

$$r = (x^2 + y^2)^{1/2}.$$

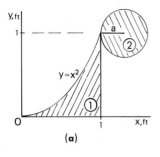

(a)

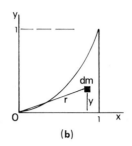

(b)

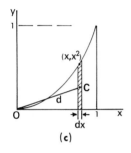

(c)

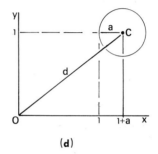

(d)

Figure 8.30

Integrating over all such elements yields 3

$$I_{zz}^0 = \mu t \int_0^1 \int_0^{x^2} (x^2 + y^2)\, dy\, dx = \mu t(\tfrac{1}{5} + \tfrac{1}{21}) \text{ slug-ft}^2.$$ (a)

Some of you may be unfamiliar with multiple integrals. If so, there is an alternative approach to the above. Consider Fig. 8.30(c). We slice the section into vertical strips of width dx. If a strip is a distance x from O, its height is x^2. The moment of inertia of the strip about its center of mass is, by (8.32′),

$$dI_{zz}^C = \tfrac{1}{12} dm\, (x^2)^2,$$

where $dm = \mu t\, (x^2)\, dx$. Thus

$$dI_{zz}^C = \tfrac{1}{12} \mu t x^6\, dx.$$

The distance d from O to the center of mass of the section is

$$d = \left[x^2 + \left(\frac{x^2}{2}\right)^2 \right]^{1/2} = \left(x^2 + \frac{x^4}{4} \right)^{1/2}.$$

Thus the parallel-axis theorem (8.35) gives

$$dI_{zz}^0 = dI_{zz}^C + (dm)\, d^2$$

or

$$dI_{zz}^O = \tfrac{1}{12}\mu t x^6 \, dx + \mu t \left(x^2 + \frac{x^4}{4}\right) x^2 \, dx$$

$$= \mu t (x^4 + \tfrac{1}{3}x^6) \, dx.$$

Finally, integrating over all the slices, we get

$$I_{zz}^O = \int_0^1 \mu t (x^4 + \tfrac{1}{3}x^6) \, dx$$

$$= \mu t (\tfrac{1}{5} + \tfrac{1}{21}) \text{ slug-ft}^2,$$

which agrees with (a).

We now compute the moment of inertia for the circular portion of the body. The mass of the section is

$$m = \mu t \pi a^2.$$

Thus, by (8.33), we have the moment of inertia of the section about its center of mass as

$$I_{zz}^C = \tfrac{1}{2}(\mu t \pi a^2)a^2 = \tfrac{1}{2}\mu t \pi a^4.$$

From Fig. 8.30(d) we see that the distance d from the center of mass to O is

$$d = [(1 + a)^2 + (1)^2]^{1/2} = [2 + 2a + a^2]^{1/2}.$$

Thus, applying the parallel-axis theorem, we have

$$I_{zz}^O = \tfrac{1}{2}\mu t \pi a^4 + \mu t \pi a^2 (2 + 2a + a^2)$$

$$= \mu t \pi a^2 (\tfrac{3}{2}a^2 + 2a + 2) \text{ slug-ft}^2. \tag{b}$$

Adding (a) and (b) gives the total moment of inertia about O:

$$I_{zz}^O = \mu t \left[\frac{26}{105} + \frac{\pi a^2}{2}(3a^2 + 4a + 4)\right] \text{ slug-ft}^2.$$

In adding these quantities to give the above result, we have used the idea of a composite body, explained below.

....

If a body can be broken into simple subsections, we can use the method of composite bodies to compute the total moment of inertia.

Suppose that, in Fig. 8.31, we wish to compute the moment of inertia of a body with two subsections, 1 and 2. The moment of inertia of the entire body about O is then

$$I_{zz}^O = \int_V r^2 \, dm.$$

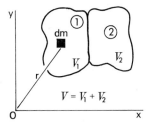

$$V = V_1 + V_2$$

Figure 8.31

Now suppose we split this integral into two parts, one over subsection 1 alone and one over subsection 2 alone. We denote these by $\int_{V_1}$ and $\int_{V_2}$. Then

$$I_{zz}^0 = \int_{V_1} r^2 \, dm + \int_{V_2} r^2 \, dm.$$

But the first sum is the moment of inertia of subsection 1 about O, $I_{zz}^0(1)$. The second sum is the moment of inertia of subsection 2 about O, $I_{zz}^0(2)$. Thus we have

$$I_{zz}^0 = I_{zz}^0(1) + I_{zz}^0(2). \tag{8.36}$$

That is, we can break the moment of inertia of the composite body into the sum of the individual moments of inertia of its constituent parts.

7. THE WORK–ENERGY RELATION

There are several ways to derive the work–energy relation for a rigid body in a rigorous way. We indicate these in Appendix 3 at the end of the book. Here we shall simply do the following:

a)   Write the kinetic energy of a rigid body in plane motion.

b)   Write the work done by a system of forces on a rigid body.

A.   Kinetic Energy of a Rigid Body in Plane Motion

For a single particle we had

$$T = \frac{p^2}{2m}$$

(where $\mathbf{p} = m\mathbf{v}$ is the linear momentum of the particle) or, if we wrote T in terms of velocities,

$$T = \tfrac{1}{2}mv^2.$$

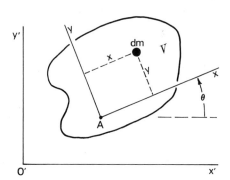

Figure 8.32

It is natural to define the energy of a rigid body as

$$T = \int_V \tfrac{1}{2} v^2 \, dm \tag{8.37}$$

Consider Fig. 8.32. From (8.10), we have the velocity of dm for plane motion,

$$\mathbf{v} = \mathbf{v}_A + \dot{\theta}(-y\mathbf{i} + x\mathbf{j}).$$

Thus

$$v^2 = \mathbf{v} \cdot \mathbf{v} = v_A^2 + 2\dot{\theta}\mathbf{v}_A \cdot (-y\mathbf{i} + x\mathbf{j}) + \dot{\theta}^2(x^2 + y^2),$$

and so we have

$$T = \tfrac{1}{2} v_A^2 \int_V dm + \dot{\theta}\mathbf{v}_A \cdot \left[\int_V (-y\mathbf{i} + x\mathbf{j}) \, dm \right] + \tfrac{1}{2}\dot{\theta}^2 \int_V (x^2 + y^2) \, dm.$$

The first term is

$$\tfrac{1}{2} m v_A^2, \tag{a}$$

where m is the total mass of the body. Noting that

$$\int_V y \, dm = m y_{C/A} \quad \text{and} \quad \int_V x \, dm = m x_{C/A},$$

we can show the middle term to be

$$m\mathbf{v}_A \cdot (\boldsymbol{\omega} \times \mathbf{r}_{C/A}), \tag{b}$$

where $\boldsymbol{\omega} = \dot{\theta}\mathbf{k}$. By (8.28), the final term is

$$\tfrac{1}{2} I_{zz}^A \dot{\theta}^2. \tag{c}$$

Thus from (a), (b), (c), we obtain

$$T = \tfrac{1}{2}mv_A^2 + mv_A \cdot (\boldsymbol{\omega} \times \mathbf{r}_{C/A}) + \tfrac{1}{2}I_{zz}^A\dot{\theta}^2.$$

There are three special points about which we commonly write T:

i) $A = O$,   fixed point, $v_O = 0$

$$T = \tfrac{1}{2}I_{zz}^O\,\dot{\theta}^2 \tag{8.38a}$$

ii) $A = IC$,   instant center,* $v_{IC} = 0$

$$T = \tfrac{1}{2}I_{zz}^{IC}\,\dot{\theta}^2 \tag{8.38b}$$

iii) $A = C$,   center of mass, $\mathbf{r}_{C/A} = 0$

$$T = \tfrac{1}{2}mv_C^2 + \tfrac{1}{2}I_{zz}^C\,\dot{\theta}^2. \tag{8.38c}$$

B. Work of the Forces on a Rigid Body

When we have a force system acting on a rigid body, we can either compute the work of the individual forces and then combine these to give the total work on the body, or we can consider the work of the force resultant $\mathbf{F}$ and the moment resultant about the center of mass, $\mathbf{M}_C$.

Suppose that forces $\mathbf{F}_1$, $\mathbf{F}_2$, . . . , $\mathbf{F}_n$ act on the rigid body (any couple acting on the body can be represented by a pair of equal and opposite forces separated by an appropriate distance). Then

$$\mathbf{F} = \sum_{i=1}^{n} \mathbf{F}_i \quad \text{and} \quad \mathbf{M}_C = \sum_{i=1}^{n} \mathbf{r}_{i/C} \times \mathbf{F}_i \tag{a}$$

are the force resultant and moment resultant about the center of mass. If v_i is the velocity of the point of application of the force $\mathbf{F}_i$, we have the *power of the force system* as

$$P = \sum_{i=1}^{n} \mathbf{F}_i \cdot \mathbf{v}_i.$$

---

* Equation (8.38b) is always valid. It is common to write $M_{ICz} = I_{zz}^{IC}\,\ddot{\theta}$ about the instant center. This can be shown to be valid if the acceleration of the instant center passes through the center of mass. This can be extremely convenient when one is analyzing a homogeneous cylinder which rolls without slipping. However, except for that problem, the difficulty in determining whether or not the equation is valid is often more trouble than using (8.27c) in the first place.

Or by (8.8), $\mathbf{v}_i = \mathbf{v}_C + \boldsymbol{\omega} \times \mathbf{r}_{i/C}$, and thus

$$P = \sum_{i=1}^{n} \mathbf{F}_i \cdot \mathbf{v}_C + \sum_{i=1}^{n} \mathbf{F}_i \cdot (\boldsymbol{\omega} \times \mathbf{r}_{i/C}). \tag{b}$$

Using (a), the first term of (b) becomes $\mathbf{F} \cdot \mathbf{v}_C$. Rewriting the second term using (A1.5), we get $\Sigma \boldsymbol{\omega} \cdot (\mathbf{r}_{i/C} \times \mathbf{F}_i)$, or by (a), $\boldsymbol{\omega} \cdot \mathbf{M}_C$. Thus (b) becomes

$$P = \mathbf{F} \cdot \mathbf{v}_C + \mathbf{M}_C \cdot \boldsymbol{\omega}. \tag{c}$$

Finally, to obtain the *work of the force system*, we integrate the power with respect to time:

$$W_{AB} = \int_{t_A}^{t_B} P \, dt = \int_A^B \mathbf{F} \cdot d\mathbf{r}_C + \int_{t_A}^{t_B} \mathbf{M}_C \cdot \boldsymbol{\omega} \, dt.$$

In the present case, $\boldsymbol{\omega} = \dot{\theta}\mathbf{k}$. Thus

$$W_{AB} = \int_A^B \mathbf{F} \cdot d\mathbf{r}_C + \int_{\theta_A}^{\theta_B} M_{Cz} \, d\theta. \tag{8.39}$$

C. The Work–Energy Relation

The work–energy relation is commonly written in one of two forms. If we evaluate the work of each force on the body separately, we write

$$E(A) + W_{AB} = E(B) \tag{8.40}$$

where $E = T + V$ and V represents all conservative forces and T is given by (8.38).

If we use (8.38c) and (8.39), we can evaluate the work by considering the force and moment resultants:

$$\int_{A_C}^{B_C} \mathbf{F} \cdot d\mathbf{r}_C = \tfrac{1}{2}mv_C^2 \Big|_A^B, \tag{8.41a}$$

$$\int_{\theta_A}^{\theta_B} M_{Cz} \, d\theta = \tfrac{1}{2}I_{zz}^C \, \dot{\theta}^2 \Big|_{\theta_A}^{\theta_B}. \tag{8.41b}$$

The proofs of (8.40), and (8.41) are indicated in Appendix 3.

The work of the external forces acting on a rigid body in plane motion can be computed individually, as in (8.40), or represented as the sum of the work of the force resultant $\mathbf{F}$ moving between two positions of the center of mass, and the work of the z component of the moment resultant about the center of mass in changing the orientation of the body from θ_A to θ_B, as in (8.41).

Example 17

A homogeneous cylinder rolls without slip on the plane in Fig. 8.33(a). Figure 8.33(b) shows the free-body diagram. In particular, we are concerned with the force f, the (static friction) force of constraint. Show that this force does no work.

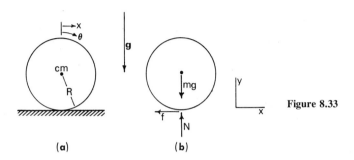

Figure 8.33

(a)   (b)

Solution

The force resultant is

$$\mathbf{F} = -f\mathbf{i} \quad \text{and} \quad d\mathbf{r}_C = dx\,\mathbf{i}$$

The moment of f about the center of mass is

$$\mathbf{M}_C = -fR\mathbf{k} \quad \text{and} \quad \boldsymbol{\omega} = -\dot{\theta}\mathbf{k}.$$

Thus (8.39) becomes

$$W = \int_{x_0}^{x_1} -f\,dx + \int_{\theta_0}^{\theta_1} fR\,d\theta.$$

If the cylinder rolls without slip,

$$\dot{x} = R\dot{\theta} \quad \text{or} \quad dx = R\,d\theta$$

and

$$W = \int_{x_0}^{x_1} -f\,dx + \int_{x_0}^{x_1} f\,dx = 0.$$

What we have shown here is not limited to homogeneous cylinders. Suppose that the center of mass lies off the geometric center of the cylinder, but the cylinder rolls without slip. Then the force f will do no net work. See Problem 8.75.

....

The above example gives a case in which we compute the work of a force which is applied to a series of points of the rigid body. We see that (8.39) enables us to compute the work of forces which are applied to various points of the body by considering the work of the force and moment resultants of those forces.

Example 18

In Fig. 8.34(a), a cylinder rolls without slip on the cart, which is moving at a constant speed v_0. Determine the work of the (static friction) constraint force f.

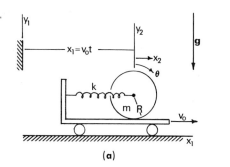

(a)

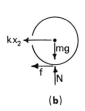

(b)

Figure 8.34

Solution

There are two immediate replies to this question. Suppose we station ourselves on the car. We note that we are in an inertial reference frame, and thus (referring to Example 17) we say that f in the free-body diagram in Fig. 8.34(b) does no work.

Now suppose we station ourselves in the x_1, y_1 reference frame. Now we see the force f move with the car, and thus we guess that f does work. Let's check. Using (8.39), we have for f:

$$\mathbf{F} = -f\mathbf{i}, \qquad d\mathbf{r}_C = (dx_1 + dx_2)\mathbf{i},$$

$$\mathbf{M}_C = -fR\mathbf{k}, \qquad \boldsymbol{\omega} = -\dot{\theta}\mathbf{k},$$

where $\dot{x}_2 = R\dot{\theta}$ by the rolling constraint. Thus the element of work of f is

$$dW = -f\,dx_1 - f\,dx_2 + fR\,d\theta$$

$$= -f\,dx_1 - f\,dx_2 + f\,dx_2.$$

Thus

$$dW = -f\,dx_1 = -fv_0\,dt.$$

In the present problem, we know that it is possible that $f \neq 0$. How then can it be that f both does and does not do work?

The answer to this simple paradox is that the computation of work and kinetic energy depends on which inertial reference we are in when we make our computation. *We must compute both work and energy in a single inertial reference frame.*

Note the computation of the kinetic energy in the two frames x_1, y_1 and x_2, y_2:

x_1, y_1: $T = \frac{1}{2}m(\dot{x}_1 + \dot{x}_2)^2 + \frac{1}{2}I^C\dot{\theta}^2$,

x_2, y_2: $T = \frac{1}{2}m(\dot{x}_2)^2 + \frac{1}{2}I^C\dot{\theta}^2$.

The expressions differ by an amount $\frac{1}{2}m(\dot{x}_1^2) = \frac{1}{2}mv_0^2$. Thus it is clear that the computation of both W and T depend on the coordinate system in which the computation is made. We now give several examples which illustrate all the techniques of this chapter.

....

Example 19

The system in Fig. 8.35(a) consists of a cylinder with a string wound about it, which passes over a weightless pulley to another mass, m_1. A force $F \sin \omega t$ is applied to the center of the cylinder.

Determine the differential equation of motion of the system, assuming that the cylinder rolls without slipping and that the string is always taut. Give criteria to show when these assumptions are violated.

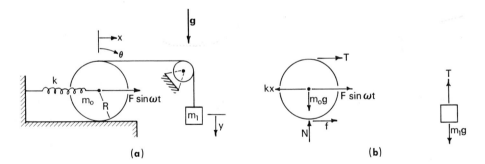

(a)   (b)

Figure 8.35

Solution

Figure 8.35(b) shows free-body diagrams for the cylinder and m_1.

The cylinder rolls without slipping. Thus

$$\dot{x} = R\dot{\theta}, \tag{a}$$

where x is the horizontal displacement of the center of the cylinder. Writing $\mathbf{F} = \dot{\mathbf{p}}_C$ and $\mathbf{M}_C = \dot{\mathbf{H}}_C$ for the cylinder, we obtain:

$$F_x = ma_x: \qquad f - kx + F \sin \omega t + T = m_0 \ddot{x},$$

$$F_y = 0: \qquad m_0 g - N = 0,$$

$$M_{Cz} = I_{zz}^C \ddot{\theta}: \qquad TR - fR = I_{zz}^C \ddot{\theta}, \tag{b}$$

where $I_{zz}^C = \frac{1}{2} m_0 R^2$. Writing $\mathbf{F} = \dot{\mathbf{p}}$ for particle m_1, we have

$$F_y = ma_y: \qquad m_1 g - T = m_1 \ddot{y}. \tag{c}$$

In addition to (a), there is another geometrical relation between variables x and y. From (a) and the fact that the point of contact is the instant center, we find that the velocity of the top of the cylinder is

$$2R\dot{\theta} = 2\dot{x}.$$

Since this is also the velocity of the string, we have

$$2R\dot{\theta} = 2\dot{x} = \dot{y}. \tag{d}$$

From (a), (b), (c), and (d), we can now eliminate the variables f, T, θ, and y to obtain

$$\ddot{x} + \left(\frac{2k}{3m_0 + 4m_1} \right) x = \frac{4m_1 g}{3m_0 + 4m_1} + \frac{2F}{3m_0 + 4m_1} \sin \omega t. \tag{e}$$

Using the methods of Chapter 6, we can easily solve this equation for the motion $x(t)$.

Let us now return to the question concerning the validity of our assumptions concerning the motion. There are two assumptions:

i) *The cylinder rolls without slip.* If we take the solution to (e), $x(t)$, and insert it into the first equation of (b), we can evaluate the force f, the force which causes the cylinder to roll without slip. If the coefficient of static friction is μ, we know that

$$f < \mu N = \mu m_0 g.$$

If this inequality is not satisfied, we know that the assumption of rolling without slip is not valid and our analysis is incorrect.

ii) *The particle m_1 moves so that $\dot{y} = 2\dot{x}$.* This will be true so long as the string is in tension. In our notation, this means that $T > 0$. If we insert $x(t)$ into (c) and find that at any time $T < 0$, our analysis will be incorrect.

··

Example 20

The rod in Fig. 8.36(a) slides without friction along the vertical wall and horizontal surface. What is the equation of motion of the rod?

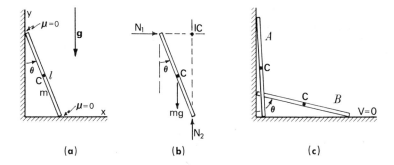

Figure 8.36

Solution

We begin by writing $\mathbf{F} = \dot{\mathbf{p}}_C$ and $\mathbf{M}_C = \dot{\mathbf{H}}_C$. The free-body diagram is shown in Fig. 8.36(b).

$$F_x = ma_{Cx}: \qquad N_1 = m\ddot{x}, \tag{a}$$

$$F_y = ma_{Cy}: \qquad mg - N_2 = m\ddot{y}, \tag{b}$$

$$M_{Cz} = I_{zz}^C \ddot{\theta}: \qquad N_2\left(\frac{l}{2}\right)\sin\theta - N_1\left(\frac{l}{2}\right)\cos\theta = I_{zz}^C \ddot{\theta} \tag{c}$$

The variables x, y (the position of the center of mass of the bar) and the angle θ are related by the geometric constraint

$$x = \frac{l}{2}\sin\theta, \qquad y = \frac{l}{2}\cos\theta. \tag{d, e}$$

For a thin rod, by (8.32′), we have

$$I_{zz}^C = \tfrac{1}{12}ml^2. \tag{f}$$

Equation (f) fixes the value of I_{zz}^C. Then equations (a), (b), (c), (d), and (e) are five equations for the five unknowns: N_1, N_2, x, y, and θ. Although these equations should be sufficient, in theory, to obtain a solution, in practice this would provide a significant challenge. Fortunately there is an alternative to the above formulation of the problem.

The simplest solution to this problem involves the use of the work–energy theorem. Again referring to Fig. 8.36(b), we see that N_1 and N_2 do no work. The weight mg is a conservative force. Thus the system is conservative.

We have the option of writing the kinetic energy T about the center of mass or the instant center. From the first formulation of this problem, it is clear that the latter is preferable.

Assume that when the inclination of the bar is $\theta = \theta_0$, $\dot\theta = 0$ (configuration A in Fig. 8.36c). Let the bar have an inclination θ in configuration B. Then we have

$$A: \qquad T = 0, \qquad V = mg\,\frac{l}{2}\cos\theta_0,$$

$$B: \qquad T = \tfrac{1}{2}(\tfrac{1}{3}ml^2)\dot\theta^2, \qquad V = mg\,\frac{l}{2}\cos\theta.$$

Equating $(T + V)$ at A and B, we get

$$\dot\theta = \sqrt{3(g/l)(\cos\theta_0 - \cos\theta)}. \tag{g}$$

If we wish $\theta(t)$, we can write

$$\int_{\theta_0}^{\theta}\frac{d\theta}{\sqrt{3(g/l)(\cos\theta_0 - \cos\theta)}} = \int_0^t dt = t. \tag{h}$$

A numerical integration of (h) will give $t = t(\theta)$. In principle this may be inverted to yield $\theta = \theta(t)$, although the inversion may have to be done graphically.

It is worth pointing out that although the energy equation leads to a quick solution to the problem, energy alone cannot answer all questions which arise in regard to this problem. There is an angle at which the left end of the bar leaves the vertical wall. To solve for this angle, we must resort to equations (a), (b), (c), (d), (e), and use (g).

....

Example 21

A weightless professor (Fig. 8.37) stands on a rotating turntable holding dumbbells, each of mass $m/2$. The turntable rotates freely, without friction. Suppose that, when the dumbbells are at a radius r_0, the angular rate of the turntable is ω. The dumbbells are then brought in to a radius r_1. Determine (i) the new angular rate ω_1, and (ii) the work done by the professor.

Solution

A free-body diagram of the system will show no net external forces or moments. Thus, writing $\mathbf{M}_C = \dot{\mathbf{H}}_C$, we get $\mathbf{H}_C = \text{const}$. Thus $I_{zz}^C \dot\theta = \text{const}$. In the first position, we have

$$(I + mr_0^2)\omega_0 \qquad \text{at} \qquad r = r_0,$$

and at the second position

$$(I + mr_1^2)\omega_1 \qquad \text{at} \qquad r = r_1,$$

where I is the moment of inertia of the turntable about its center. Equating these two values yields:

$$\omega_1 = \left(\frac{I + mr_0^2}{I + mr_1^2}\right)\omega_0. \tag{a}$$

Thus, as r decreases, ω increases.

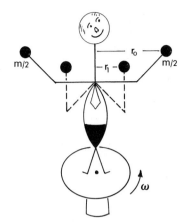

Figure 8.37

Now let us compute the work done by the professor. Since the external forces do no work, any change in the kinetic energy of the system must be equal to the work of the internal forces, i.e., the work done by the professor. Computing the kinetic energy at the two positions gives

$$T = \tfrac{1}{2}(I + mr_0^2)\omega_0^2 \qquad \text{at } r = r_0$$

or

$$T = \tfrac{1}{2}(I + mr_1^2)\omega_1^2$$
$$T = \frac{1}{2}\frac{(I + mr_0^2)^2}{I + mr_1^2}\,\omega_0^2$$

$$\text{at } r = r_1$$

Subtracting the first from the second gives the work:

$$\text{Work} = \frac{1}{2}\left(\frac{I + mr_0^2}{I + mr_1^2}\right)m(r_0^2 - r_1^2)\omega_0^2.$$

· ·

Example 22

Consider the system in Fig. 8.38(a), in which the car has constant acceleration a_0. A thin bar of length l and mass m is pinned at O and rests on the car.

i)   Calculate the acceleration a_0 such that the bar just begins to rotate.

ii) If a_0 is twice the value determined above, write the equation for the angular motion of the bar.

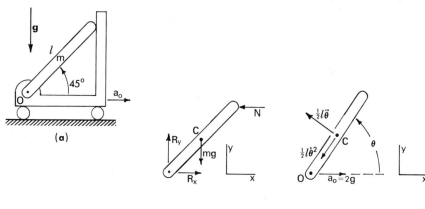

Figure 8.38   (b)   (c)

Solution

i) So long as the bar rests on the car, the acceleration of the center of mass of the bar is $a_0\mathbf{i}$. Noting the free-body diagram in Fig. 8.38(b), we can write $\mathbf{F} = \dot{\mathbf{p}}_C$ for the bar.

$$F_x = ma_{Cx}: \qquad R_x - N = ma_0, \tag{a}$$

$$F_y = ma_{Cy}: \qquad R_y - mg = 0. \tag{b}$$

And since the bar does not rotate, $\mathbf{M}_C = 0$. Thus

$$\mathbf{M}_C = 0: \qquad \left(R_x \frac{l}{2} \sin 45° - R_y \frac{l}{2} \cos 45° + N \frac{l}{2} \sin 45°\right)\mathbf{k} = 0$$

or

$$R_x - R_y + N = 0. \tag{c}$$

We wish to determine a_0 for the case $N \to 0$. Thus, from (a), (b), and (c), we get

$$a_0 = g. \tag{d}$$

ii) We now suppose that $a_0 = 2g$. The bar rotates, so that we now have to determine the acceleration of the center of mass. The point O has acceleration a_0. Thus, from Fig. 8.38(c), we obtain

$$a_{Cx} = 2g - \frac{l}{2}\dot{\theta}^2 \cos \theta - \frac{l}{2}\ddot{\theta} \sin \theta,$$

$$a_{Cy} = -\frac{l}{2}\dot{\theta}^2 \sin \theta + \frac{l}{2}\ddot{\theta} \cos \theta.$$

Writing $\mathbf{F} = \dot{\mathbf{p}}_C$, we now have

$$F_x = ma_{Cx}: \qquad R_x = m\left(2g - \frac{l}{2}\dot{\theta}^2\cos\theta - \frac{l}{2}\ddot{\theta}\sin\theta\right), \tag{a'}$$

$$F_y = ma_{Cy}: \qquad R_y - mg = m\left(-\frac{l}{2}\dot{\theta}^2\sin\theta + \frac{l}{2}\ddot{\theta}\cos\theta\right). \tag{b'}$$

And from $\mathbf{M}_C = \dot{\mathbf{H}}_C$, noting that $I_{zz}^C = \frac{1}{12}ml^2$, we get

$$M_{C2} = I_{zz}^C\ddot{\theta}: \qquad R_x\frac{l}{2}\sin\theta - R_y\frac{l}{2}\cos\theta = \frac{1}{12}ml^2\ddot{\theta}$$

or

$$R_x\sin\theta - R_y\cos\theta = \tfrac{1}{6}ml\ddot{\theta}. \tag{c}$$

Multiplying (a') by $\sin\theta$ and (b') by $\cos\theta$ and subtracting gives

$$R_x\sin\theta - R_y\cos\theta + mg\cos\theta = 2mg\sin\theta - m\frac{l}{2}\ddot{\theta}. \tag{d}$$

Subtracting (c) from (d) yields

$$mg\cos\theta = 2mg\sin\theta - \tfrac{2}{3}ml\ddot{\theta}$$

or

$$\ddot{\theta} = \tfrac{3}{2}(g/l)(2\sin\theta - \cos\theta).$$

This equation can be integrated one time to give $\dot{\theta} = \dot{\theta}(\theta)$. Let $\dot{\theta} = \omega$. Then

$$\ddot{\theta} = \frac{d\omega}{d\theta}\omega \qquad \text{(why?)}.$$

See Problem (8.109).

....

8. VIBRATION PROBLEMS INVOLVING RIGID BODIES

In Chapter 6, we discussed simple vibrations problems. We said that we had a linear vibration problem if the differential equation of motion for the system was reducible to the form

$$\ddot{x} + 2\zeta\omega_n\dot{x} + \omega_n^2 x = F_0(t). \tag{8.42}$$

Although all the problems in Chapter 6 involved the motion of a single particle, rigid bodies are often involved in vibration problems. Recall Examples 9, 10, and 19 of this chapter. In each of these examples, the final equation of motion for the system was essentially (8.42). Thus all the results of Chapter 6 pertinent to (8.42) apply to these rigid-body problems.

Example 23

A cylinder of mass m and radius r rolls without slipping inside another cylinder of radius R (Fig. 8.39). Calculate the natural frequency of small motions of the rolling cylinder.

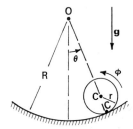

Figure 8.39

Solution

Let θ measure the angle which OC makes with the vertical. The point of contact of the rolling cylinder is its instant center. Thus, if ϕ measures the rotation of the cylinder, we have

$$(R - r)\dot{\theta} = r\dot{\phi}.$$

Suppose that the cylinder oscillates between θ_0 and $-\theta_0$. Then, noting that this is a conservative system, we obtain

$$\theta = \theta_0: \qquad T = 0, \qquad V = mg(R - r)(1 - \cos\theta_0)$$

$$\theta = 0: \qquad T = \tfrac{1}{2}(\tfrac{3}{2}mr^2)\dot{\phi}^2 = \tfrac{3}{4}m(R - r)^2\dot{\theta}^2, \qquad V = 0.$$

We let

$$\cos\theta_0 \approx 1 - \frac{\theta_0^2}{2} \qquad \text{and} \qquad \theta = \theta_0 \cos\omega_n t.$$

Then, equating $V_{\max} = T_{\max}$, we obtain

$$mg(R - r)\frac{\theta_0^2}{2} = \tfrac{3}{4}m(R - r)^2\omega_n^2\theta_0^2$$

or

$$\omega_n^2 = \frac{2}{3}\left(\frac{g}{R - r}\right).$$

Thus

$$f = \frac{1}{2\pi} \sqrt{\frac{2}{3}\left(\frac{g}{R-r}\right)} \text{ hertz.}$$

....

9. SUMMARY

Kinematics of a Rigid Body in Plane Motion

Relative Motion

$$\mathbf{r}_P = \mathbf{r}_A + \mathbf{r}_{P/A}, \qquad \mathbf{v}_P = \mathbf{v}_A + \mathbf{v}_{P/A}, \qquad \mathbf{a}_P = \mathbf{a}_A + \mathbf{a}_{P/A}.$$

If A and P are points of a rigid body in plane motion, $\mathbf{v}_{P/A}$ and $\mathbf{a}_{P/A}$ are the velocity and acceleration of a point in circular motion.

If z is an axis perpendicular to the plane of motion and the angular velocity of the body is

$$\boldsymbol{\omega} = \dot{\theta}\mathbf{k},$$

then the relative velocity $\mathbf{v}_{P/A}$ can be expressed

$$\mathbf{v}_{P/A} = \boldsymbol{\omega} \times \mathbf{r}_{P/A}.$$

Instant Center

At each instant for which $\dot{\theta} \neq 0$, there is one point of the rigid body for which $\mathbf{v} = 0$. This point is called the instant center. Then

$$\mathbf{v}_P = \boldsymbol{\omega} \times \mathbf{r}_{P/IC}.$$

Equations of Motion

$\mathbf{F} = \dot{\mathbf{p}}_C$: $F_x = ma_{Cx}, \qquad F_y = ma_{Cy}$

$\mathbf{M}_O = \dot{\mathbf{H}}_O$:   i) If O is a fixed point of the body, $M_{Oz} = I_{zz}^O\, \ddot{\theta}$.

ii) If there is no fixed point in the body, the most convenient equation is written about the center of mass $M_{Cz} = I_{zz}^C\, \ddot{\theta}$.

Work–Energy: $T = \frac{1}{2} I_{zz}^O\, \dot{\theta}^2$ O, a fixed point of the body

$T = \frac{1}{2} I_{zz}^{IC}\, \dot{\theta}^2$ IC, the instant center

$T = \frac{1}{2} m v_C^2 + \frac{1}{2} I_{zz}^C\, \dot{\theta}^2$ C, the center of mass.

Moment of Inertia

$$I_{zz}^A = \int_V (x^2 + y^2)\ dm$$

where x, y are measured from the point A. Also

$$I_{zz}^O = I_{zz}^C + md^2 \qquad \text{parallel-axis theorem.}$$

The moment of inertia I_{zz} can be evaluated experimentally, numerically, or from a table of inertias. See Appendix 2.

PROBLEMS

Section 8.2

Problem 8.1

The cart in the figure has velocity $\mathbf{v} = 10\mathbf{i}$ ft/sec and acceleration $\mathbf{a} = -5\mathbf{i}$ ft/sec^2 at the instant shown. The angle $\theta = 30°$, $\dot\theta = 2$ rad/sec, and $\ddot\theta = 1$ rad/sec^2. Calculate: (a) $\mathbf{v}_P$, (b) $\mathbf{a}_P$, at the instant shown.

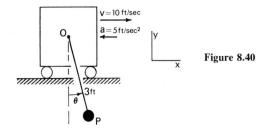

Figure 8.40

Problem 8.2

At the instant shown in the figure, the velocity of point B is zero and the acceleration is $\mathbf{a}_B = 10\mathbf{i}$ ft/sec^2. Assuming that the point A remains on the vertical wall, calculate: (a) $\dot\theta$ and $\ddot\theta$ of the bar, (b) $\mathbf{v}_A$ and $\mathbf{a}_A$.

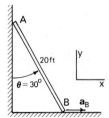

Figure 8.41

Problem 8.3

Point A of the right triangular plate shown in the figure slides on the vertical wall, and point B slides on the horizontal surface. At the instant shown, $\mathbf{v}_A = -20\mathbf{j}$ ft/sec and $\mathbf{a}_B = -10\mathbf{i}$ ft/sec^2. Calculate (a) $\dot\theta$ and $\ddot\theta$ of the plate, (b) $\mathbf{v}_P$ and $\mathbf{a}_P$.

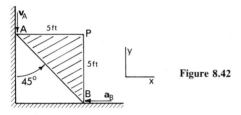

Figure 8.42

Problem 8.4

The bar in the figure has the velocities shown at points A and B. Determine $\dot\theta$ of the bar.

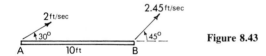

Figure 8.43

Problem 8.5

The bar in the figure passes over point A and has one end which follows the circular path. In the position shown, $\mathbf{v}_B = 10\mathbf{i}$ ft/sec. Determine: (a) $\dot\theta$ of the bar, (b) the velocity of that point of the bar which is directly over point A.

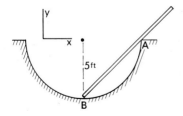

Figure 8.44

Problem 8.6

A bar of length $\sqrt{2}\,R$ slides inside the circular surface as shown in the figure. When the bar has the position shown in (a), $\mathbf{v}_A = -2\mathbf{j}$ ft/sec and $\mathbf{a}_A = \mathbf{i} - 8\mathbf{j}$ ft/sec^2.

a) Prove that $\theta = \phi - \frac{1}{4}\pi$ and thus that $\dot{\theta} = \dot{\phi}$ and $\ddot{\theta} = \ddot{\phi}$.

b) Determine the radius R of the circle.

c) Determine $\dot{\theta}$ and $\ddot{\theta}$ of the bar.

d) Determine $\mathbf{v}_B$, $\mathbf{a}_B$, and $\mathbf{v}_{B/A}$, $\mathbf{a}_{B/A}$.

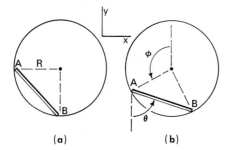

(a)   (b)   **Figure 8.45**

Problem 8.7

The string in the figure has a total length l. The support pulley moves to the right with the constant speed v_0, starting at the wall when $t = 0$. The angle θ has the derivatives $\dot{\theta}$ and $\ddot{\theta}$ at time t. Calculate (a) $\mathbf{v}_P$, (b) $\mathbf{a}_P$.

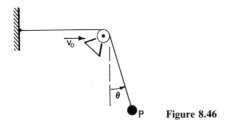

Figure 8.46

Problem 8.8

Consider again the system of Problem 8.7. Suppose now that the support moves to the right with a constant acceleration a_0. Assume that the support starts from rest at the wall at $t = 0$. Compute $\mathbf{v}_P$ and $\mathbf{a}_P$.

Problem 8.9

Consider the system in the figure. The angles θ and ϕ are measured as shown, and have derivatives $\dot{\theta}$, $\ddot{\theta}$ and $\dot{\phi}$, $\ddot{\phi}$. Find the horizontal and vertical (that is, x and y) components of the following vectors: (a) $\mathbf{v}_{A/O}$, $\mathbf{v}_{P/A}$, and $\mathbf{v}_{P/O}$, (b) $\mathbf{a}_{A/O}$, $\mathbf{a}_{P/A}$, and $\mathbf{a}_{P/O}$.

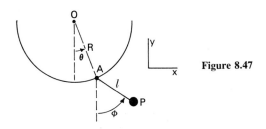

Figure 8.47

Problem 8.10

Consider the system in the figure. The angles θ and ψ, which describe the motion, have the derivatives $\dot{\theta}$, $\ddot{\theta}$ and $\dot{\psi}$, $\ddot{\psi}$. Write the $\mathbf{e}_1$ and $\mathbf{e}_2$ components of the following vectors: (a) $\mathbf{v}_{P/O}$, (b) $\mathbf{a}_{P/O}$.

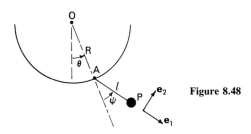

Figure 8.48

Section 8.3

Problem 8.11

For the system shown in Fig. 8.49, the cylinder rolls without slip. Show that the instant center of the cylinder is I_1 and the instant center of the bar is I_2 in the position shown.

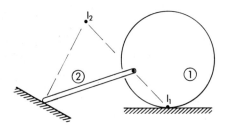

Figure 8.49

Problem 8.12

Consider the bar in the figure; its ends remain on the surfaces indicated. Show that the instant center has the position shown.

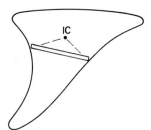

Figure 8.50

Problem 8.13

For the system in the figure, determine (a) the angular speed ω of each bar, (b) the speed of point A at the instant indicated.

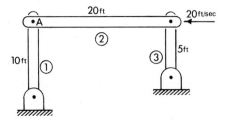

Figure 8.51

Problem 8.14

For the system in the figure, determine (a) the angular speed ω of each bar, (b) the position of the instant center of bar (2), (c) the speed of point A.

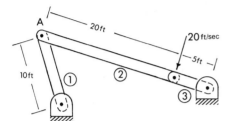

Figure 8.52

Problem 8.15

The cart in Fig. 8.53 moves with a speed of $v_0 = 20$ ft/sec. The cylinder moves so that there is no slip with respect to either the cart or the fixed surface above it. Calculate: (a) the angular speed ω of the disk, (b) the angular speed ω of the bar, (c) the speed of point A of the bar.

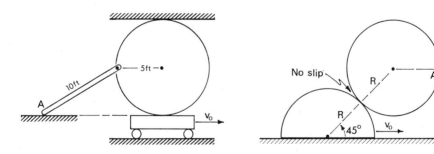

Figure 8.53   Figure 8.54

Problem 8.16

The semicircular disk of radius R in Fig. 8.54 moves to the right with a speed of $v_0 = 10$ ft/sec, while the circular disk of radius R rolls without slip on the semicircular disk and slides on the wall. For the position shown: (a) calculate the angular speed of the circular disk, (b) calculate the speed of point A of the circular disk.

Problem 8.17

The cart in the figure has a speed v_0 to the right. When the angle of the bar is $\theta = 0$, suppose that $\dot{\theta} = 2v_0/l$. What is the position of the instant center of the bar at this instant?

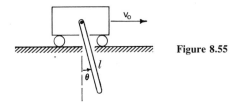

Figure 8.55

Problem 8.18

The cart in the figure moves with the constant speed v.

a) Find the position of the instant center of bar (2) as a function of the angle θ.

b) Find $\dot{\theta}$ as a function of θ and v.

c) Find the speed of the point P as a function of θ and v.

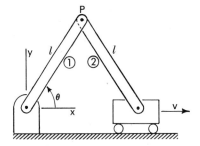

Figure 8.56

Problem 8.19

Consider the system of Problem 8.5.

a) Calculate the position of the instant center of the bar: (i) when the bar is horizontal, (ii) when the left end of the bar is at the lowest point of the circular path.

b) Suppose that the speed of the left end of the bar is 10 ft/sec. Determine the angular speed of the bar ω: (i) when the bar is horizontal, (ii) when the left end of the bar is at the lowest point of the circular path.

Problem 8.20

A bar of *unit length* moves so that its ends lie on the surfaces shown in the figure, with the right end higher than the left. Show that the position of the instant center is

$$x_{IC} = x + 2x \sqrt{1 - x^2}, \qquad y_{IC} = x^2 - \sqrt{1 - x^2},$$

where x is the horizontal displacement of the right end.

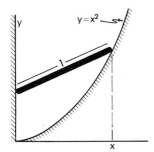

Figure 8.57

Problem 8.21

A cylinder 10 ft in diameter rolls without slip on a cart, as shown in the figure. The angular rate of the cylinder is $\omega = 1.0$ rad/sec, the speed of the cart is $v = 5$ ft/sec, and the acceleration is $a = 6294$ ft/sec^2.

a) Calculate the position of the instant center of the cylinder.

b) Suppose that the direction of the rotation of the cylinder were reversed. Now compute the position of the instant center.

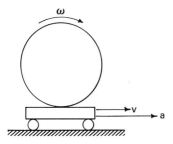

Figure 8.58

Section 8.4

Problem 8.22

For the system in Fig. 8.59, suppose that the angular speed ω_0 and the angular acceleration α_0 of bar BC are given in the senses shown.

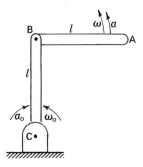

Figure 8.59

a) Calculate ω and α of bar AB such that $\mathbf{a}_A = 0$ in the position shown.

b) Under the conditions of part (a), compute $\mathbf{v}_A$.

Problem 8.23

For the system shown in the figure, bar AB has angular speed ω_2 and angular accelera-
tion α_2, and bar BC has ω_1 and α_1. For the position shown, calculate: (a) $\mathbf{v}_B$ and $\mathbf{a}_B$,
(b) $\mathbf{v}_A$ and $\mathbf{a}_A$, (c) the conditions under which $\mathbf{a}_A = 0$ in the position shown.

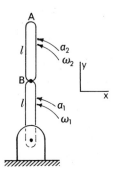

Figure 8.60

Problem 8.24

The cylinder in the figure rolls without slip. The bar of length $2R$ is pinned to the cylinder at A and slides on the horizontal surface, and point C of the cylinder has constant
speed v_0. In the position shown, calculate (a) ω of the bar, (b) α of the bar, (c) $\mathbf{a}_B$.

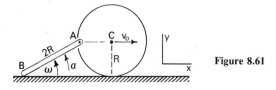

Figure 8.61

Problem 8.25

The cylinder of radius r shown in the figure rolls without slip inside a cylindrical surface of radius R. Its motion is described by the rotation angles θ and ϕ.

a) Show that, if there is no slipping,

$$\dot{\theta} = \left(\frac{R - r}{r}\right) \dot{\phi}.$$

b) Suppose that the angle ϕ has derivatives $\dot{\phi}$ and $\ddot{\phi}$. What is the acceleration of
the instant center of the cylinder?

c) The speed of the center of the cylinder is v_C. Show that $\mathbf{a}_{IC} \to -(v_C^2/r)\mathbf{e}_1$ as
$R \to \infty$. [*Hint:* Write $\mathbf{a}_{IC}$ in terms of $\dot{\theta}$ and $\ddot{\theta}$, then let $R \to \infty$.]

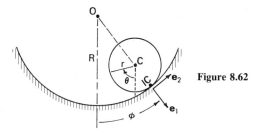

Figure 8.62

Problem 8.26

The cylinder of radius r shown in the figure rolls without slip on the cylindrical surface
of radius R, and the rotation angles θ and ϕ describe the motion.

a) Show that, if there is no slipping,

$$\dot{\theta} = \left(\frac{R + r}{r}\right) \dot{\phi}.$$

b) The angle ϕ has derivatives $\dot{\phi}$ and $\ddot{\phi}$. What is the acceleration of the instant
center of the cylinder?

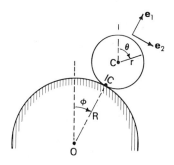

Figure 8.63

c) The speed of point C is v_C. Show that as $R \to \infty$, $\mathbf{a}_{IC} \to (v_C^2/r)\mathbf{e}_1$.

d) In part (c), show that, as $R \to 0$, $\mathbf{a}_{IC} \to 0$.

★ Problem 8.27

The system shown in the figure is released from rest at the angle $\theta_0 > 61°$. It is given that

$$\ddot{\theta} = \frac{3}{2}\left(\frac{g}{l}\right)\sin\theta, \qquad \dot{\theta}^2 = 3\left(\frac{g}{l}\right)(\cos\theta_0 - \cos\theta).$$

The acceleration of the mass m is $\ddot{y} = -g$. Show that the mass m will leave contact with the bar upon release of the system, and that the mass will not catch up to the bar until after the bar stops.

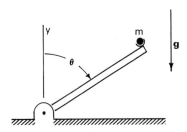

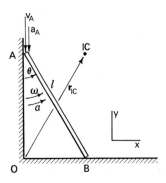

Figure 8.64

Problem 8.28

The bar in the figure slides so that point A remains on the vertical wall and point B remains on the horizontal surface. The speed and acceleration of point A are given to be v_A and a_A, and the inclination angle of the bar is θ.

Figure 8.65

a) Compute ω of the bar.

b) Calculate α of the bar. [*Hint:* Use the condition that $a_{By} = 0$.]

c) Show that the acceleration of the instant center *IC* of the bar can be written

$$\mathbf{a}_{IC} = -\left[\frac{v_A^2}{(l \sin \theta)^2}\right] \mathbf{r}_{IC},$$

where $\mathbf{r}_{IC}$ is the vector from *O* to the instant center.

Problem 8.29

Derive equations (8.15).

Problem 8.30

For a rigid body in plane motion, the angle θ identifies the rotation of the body in the plane; see the figure. Suppose that either $\dot{\theta} \neq 0$ or $\ddot{\theta} \neq 0$ (or both). Show that at each instant, some point *P* of the body has a zero acceleration, $\mathbf{a}_P = 0$. [*Hint:* Use Eqs. (8.15) and solve for the position of the point *P*.]

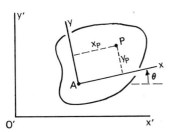

Figure 8.66

Problem 8.31

Consider a rigid body in plane motion. At some instant, the acceleration of point *P* of the body is zero, $\mathbf{a}_P = 0$. Consider any other point of the body, P_1, at the same instant (see the figure).

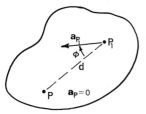

Figure 8.67

a) Show that $|\mathbf{a}_{P1}|$ is proportional to the distance d between points P and P_1.

b) Show that the angle which $\mathbf{a}_{P1}$ makes with the line from P to P_1 is the same for all points P_1. Evaluate the angle in terms of $\dot{\theta}$ and $\ddot{\theta}$.

Section 8.5

Problem 8.32

A homogeneous cylinder of radius R is suspended at its edge, as shown in the figure. The natural frequency of small amplitude motions of the cylinder is $\omega_n = \sqrt{2g/3R}$ rad/sec. Show that the moment of inertia of the cylinder about the point on its edge is $I_{zz}^0 = \frac{3}{2}mR^2$, where m is the mass of the cylinder.

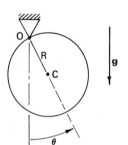

Figure 8.68

Problem 8.33

For the system of Problem 8.32, write $\mathbf{F} = \dot{\mathbf{p}}_C$ and $\mathbf{M}_C = \dot{\mathbf{H}}_C$, and use $I_{zz}^C = \frac{1}{2}mR^2$. Show that the differential equation of motion of the cylinder is

$$\ddot{\theta} + \left(\frac{2g}{3R}\right)\sin\theta = 0.$$

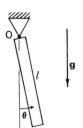

Figure 8.69

Problem 8.34

A thin homogeneous bar is suspended at one end, as shown in the figure. The period of small oscillations of the bar is

$$\tau = 2\pi \sqrt{2l/3g} \text{ sec.}$$

Show that the moment of inertia of the bar about an endpoint is $I_{zz}^0 = \frac{1}{3}ml^2$, where m is the mass of the bar.

Problem 8.35

The thin homogeneous bar of length l and mass m shown in the figure is pinned at its center of mass and restrained by a spring k at one end. The natural frequency of small motions of the system is $\omega_n = \sqrt{3k/m}$ rad/sec. Show that the moment of inertia of the bar about its center of mass is

$$I_{zz}^C = \frac{1}{12}ml^2.$$

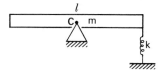

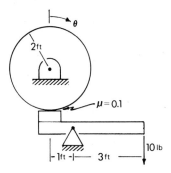

Figure 8.70

Problem 8.36

The homogeneous disk shown in the figure weighs 64 lb, and is rotating at an angular speed of 2.0 rad/sec when the brake is applied.

a) Determine the time required to stop the disk.

b) Determine the total angular rotation of the disk before it comes to a stop. [Hint: $I_{zz}^C = \frac{1}{2}mR^2$.]

Figure 8.71

Problem 8.37

The cylinder in the figure rolls and slips on the horizontal surface. At $t = 0$, $x = 0$, $\dot{x} = v_0$, $\theta = \dot{\theta} = 0$. Calculate the time at which the cylinder begins to roll without slipping. [*Hint*: $I^C_{zz} = \frac{1}{2}mR^2$ for the cylinder.]

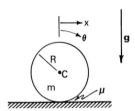

Figure 8.72

Problem 8.38

The homogeneous cylinder of mass m and radius R shown in the figure is subjected to a constant force F. The coefficient of static friction between the cylinder and the plane is μ.

a) What is the maximum value of the force F such that the cylinder still rolls without slip?

b) Suppose that F is larger than the maximum determined in part (a). At $t = 0$, $x = \dot{x} = \theta = \dot{\theta} = 0$. Calculate $x(t)$ and $\theta(t)$. [*Hint*: $I^C_{zz} = \frac{1}{2}mR^2$ for the cylinder.]

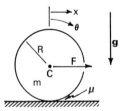

Figure 8.73

Problem 8.39

The hula hoop shown in Fig. 8.74 is considered to be a thin ring of mass m and radius R. The hoop is given a backward spin and a forward speed.

At $t = 0$: $x = 0$, $\dot{x} = 10$ ft/sec,

$\theta = 0$, $\dot{\theta} = -6$ rad/sec.

Given that $mg = 2$ lb, $R = 2$ ft, $I_{zz}^C = mR^2$, and the coefficient of sliding friction is $\mu = 0.3$, determine $x(t)$ until the hoop stops slipping.

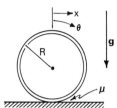

Figure 8.74

Problem 8.40

The homogeneous cylinder of mass m and radius R rolls without slip on the inclined plane shown in the figure.

a) What is the acceleration of the center of mass of the cylinder as a function of the angle θ?

b) What is the force of static friction acting on the cylinder as a function of the angle θ?

c) Given that μ is the coefficient of static friction between the cylinder and the incline, compute the range of angles θ for which rolling without slipping is possible. [*Hint:* $I_{zz}^C = \frac{1}{2}mR^2$.]

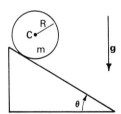

Figure 8.75

Problem 8.41

The cylinder shown in Fig. 8.76 has a groove cut in it to an inner radius r, and the outer radius is R. The cylinder rolls without slip on its inner radius down an inclined plane, and $I_{zz}^C \approx \frac{1}{2}mR^2$. Calculate:

a) the acceleration of the center of mass of the cylinder,

b) the force of static friction acting on the cylinder.

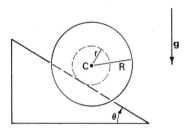

Figure 8.76

Problem 8.42

The homogeneous cylinder in the figure, which has a weightless string wound around it, slides down the smooth inclined plane. What is the acceleration of the center of mass of the cylinder?

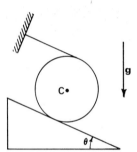

Figure 8.77

Problem 8.43

Consider again the system of Problem 8.42, and suppose now that the coefficients of static and sliding friction between the cylinder and the incline are both μ.

a) The cylinder is released from rest. Calculate the range of angles θ for which the cylinder will remain at rest.

b) Suppose that θ is larger than the values determined in part (a). Determine the acceleration of the center of mass of the cylinder.

Section 8.6

Problem 8.44

A thin bar of length l has a variable density (mass/length) of

$$\mu(x) = \mu_1 \left(1 - \frac{x}{l}\right) + \mu_2 \left(\frac{x}{l}\right).$$

a) What is the total mass of the bar?

b) What is the moment of inertia I^O_{zz} as a function of μ_1, μ_2, and l? Here O is the point $x = 0$, the end of the bar.

Problem 8.45

The rigid body in the figure is suspended from the point P, which is a distance y from the center of mass of the body. It is found that the period of small oscillations of the body is

$$\tau(y) = 2\pi \sqrt{\frac{d^2 + y^2}{gy}} \, .$$

Given that the total mass of the body is m:

a) show that $I^C_{zz} = md^2$,

b) show that I^P_{zz} can be written $I^P_{zz} = I^C_{zz} + my^2$.

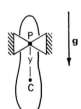

Figure 8.78

Problem 8.46

The triangular plate shown in the figure has a constant density and a total mass of m.

a) Compute the moment of inertia I^B_{zz}, where B is the point $(0,b)$.

b) Find the minimum value of I^B_{zz} over the range of possible values of the parameter b. What is the significance of the point B which is determined?

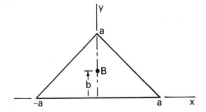

Figure 8.79

Problem 8.47

Calculate the moment of inertia I_{zz}^C of the constant-density, rectangular plate shown in the figure. Its total mass is m.

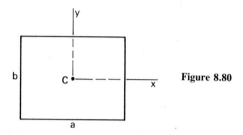

Figure 8.80

Problem 8.48

The rectangular plate shown in the figure has a hole of radius R cut from its center. The mass of the plate is m. Compute I_{zz}^C.

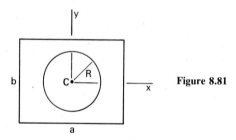

Figure 8.81

Problem 8.49

Calculate the moment of inertia I_{zz}^O of the plate shown in the figure, which has a constant density and a hole with a radius R. The total mass of the plate is m.

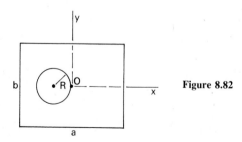

Figure 8.82

Problem 8.50

Determine I^0_{yy} for the right circular cone of mass m shown in the figure.

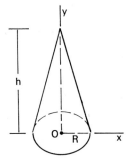

Figure 8.83

Problem 8.51

Compute the moment of inertia of a homogeneous sphere of mass m and radius R about an axis through the center of the sphere.

Problem 8.52

Determine the moment of inertia of a thin homogeneous disk of radius R and mass m about an axis l through a diameter of the disk.

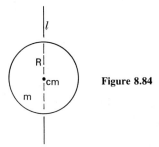

Figure 8.84

Problem 8.53

A plate of constant density is bounded by the curves $y = \frac{1}{2}b\sqrt{x/a}$, $y = -\frac{1}{2}b\sqrt{x/a}$, and the line $x = a$. (See Fig. 8.85.) Determine I^0_{zz} for the plate, and write the result in terms of the total mass m and the dimensions a and b.

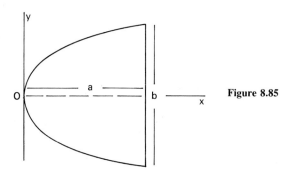

Figure 8.85

Problem 8.54

A cube with sides of length a has a right circular conical hole of base radius R and height a cut from it, as shown in the figure. The mass of the block is m. Calculate I^0_{zz}.

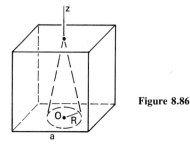

Figure 8.86

Section 8.7

Problem 8.55

The plate in Problem 8.3 weighs 16 lb. Determine the kinetic energy of the plate for the conditions stated in Problem 8.3. [*Hint:* Determine the moment of inertia of the plate from the Table of Inertias in Appendix 2.]

Problem 8.56

The bar in Problem 8.4 weighs 32 lb. Determine the kinetic energy of the bar for the conditions indicated in Problem 8.4.

Problem 8.57

The bar in Problem 8.5 weighs 16 lb and is 14.14 ft long so that, at the instant indicated, the center of mass is directly above the point A. Calculate the kinetic energy of the bar for the conditions indicated in Problem 8.5.

Problem 8.58

Suppose that in figure (a) of Problem 8.6, $R = 5\sqrt{2}$ ft $= 7.07$ ft. The bar weighs 8 lb and $v_A = -2\mathbf{j}$ ft/sec. Determine the kinetic energy of the bar.

Problem 8.59

For the system of Problem 8.13, suppose that each bar has a density of 0.1 slug/ft. (Neglect the overhang of the bars beyond the pins.) Determine the total kinetic energy of the system for the instant indicated in Problem 8.13.

Problem 8.60

For the system of Problem 8.14, suppose that each bar has a density of 0.1 slug/ft. Determine the kinetic energy of the system for the conditions indicated in Problem 8.14.

Problem 8.61

Consider the system of Problem 8.15, and suppose that the disk weighs 64 lb, the bar weighs 8 lb, and the cart weighs 160 lb. Determine the kinetic energy of the entire system for the conditions indicated in Problem 8.15.

Problem 8.62

Consider the system of Problem 8.16, and suppose that the semicircular disk weighs 32 lb, the circular disk weighs 64 lb, and $R = 2$ ft. Determine the total kinetic energy of the system for the conditions indicated in Problem 8.16.

Problem 8.63

For the system of Problem 8.17, let the mass of the bar be m and the mass of the cart be $10m$. Write the total kinetic energy of the system in terms of the parameters m, l, and v_0.

Problem 8.64

Suppose that the disk in Problem 8.21 weighs 1600 lb.

   a) Determine the kinetic energy of the *disk* for part (a) of Problem 8.21.

   b) Determine the kinetic energy of the *disk* for part (b) of Problem 8.21.

Problem 8.65

Suppose that the mass of the disk in Problem 8.25 is m. Show that the kinetic energy of the disk can be written

$$T = \tfrac{3}{4}m(R-r)^2\dot{\phi}^2.$$

Problem 8.66

Suppose that the mass of the disk in Problem 8.26 is m. Show that the kinetic energy of the disk can be written

$$T = \tfrac{3}{4}m(R+r)^2\dot{\phi}^2.$$

Problem 8.67

Consider the system of Problem 8.35, and let θ be the angle of rotation of the bar. Suppose that when $\theta = 0$, $\dot{\theta} = \omega_0$. Calculate the total energy of the system, $T + V$, and hence determine an expression for $\dot{\theta}$ as a function of θ. Assume that θ remains small during the motion of the system.

Problem 8.68

In Problem 8.36, write the work–energy relation to determine the total angular rotation of the disk [part (b)].

Problem 8.69

The mass of the cylinder in Problem 8.43 is m and the radius is R. Compute the work of the force of friction when the point C moves a distance Δ down the incline.

Problem 8.70

Suppose that, in Problem 8.43, when $x_C = 0$, $\dot{x}_C = 0$, where x_C is the displacement of point C along the incline. Use the result of Problem 8.69 to determine $\dot{x}_C$ when $x_C = \Delta$.

Problem 8.71

In the system shown in the figure, the constant forces P and Q are applied to the center of the disk. Suppose that the disk is released from rest at $\phi = 0$.

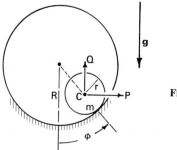

Figure 8.87

a) Compute the total work done by the forces P and Q when the disk moves from $\phi = 0$ to ϕ.

b) Given that P and Q are such that the disk rolls without slipping, calculate the maximum angular displacement which the disk will attain.

Problem 8.72

In Problem 8.71, suppose that $P = mg$ and $Q = 0$.

a) Calculate $\dot{\phi}$ as a function of ϕ.

b) Calculate the maximum value of $\dot{\phi}$ and the angle at which it occurs.

c) Calculate the angle at which $\dot{\phi} = 0$.

Problem 8.73

For the system shown in the figure, the force P is always applied perpendicular to the line OC and has a constant magnitude. At $\phi = 0$, the disk is released from rest. At $\phi = \frac{1}{2}\pi$, the rate $\dot{\phi}$ again becomes zero. Assuming that the cylinder rolls without slipping, compute the magnitude of the force P.

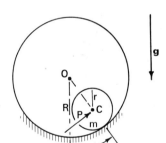

Figure 8.88

Problem 8.74

For the system of Problem 8.73:

a) Calculate $\dot{\phi}$ as a function of ϕ.

b) Calculate the maximum value of $\dot{\phi}$ and the angle at which it occurs.

c) If $P = \frac{1}{2}mg$, determine (i) the angle at which $\dot{\phi}$ is maximum, (ii) the angle at which $\dot{\phi} = 0$.

Problem 8.75

The nonhomogeneous cylinder in Fig. 8.89 rolls without slipping. Show that the force of static friction f which acts on the cylinder does no work.

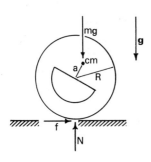

Figure 8.89

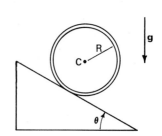

Figure 8.90

Problem 8.76

The thin ring in Fig. 8.90 rolls without slip down the inclined plane. Determine the acceleration of the center of mass of the ring.

Problem 8.77

The disk of mass m and radius R shown in the figure rolls without slip on an inclined plane with angle θ. The particle m slides down another inclined plane with angle θ. The coefficient of sliding friction between the particle and its inclined plane is μ, and the disk and particle are released from rest at the tops of the inclines.

    a) Determine μ so that the disk and the particle arrive at the bottoms of the inclines at the same time.

    b) Determine the kinetic energy at the bottoms of the inclines of (i) the disk, (ii) the particle.

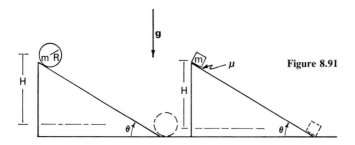

Figure 8.91

Problem 8.78

The disk in Fig. 8.92 rolls and slips on the horizontal surface, and μ is the coefficient of sliding friction between the disk and the horizontal surface. At $t = 0$:

$$\dot{x} = v_0 \quad \text{and} \quad \dot{\theta} = -\omega_0 \quad (v_0 > 0, \, \omega_0 > 0).$$

a) Determine the time t_1 at which the disk starts rolling without slip.
b) Determine v_0 and ω_0 such that the disk stops at the time t_1 determined in part (a).

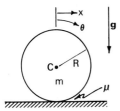

Figure 8.92

Problem 8.79

Consider again the system of Problem 8.78. At $t = 0$, suppose that

$$\dot{x} = v_0 \quad \text{and} \quad \dot{\theta} = \omega_0, \quad \text{where } R\omega_0 > v_0 \quad (v_0 > 0, \ \omega_0 > 0).$$

a) Determine the time t_1 at which the disk begins to roll without slip.
b) Determine the energy of the disk at $t = 0$ and at $t = t_1$.

Problem 8.80

For the spool shown in the figure, P is a constant force. Let μ be the coefficient of static and sliding friction, m be the mass, and I^C be the moment of inertia of the cylinder.

a) Calculate $\ddot{x}$, assuming that the cylinder rolls without slip.
b) Calculate the force of static friction.

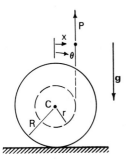

Figure 8.93

Problem 8.81

In Problem 8.80:

    a) Calculate the minimum value of P so that the cylinder will slip.

    b) Given that P is large enough to cause slipping, determine $\ddot{x}$ and $\ddot{\theta}$ of the cylinder.

Problem 8.82

The spool in the figure has an inner radius r and an outer radius R. The mass is m, the moment of inertia about the center of mass is I^c, and μ is the coefficient of static and sliding friction between the disk and the surface.

    a) Given that the spool rolls without slip, calculate $\ddot{x}$ of the disk, assuming that P is a constant force.

    b) Calculate $\ddot{x}_1$, the acceleration of the string.

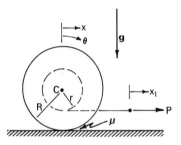

Figure 8.94

Problem 8.83

In Problem 8.82:

    a) Calculate the force of friction.

    b) Calculate the minimum value of P so that the cylinder will slip.

    c) Suppose that P is large enough so that the cylinder will slip. Determine $\ddot{x}$ and $\ddot{\theta}$.

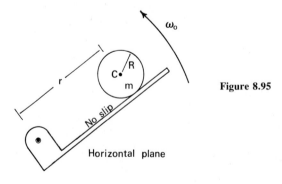

Figure 8.95

Problem 8.84

The disk in the figure rolls without slip on the bar, which rotates in the smooth horizontal plane with constant angular speed ω_0.

    a) Determine the differential equation of motion for $r(t)$, the motion of the center of the disk along the bar.

    b) Of the forces acting on the disk, identify those which do work on the disk.

Problem 8.85

A bar of mass m is pinned at the point A and suspended by a string at point B. At $t = 0$, the string is cut. Find the horizontal and vertical pin reactions (a) before $t = 0$, (b) just after $t = 0$.

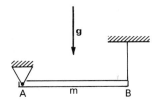

Problem 8.86

In Problem 8.85, calculate the horizontal and vertical pin reactions at point A when the bar has rotated through an angle of 45°.

Problem 8.87

The bar in the figure is suspended by strings at A and B. At $t = 0$, the string at B is cut. Find the tension in the string at A, (a) before $t = 0$, (b) just after $t = 0$.

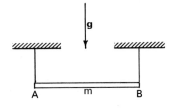

Figure 8.97

Problem 8.88

The bar in the figure slides without friction so that A remains on the vertical wall and B remains on the horizontal surface. The bar is released from rest in the vertical position. Calculate the angular speed of the bar at the moment it reaches the horizontal position.

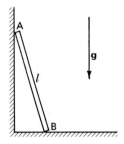

Figure 8.98

Problem 8.89

The bar in Fig. 8.99 slides without friction on the two walls. The spring is at its free length when the bar is vertical. The bar is released from rest at this position.

    a) Compute $\dot{\theta}$ of the bar when $\theta \neq 0$.

    b) Calculate the angles of equilibrium positions of the bar.

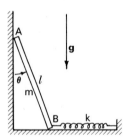

Figure 8.99

Problem 8.90

The disk in Fig. 8.100 rolls without slip inside a cylindrical surface of radius R. The disk is released from rest at the position shown.

    a) Calculate the kinetic energy of the disk at the lowest point of its trajectory.

    b) Calculate the angular speed of the disk at the lowest point of its trajectory.

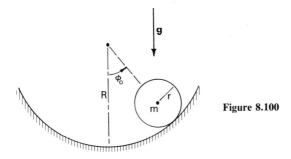

Figure 8.100

Problem 8.91

A disk of radius r rolls without slip inside a cylindrical surface of radius R, as shown in Fig. 8.101. Let v be the speed of the center of mass of the disk. Consider the disk at the position shown.

   a) Calculate the minimum speed v so that the disk will remain on the cylindrical surface.

   b) Suppose that the disk is slipping on the cylindrical surface. Compute the minimum value of v so that the disk will remain on the surface at the position shown.

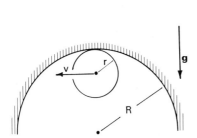

Figure 8.101

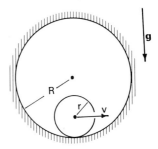

Figure 8.102

Problem 8.92

A disk of radius r rolls without slip inside a cylindrical surface of radius R (Fig. 8.102). What is the minimum speed of the center of mass of the disk at the bottom if the disk is to remain on the cylindrical surface? [*Hint:* Use the result of Problem 8.91.]

★ Problem 8.93

Suppose that the coefficient of static friction in Problem 8.92 is μ. Is the analysis in the problem valid?

Problem 8.94

The disk of radius r rolls without slip on the fixed cylinder of radius R shown in the figure. The coefficient of static friction between the disk and the cylinder is μ. Suppose that the disk is released from rest at $\theta = 0$.

a) Calculate the angle ϕ at which the disk begins to slip. (Write an equation from which the angle could be determined.)

b) Suppose that the surfaces are such that the disk rolls without slip until it loses contact with the cylinder. At what angle would this occur?

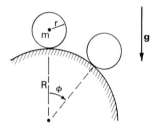

Figure 8.103

Problem 8.95

The center of mass of the disk in the figure moves on the (dashed) elliptical path

$$\frac{x^2}{a^2} + \frac{y^2}{b^2} = 1, \text{ assume } a > b.$$

Suppose that the disk rolls without slip. Calculate the minimum speed of the center of mass at the bottom of the path in order that the disk will remain in contact with the surface at the top of the path.

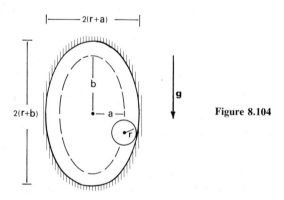

Figure 8.104

Problem 8.96

In Problem 8.95, suppose that $a = 8$ ft and $b = 4$ ft. Calculate the minimum speed of the center of mass at the bottom of the path in order that the disk will remain in contact with the surface at the top of the path.

Problem 8.97

The disk in the figure is free to rotate in the *horizontal plane*. The moment of inertia of the disk about O is I^0. A particle m can slide in a smooth groove cut in the disk. When m is at $r = a \neq 0$,

$$\dot{r} = 0 \quad \text{and} \quad \dot{\theta} = \omega_0.$$

a) Calculate the angular speed $\dot{\theta}$ when m has position r.

b) Compute $\dot{r}$ when the position of m is r.

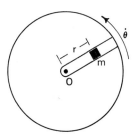

Figure 8.105

★ Problem 8.98

The system in the figure is released from rest at the position shown. Show that the ball m will appear to hop from cup 1 to cup 2. [*Hint:* Calculate the acceleration of m compared to the acceleration of the set of points of the bar which are beneath m. Assume that $\theta_0 > 61°$.]

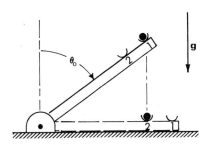

Figure 8.106

Problem 8.99

A bullet with speed v_0 becomes lodged in the rim of a bar stool (see the figure), which is initially at rest. Compute the angular speed of the stool after the collision.

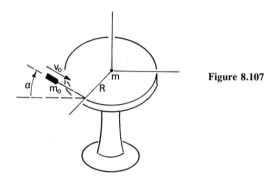

Figure 8.107

Problem 8.100

Point A of the bar shown in the figure can slide in the smooth horizontal track.

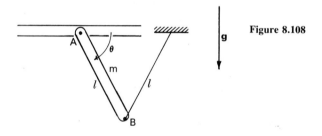

Figure 8.108

a) Show that the moment of inertia of the bar about its instant center is
$$I^{IC} = \frac{ml^2}{3}(1 + 6\sin^2\theta).$$

b) Given that the bar is released from rest with $\theta = 0$, compute the rate $\dot{\theta}$ when $\theta \neq 0$.

c) Determine the speed of point A when $\theta \neq 0$.

Problem 8.101

The bar shown in the figure can slide without friction against the end of the cart. The entire system is at rest when $\theta = 0$.

a) Determine the speed of the cart when the angle of the bar is θ.

b) Determine the force exerted by the bar on the cart as a function of the angle θ.

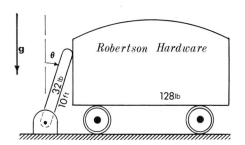

Figure 8.109

Problem 8.102

An impulsive force has a very large magnitude over a very brief time interval. We define the *linear impulse* of such a force as

$$\hat{\mathbf{F}} = \int_{t}^{t+\Delta t} \mathbf{F} \, dt.$$

For an impulsive moment, we define the *angular impulse* as

$$\hat{\mathbf{M}}_A = \int_{t}^{t+\Delta t} \mathbf{M}_A \, dt.$$

Consider a rigid body, initially at rest, subject to a plane, impulsive force system at time $t = 0$.

a) Show that the resultant linear impulse equals $m\mathbf{v}_C$, where $\mathbf{v}_C$ is the velocity of the center of mass just after $t = 0$.

b) Suppose that the rigid body has a fixed point O. Show that the resultant angular impulse $\hat{\mathbf{M}}_O$ equals $(I_{zz}^O \dot{\theta})\mathbf{k}$, where $\dot{\theta}$ is the angular velocity of the body just after $t = 0$.

c) Show that the resultant angular impulse about the center of mass, $\hat{\mathbf{M}}_C$, equals $(I_{zz}^C \dot{\theta})\mathbf{k}$, where $\dot{\theta}$ is the angular speed just after $t = 0$.

Problem 8.103

The rigid body shown in the figure is pinned through the point O. Let the mass of the body be m and the moment of inertia about the point O be I^O. A linear impulse is applied to the point O'. Find the distance l such that there is no horizontal pin reaction at O. The point O' is called the *center of percussion* with respect to O.

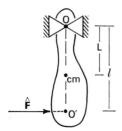

Figure 8.110

Problem 8.104

Consider the rigid body shown in the figure, and suppose that O' is the center of percussion of the body with respect to O. (See Problem 8.103.)

a) Show that O is the center of percussion with respect to O'.

b) Show that the points O and O' are related by

$$ab = \frac{I^C}{m},$$

where m is the mass of the body and I^C is the moment of inertia of the body about the center of mass.

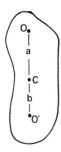

Figure 8.111

Problem 8.105

The baseball bat in the figure is so designed that the center of percussion with respect to the point A_0 is P_0 (refer to Problem 8.104). Show that a batter who "chokes up" to the point A_1 should try to hit the ball at the point P_1, where

$$a_0 b_0 = \frac{I^C}{m} = a_1 b_1.$$

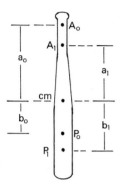

Figure 8.112

Problem 8.106

A very large force of short duration, P, is applied to the cylinder shown in the figure. Find the height h so that the force of static friction acting on the cylinder is zero. [*Hint:* If the static friction is zero, the cylinder will roll without slip after the application of P.]

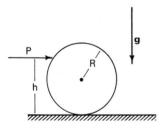

Figure 8.113

Problem 8.107

Repeat the analysis of Problem 8.106, considering the body to be a sphere of radius R and mass m. (This is the case of a billiard ball.)

Problem 8.108

Repeat the analysis of Problem 8.106, considering the body to be a thin ring of radius R and mass m.

Problem 8.109

The final equation of motion in Example 22 was found to be

$$\ddot{\theta} = \frac{3}{2}\left(\frac{g}{l}\right)(2 \sin \theta - \cos \theta).$$

Differential equations of this form may be integrated to give $\dot{\theta}$ as a function of θ by the following trick: Let $\dot{\theta} = \omega$. Then

$$\ddot{\theta} = \omega \frac{d\omega}{d\theta}.$$

If this expression is inserted in the differential equation, the variables ω and θ can be separated, and the relation $\dot{\theta} = f(\theta)$ can be found by integration.

a) Prove that $\ddot{\theta} = \omega(d\omega/d\theta)$.

b) For the system of Example 22, suppose that $\dot{\theta} = 0$ when $\theta = 45°$. Determine the relation $\dot{\theta} = \omega = f(\theta)$.

Problem 8.110

Suppose that by some means (e.g., an energy expression, or the method of Problem 8.109) we have determined $\dot{\theta}$ as a function of θ:

$$\dot{\theta} = f(\theta).$$

Given that $\theta = \theta_0$ at $t = t_0$, show that $\theta(t)$ can be obtained from

$$t = t_0 + \int_{\theta_0}^{\theta} \frac{d\theta}{f(\theta)}.$$

(The integral may require numerical evaluation and the relation $t = t(\theta)$ may have to be inverted graphically.)

Section 8.8

Problem 8.111

a) Determine the natural frequency of the bar shown in the figure.

b) Calculate the length l_1 of a simple pendulum which has the same natural frequency as the bar of part (a).

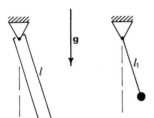

Figure 8.114

Problem 8.112

Calculate the natural frequency of the thin bar in the figure which is suspended at a point O a distance d from the center of mass. Let the length of the bar be l.

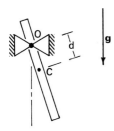

Figure 8.115

Problem 8.113

Determine the natural frequency of the homogeneous ring in the figure.

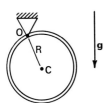

Figure 8.116

Problem 8.114

For the system shown in the figure:
  a) Show that $I^0 = \frac{1}{2}mR^2$.
  b) Calculate the natural frequency of the system. [*Hint:* The distance $OC = 4R/3\pi$.]

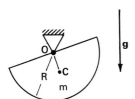

Figure 8.117

Problem 8.115

Assume that the string which passes under the disk in the figure does not slip with respect to the disk. Calculate the natural frequency of the vertical motion of the system.

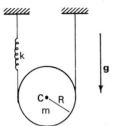

Figure 8.118

Problem 8.116

Two pendulum systems are shown in the figure. In the first, the disk is free to rotate about the pin at C. In the second, the disk has been welded to the bar. Determine the natural frequencies of the two systems, given that the mass of the bar is negligible in each case.

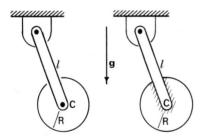

Figure 8.119

Problem 8.117

Suppose that the rigid body in the figure has the same natural frequency when it is suspended from either O or O'. Then O' is called the *center of oscillation of the body*

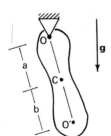

Figure 8.120

with respect to O. Show that the points O and O' are related by

$$ab = \frac{I^C}{m},$$

where m is the mass of the body and I^C is the moment of inertia about the center of mass.

Problem 8.118

A rigid body has the center of oscillation (see Problem 8.117) O' with respect to O. Show that the natural frequency of the body when suspended from O or O' is the same as that of a simple pendulum of length OO'. (See the figure.)

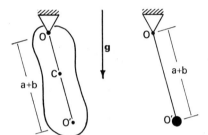

Figure 8.121

Problem 8.119

Suppose that the center of oscillation of a rigid body is O' with respect to O (see Problem 8.117). Show that O' is the center of percussion of the body with respect to O (see Problem 8.103).

Problem 8.120

Find the natural frequency of the ring in the figure, assuming that the ring rolls without slip inside the cylindrical surface.

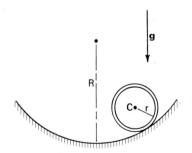

Figure 8.122

Problem 8.121

Determine the natural frequency of the system shown in the figure. [*Hint:* Equate T_{max} with V_{max} for the system.]

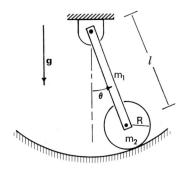

Figure 8.123

Problem 8.122

Determine the natural frequency of the system shown in the figure, assuming that the disk rolls without slipping.

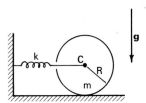

Figure 8.124

Problem 8.123

The disk in the figure rolls without slip on a cart which has the motion $x_0(t) = A_0 \sin \omega t$. Let x measure the displacement of the disk relative to the cart. At $t = 0$: $x = x_0$, $\dot{x} = 0$.

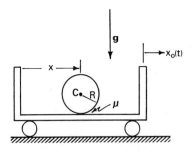

Figure 8.125

a) Calculate $x(t)$.

b) Given that μ is the coefficient of static friction between the disk and the cart, compute the frequency ω at which the disk will slip if A_0 is a given constant.

Problem 8.124

The disk in the figure rolls without slip on the cart. The motion of the cart is specified as $x_0 = A_0 \sin \omega t$. Given that x is the displacement of the disk measured relative to the cart:

a) Determine the differential equation of motion of the disk.

b) Determine the amplitude of the forced motion of the disk.

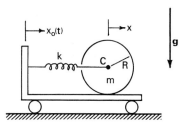

Figure 8.126

★ Problem 8.125

A homogeneous disk of mass m has a particle, also of mass m, welded to its edge (see the figure). The disk rolls without slip, and is released from rest at $\theta = \theta_0$, which is *not* assumed to be a small angle.

a) Calculate the differential equation of motion for the disk, using the angle θ as the variable.

b) Linearize the differential equation of part (a), and determine the natural frequency of small motions of the disk.

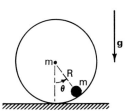

Figure 8.127

Problem 8.126

In Problem 8.125, write expressions for T_{max} and V_{max}. Let θ_0 become small, and then determine the natural frequency of the disk.

APPENDIXES

Appendix 1

MATHEMATICS

A. VECTOR ALGEBRA

Scalar Quantities

Any quantity which is completely defined by a single number (e.g., area, volume, work, power, energy) is called a *scalar*. We are concerned with scalars which are real (and occasionally complex) numbers.

Vector Quantities

A quantity which has (i) magnitude (or length), (ii) direction (or orientation), and (iii) sense is called a *vector*. Examples of vectors are velocity, acceleration, force, moment. Vectors are denoted by boldface type in print, $\mathbf{v}$, $\mathbf{a}$, $\mathbf{F}$, $\mathbf{M}$. A vector is indicated graphically by means of an arrow, as in Fig. A1.1.

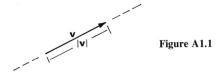

Figure A1.1

The relevant defining characteristics of the vector $\mathbf{v}$ are indicated as follows.

   i) Magnitude: the length of the arrow, denoted by $|\mathbf{v}|$

   ii) Direction: the orientation of the arrow

   iii) Sense: the direction indicated by the arrowhead

Addition of Two Vectors*

The sum of two vectors $\mathbf{v}_1$ and $\mathbf{v}_2$ is defined by the parallelogram rule (Fig. A1.2). The vectors $\mathbf{v}_1$ and $\mathbf{v}_2$ are moved to form two sides of a parallelogram.

---

  * See also the section concerning Cartesian coordinates in this Appendix.

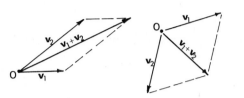

Figure A1.2

The vector from the common origin O of v_1 and v_2 along the diagonal of the parallelogram is taken to be the sum $(v_1 + v_2)$. Thus

$$v_1 + v_2 = v_2 + v_1. \tag{A1.1}$$

Multiplication of a Vector by a Real Scalar*

Suppose we have a vector v and a real scalar c. The product cv is defined as follows.

i)   Magnitude: $|c||v|$; i.e., the magnitude of v multiplied by the absolute value of c

ii)   Direction: the direction of v

iii) Sense: (a) If $c > 0$, the sense of cv is the same as that of v.
        (b) If $c < 0$, the sense of cv is opposite that of v.

An illustration of the above is shown in Fig. A1.3.

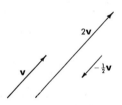

Figure A1.3

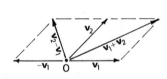

Figure A1.4

Subtraction of Two Vectors*

The difference $(v_2 - v_1)$ can be defined through multiplication by a scalar, -1, and addition of two vectors $[v_2 + (-1)v_1]$. Thus we have $(v_2 - v_1)$ and $(v_1 + v_2)$, as shown in Fig. A1.4.

_____

* See also the section on Cartesian coordinates in this Appendix.

Dot Product of Two Vectors*

The *dot product* (inner product or scalar product are alternative names) of two vectors v_1 and v_2 is defined to be the real scalar

$$v_1 \cdot v_2 = |v_1||v_2| \cos \theta, \tag{A1.2}$$

where θ is the angle between v_1 and v_2 which is less than or equal to 180° (π radians). (See Fig. A1.5.) In particular,

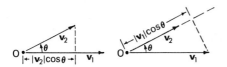

Figure A1.5

$$v \cdot v = |v||v| = v^2. \tag{A1.2'}$$

There are two interpretations of $v_1 \cdot v_2$:

a)   The projection of v_2 on v_1, ($|v_2| \cos \theta$), times the magnitude of v_1 (Fig. A1.5a).

b)   The projection of v_1 on v_2, ($|v_1| \cos \theta$), times the magnitude of v_2 (Fig. A1.5b).

The dot product $v_1 \cdot v_2$ is a (positive, negative, or zero) real number.
    If the vectors v_1 and v_2 are such that

$$v_1 \cdot v_2 = 0 \qquad \text{which implies} \qquad \theta = \left(\frac{\pi}{2}\right) \text{ radians,}$$

the vectors v_1 and v_2 are perpendicular (or orthogonal).

Cross Product of Two Vectors*

The *cross product* (or vector product) of two vectors v_1 and v_2 is defined to be another vector v such that

$$v = v_1 \times v_2, \tag{A1.3}$$

where v has the following properties

    i) Magnitude: $|v| = |v_1||v_2| \sin \theta$

---

* See also the section on Cartesian coordinates in this Appendix.

$V_1 \times V_2 \approx 0 \implies V_1 \stackrel{?}{\sim} V_2$ are parallel.

ii) Direction: The direction of **v** is perpendicular to the plane formed by $\mathbf{v}_1$ and $\mathbf{v}_2$.

iii) Sense: The sense of **v** is obtained by the right-hand rule rotating $\mathbf{v}_1$ into $\mathbf{v}_2$ through the angle θ. See Fig. A1.6.

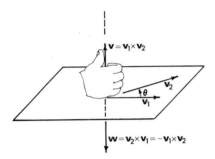

Figure A1.6

Again θ is the angle between $\mathbf{v}_1$ and $\mathbf{v}_2$ which is less than or equal to $180°$ (π radians).

Note that if we form $\mathbf{w} = \mathbf{v}_2 \times \mathbf{v}_1$, the magnitude and direction of **w** is the same as for $\mathbf{v} = \mathbf{v}_1 \times \mathbf{v}_2$. But the sense of **w** is opposite to that of **v**. Thus

$$\mathbf{v}_1 \times \mathbf{v}_2 = -\mathbf{v}_2 \times \mathbf{v}_1. \tag{A1.4}$$

Scalar and Vector Triple Products*

The *scalar triple product* of three vectors **a**, **b**, and **c** results in a scalar:

$\mathbf{a} \cdot (\mathbf{b} \times \mathbf{c})$.

We have the following identities

$$\mathbf{a} \cdot (\mathbf{b} \times \mathbf{c}) = \mathbf{b} \cdot (\mathbf{c} \times \mathbf{a}) = \mathbf{c} \cdot (\mathbf{a} \times \mathbf{b})$$
$$= -\mathbf{a} \cdot (\mathbf{c} \times \mathbf{b}) = -\mathbf{b} \cdot (\mathbf{a} \times \mathbf{c}) = -\mathbf{c} \cdot (\mathbf{b} \times \mathbf{a}). \tag{A1.5}$$

The *vector triple product* of three vectors **a**, **b**, **c** results in a vector:

$$\mathbf{a} \times (\mathbf{b} \times \mathbf{c}) = (\mathbf{a} \cdot \mathbf{c})\mathbf{b} - (\mathbf{b} \cdot \mathbf{a})\mathbf{c}$$

or $$\tag{A1.6}$$

$$(\mathbf{a} \times \mathbf{b}) \times \mathbf{c} = (\mathbf{a} \cdot \mathbf{c})\mathbf{b} - (\mathbf{b} \cdot \mathbf{c})\mathbf{a}.$$

---

* See also the section on Cartesian coordinates in this Appendix.

Elementary Vector Operations in Cartesian Coordinates

Let $\mathbf{i}$, $\mathbf{j}$, $\mathbf{k}$ be unit vectors in a Cartesian coordinate system (Fig. A1.7).

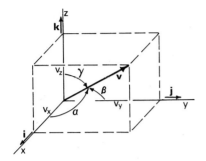

Figure A1.7

The projections

$$v_x = \mathbf{v} \cdot \mathbf{i} = |\mathbf{v}| \cos \alpha,$$
$$v_y = \mathbf{v} \cdot \mathbf{j} = |\mathbf{v}| \cos \beta, \tag{A1.7}$$
$$v_z = \mathbf{v} \cdot \mathbf{k} = |\mathbf{v}| \cos \gamma$$

are called the components of a vector $\mathbf{v}$ in the Cartesian coordinate system. The terms $\cos \alpha$, $\cos \beta$, $\cos \gamma$ (occasionally denoted simply l, m, n) are the *direction cosines* of the line of $\mathbf{v}$.

Using the components of $\mathbf{v}$, we can write

$$\mathbf{v} = v_x \mathbf{i} + v_y \mathbf{j} + v_z \mathbf{k}. \tag{A1.8}$$

Elementary Operations

In the following, let $\mathbf{a}$, $\mathbf{b}$, $\mathbf{c}$ be vectors with components

$$\mathbf{a} = a_x \mathbf{i} + a_y \mathbf{j} + a_z \mathbf{k}, \qquad \mathbf{b} = b_x \mathbf{i} + b_y \mathbf{j} + b_z \mathbf{k}, \qquad \mathbf{c} = c_x \mathbf{i} + c_y \mathbf{j} + c_z \mathbf{k}.$$

Addition

$$\mathbf{a} + \mathbf{b} = (a_x + b_x)\mathbf{i} + (a_y + b_y)\mathbf{j} + (a_z + b_z)\mathbf{k}. \tag{A1.9}$$

Multiplication by a Scalar

$$k\mathbf{a} = ka_x \mathbf{i} + ka_y \mathbf{j} + ka_z \mathbf{k}. \tag{A1.10}$$

Subtraction

$$\mathbf{a} - \mathbf{b} = (a_x - b_x)\mathbf{i} + (a_y - b_y)\mathbf{j} + (a_z - b_z)\mathbf{k}. \tag{A1.11}$$

Dot Product

$$\mathbf{a} \cdot \mathbf{b} = a_x b_x + a_y b_y + a_z b_z. \tag{A1.12}$$

The *magnitude* of a vector:

$$|\mathbf{a}| = \sqrt{\mathbf{a} \cdot \mathbf{a}} = \sqrt{a_x^2 + a_y^2 + a_z^2}. \tag{A1.13}$$

Vector Product

$$\mathbf{a} \times \mathbf{b} = \mathbf{i}(a_y b_z - a_z b_y) + \mathbf{j}(a_z b_x - a_x b_z) + \mathbf{k}(a_x b_y - a_y b_x). \tag{A1.14}$$

A useful device for remembering (A1.14) is

$$\mathbf{a} \times \mathbf{b} = \begin{vmatrix} \mathbf{i} & \mathbf{j} & \mathbf{k} \\ a_x & a_y & a_z \\ b_x & b_y & b_z \end{vmatrix} \tag{A1.14'}$$

In fact, if $\mathbf{e}_1$, $\mathbf{e}_2$, $\mathbf{e}_3$ is any *right-handed* set of mutually perpendicular unit vectors (a set of unit vectors is right-handed if $\mathbf{e}_1 \times \mathbf{e}_2 = \mathbf{e}_3$), then

$$\mathbf{a} \times \mathbf{b} = \begin{vmatrix} \mathbf{e}_1 & \mathbf{e}_2 & \mathbf{e}_3 \\ a_1 & a_2 & a_3 \\ b_1 & b_2 & b_3 \end{vmatrix}$$

where a_1, a_2, a_3 are the components of $\mathbf{a}$ relative to $\mathbf{e}_1$, $\mathbf{e}_2$, and $\mathbf{e}_3$:

$$\mathbf{a} = a_1 \mathbf{e}_1 + a_2 \mathbf{e}_2 + a_3 \mathbf{e}_3, \qquad \text{etc.}$$

For example, in spherical coordinates, we have (see Fig. A1.8) the three unit vectors

$$\mathbf{e}_\rho, \ \mathbf{e}_\phi, \ \mathbf{e}_\theta.$$

From the figure, it is clear that

$$\mathbf{e}_\phi \times \mathbf{e}_\theta = \mathbf{e}_\rho, \qquad \mathbf{e}_\theta \times \mathbf{e}_\rho = \mathbf{e}_\phi, \qquad \mathbf{e}_\rho \times \mathbf{e}_\phi = \mathbf{e}_\theta.$$

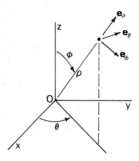

Figure A1.8

Thus the following sets are *right-handed:*

$$(\mathbf{e}_\phi, \mathbf{e}_\theta, \mathbf{e}_\rho), \qquad (\mathbf{e}_\theta, \mathbf{e}_\rho, \mathbf{e}_\phi), \qquad (\mathbf{e}_\rho, \mathbf{e}_\phi, \mathbf{e}_\theta).$$

And so, if two vectors have components

$$\mathbf{a} = a_\phi \mathbf{e}_\phi + a_\theta \mathbf{e}_\theta + a_\rho \mathbf{e}_\rho, \qquad \mathbf{b} = b_\phi \mathbf{e}_\phi + b_\theta \mathbf{e}_\theta + b_\rho \mathbf{e}_\rho,$$

we can write

$$\mathbf{a} \times \mathbf{b} = \begin{vmatrix} \mathbf{e}_\phi & \mathbf{e}_\theta & \mathbf{e}_\rho \\ a_\phi & a_\theta & a_\rho \\ b_\phi & b_\theta & b_\rho \end{vmatrix} = \begin{vmatrix} \mathbf{e}_\theta & \mathbf{e}_\rho & \mathbf{e}_\phi \\ a_\theta & a_\rho & a_\phi \\ b_\theta & b_\rho & b_\phi \end{vmatrix} = \begin{vmatrix} \mathbf{e}_\rho & \mathbf{e}_\phi & \mathbf{e}_\theta \\ a_\rho & a_\phi & a_\theta \\ b_\rho & b_\phi & b_\theta \end{vmatrix}.$$

But, for example,

$$\mathbf{a} \times \mathbf{b} \neq \begin{vmatrix} \mathbf{e}_\rho & \mathbf{e}_\theta & \mathbf{e}_\phi \\ a_\rho & a_\theta & a_\phi \\ b_\rho & b_\theta & b_\phi \end{vmatrix}.$$

Scalar Triple Product

$$\mathbf{a} \cdot (\mathbf{b} \times \mathbf{c}) = \begin{vmatrix} a_x & a_y & a_z \\ b_x & b_y & b_z \\ c_x & c_y & c_z \end{vmatrix} \tag{A1.15}$$

Vector Triple Product

No advantage is gained over the form (A1.6) by writing the expressions explicitly in terms of Cartesian coordinates.

B. VECTOR CALCULUS

Directional Derivatives

Suppose that ψ is a function of the spatial coordinates. For example, suppose that

$$\psi = \psi(x, y, z).$$

If we wish the *derivative* of ψ in the direction of $d\mathbf{r}$, where

$$d\mathbf{r} = dx\mathbf{i} + dy\mathbf{j} + dz\mathbf{k},$$

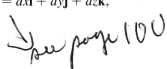

we obtain

$$\frac{d\psi}{ds} = \nabla\psi \cdot \mathbf{n}, \tag{A1.16}$$

where $\nabla\psi$ = gradient of ψ, $\mathbf{n}$ = unit vector in the direction of $d\mathbf{r}$.

The Gradient in Various Coordinate Systems

Cartesian Coordinates (See Fig. A1.9.)

$$\nabla\psi = \frac{\partial\psi}{\partial x}\mathbf{i} + \frac{\partial\psi}{\partial y}\mathbf{j} + \frac{\partial\psi}{\partial z}\mathbf{k}, \tag{A1.17a}$$

$$\mathbf{r} = x\mathbf{i} + y\mathbf{j} + z\mathbf{k}, \qquad d\mathbf{r} = dx\mathbf{i} + dy\mathbf{j} + dz\mathbf{k}$$

$$ds = \sqrt{(dx)^2 + (dy)^2 + (dz)^2}$$

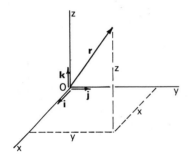

Figure A1.9

Cylindrical Coordinates (See Fig. A1.10.)

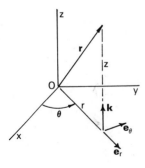

Figure A1.10

$$\nabla \psi = \frac{\partial \psi}{\partial r} \, \mathbf{e}_r + \frac{1}{r} \frac{\partial \psi}{\partial \theta} \, \mathbf{e}_\theta + \frac{\partial \psi}{\partial z} \, \mathbf{k} \qquad \text{(A1.17b)}$$

$$\mathbf{r} = r\mathbf{e}_r + z\mathbf{k}, \qquad d\mathbf{r} = dr\mathbf{e}_r + r \, d\theta \mathbf{e}_\theta + dz\mathbf{k}$$

$$ds = \sqrt{(dr)^2 + (r \, d\theta)^2 + (dz)^2}$$

Spherical Coordinates (See Fig. A1.11.)

$$\nabla \psi = \frac{\partial \psi}{\partial \rho} \, \mathbf{e}_\rho + \frac{1}{\rho} \frac{\partial \psi}{\partial \phi} \, \mathbf{e}_\phi + \frac{1}{\rho \sin \phi} \frac{\partial \psi}{\partial \theta} \, \mathbf{e}_\theta, \qquad \text{(A1.17c)}$$

$$\mathbf{r} = \rho \mathbf{e}_\rho, \qquad d\mathbf{r} = d\rho \, \mathbf{e}_\rho + \rho \, d\phi \, \mathbf{e}_\phi + \rho \sin \phi \, d\theta \, \mathbf{e}_\theta,$$

$$ds = \sqrt{(d\rho)^2 + (\rho \, d\phi)^2 + (\rho \sin \phi \, d\theta)^2}.$$

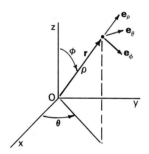

Figure A1.11

Line Integrals (See Fig. A1.12.)

Suppose that $\mathbf{F}$ is a vector function of spatial coordinates. For example,

$$\mathbf{F} = \mathbf{F}(x,y,z) = F_x(x,y,z)\mathbf{i} + F_y(x,y,z)\mathbf{j} + F_z(x,y,z)\mathbf{k}.$$

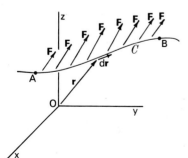

Figure A1.12

Suppose that C is a path in space, running from point A to point B. Let $\mathbf{r}$ be the vector from some origin O to a point on the curve. The integral

$$\int_C \mathbf{F} \cdot d\mathbf{r}$$

is called the *line integral of F along C*.

Cartesian Coordinates

$$\mathbf{F} = F_x \mathbf{i} + F_y \mathbf{j} + F_z \mathbf{k},$$

(A1.18a)

$$\int_C \mathbf{F} \cdot d\mathbf{r} = \int_C [F_x \, dx + F_y \, dy + F_z \, dz].$$

Cylindrical Coordinates

$$\mathbf{F} = F_r \mathbf{e}_r + F_\theta \mathbf{e}_\theta + F_z \mathbf{k},$$

(A1.18b)

$$\int_C \mathbf{F} \cdot d\mathbf{r} = \int_C [F_r \, dr + r F_\theta \, d\theta + F_z \, dz].$$

Spherical Coordinates

$$\mathbf{F} = F_\rho \mathbf{e}_\rho + F_\phi \mathbf{e}_\phi + F_\theta \mathbf{e}_\theta,$$

(A1.18c)

$$\int_C \mathbf{F} \cdot d\mathbf{r} = \int_C [F_\rho \, d\rho + \rho F_\phi \, d\phi + \rho \sin \phi \, F_\theta \, d\theta].$$

With respect to Eqs. (A1.18), it is worth pointing out that the specification of curve C will mean that the differentials of the problem (dx, dy, dz), $(dr, d\theta, dz)$, or $(d\rho, d\phi, d\theta)$ are related. There are two common ways in which this is done. We illustrate this in terms of Cartesian coordinates.

The two common ways to specify a space curve C are as follows.

i)   *An independent variable t*

$$x = x(t), \qquad y = y(t), \qquad z = z(t).$$

Then

$$dx = \frac{dx}{dt} \, dt, \qquad dy = \frac{dy}{dt} \, dt, \qquad dz = \frac{dz}{dt} \, dt.$$

And if t_0 is the value of t which corresponds to point A of C and t_1 is the value of t which corresponds to point B of C, then Eq. (A1.18a) becomes

$$\int_C \mathbf{F} \cdot d\mathbf{r} = \int_{t_0}^{t_1} [F_x \dot{x} + F_y \dot{y} + F_z \dot{z}] \, dt.$$

ii)   *Purely spatial description*

Suppose that we write

$$y = y(x), \qquad z = z(x).$$

[We could obtain the above by eliminating t in case (i).] Then we have

$$dy = \frac{dy}{dx} \, dx, \qquad dz = \frac{dz}{dx} \, dx,$$

and (A1.18a) becomes

$$\int_C \mathbf{F} \cdot d\mathbf{r} = \int_{x_0}^{x_1} \left[F_x + F_y \frac{dy}{dx} + F_z \frac{dz}{dx} \right] dx,$$

where x_0 and x_1 are the values of x which correspond to points A and B of curve C.

Conservative Force Field

The line integral

$$\int_C \mathbf{F} \cdot d\mathbf{r}$$

can be regarded as a function *only* of the endpoints A and B of curve C if and only if the following conditions are met.

i)   The spatial region R of possible paths C is *simply connected*. This means that any closed path in R can be continuously shrunk to a point. In Fig. A1.13(a), the region R_1 is simply connected. The closed curve C_1 (or any other closed curve in R_1) could be shrunk to a point within R_1. In Fig. A1.13(b), however, the region R_2 is *not* simply connected. The closed curve C_2 cannot be shrunk to a point.

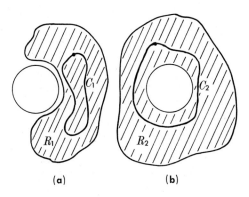

(a)   (b)   Figure A1.13

ii)   The vector field **F** must be continuously differentiable in R and

$\nabla \times \mathbf{F} = 0$ in R.

The condition $\nabla \times \mathbf{F} = 0$ is read "the curl of **F** is zero in R." If the conditions (i) and (ii) are met, there exists a scalar function ψ such that $\mathbf{F} = \nabla \psi$ and

$$\int_C \mathbf{F} \cdot d\mathbf{r} = \psi(B) - \psi(A). \tag{A1.19}$$

The Curl in Various Coordinate Systems

Cartesian Coordinates (Refer to Fig. A1.9.)

$$\nabla \times \mathbf{F} = \begin{vmatrix} \mathbf{i} & \mathbf{j} & \mathbf{k} \\ \dfrac{\partial}{\partial x} & \dfrac{\partial}{\partial y} & \dfrac{\partial}{\partial z} \\ F_x & F_y & F_z \end{vmatrix} \tag{A1.20a}$$

Cylindrical Coordinates (Refer to Fig. A1.10.)

$$\nabla \times \mathbf{F} = \left(\frac{1}{r}\right) \begin{vmatrix} \mathbf{e}_r & r\mathbf{e}_\theta & \mathbf{k} \\ \dfrac{\partial}{\partial r} & \dfrac{\partial}{\partial \theta} & \dfrac{\partial}{\partial z} \\ F_r & rF_\theta & F_z \end{vmatrix} \tag{A1.20b}$$

Spherical Coordinates (Refer to Fig. A1.11.)

$$\nabla \times \mathbf{F} = \frac{1}{\rho^2 \sin \phi} \begin{vmatrix} \mathbf{e}_\rho & \rho\mathbf{e}_\phi & \rho \sin \phi \mathbf{e}_\theta \\ \dfrac{\partial}{\partial \rho} & \dfrac{\partial}{\partial \phi} & \dfrac{\partial}{\partial \theta} \\ F_\rho & \rho F_\phi & \rho \sin \phi F_\theta \end{vmatrix} \tag{A1.20c}$$

C. MISCELLANEOUS ELEMENTARY MATHEMATICS

Some Common Trigonometric Identities

$\sin^2 \theta + \cos^2 \theta = 1$

$1 + \tan^2 \theta = \sec^2 \theta$

$\sin 2\theta = 2 \sin \theta \cos \theta$

$\cos 2\theta = 1 - 2 \sin^2 \theta$

$\sin (\theta_1 \pm \theta_2) = \sin \theta_1 \cos \theta_2 \pm \cos \theta_1 \sin \theta_2$

$$\tag{A1.21}$$

$$\cos(\theta_1 \pm \theta_2) = \cos\theta_1 \cos\theta_2 \mp \sin\theta_1 \sin\theta_2$$

$$\sin^2\theta = \tfrac{1}{2}(1 - \cos 2\theta)$$

$$\cos^2\theta = \tfrac{1}{2}(1 + \cos 2\theta)$$

Complex Numbers

$$i = \sqrt{-1}, \qquad i^2 = -1,$$

$$c = a + ib = re^{i\theta},$$

where $r = \sqrt{a^2 + b^2} = $ modulus

$\qquad\quad \theta = \tan^{-1}(b/a) = $ argument.

Addition

$$(a_1 + ib_1) + (a_2 + ib_2) = (a_1 + a_2) + i(b_1 + b_2) \tag{A1.22}$$

Multiplication

$$(a_1 + ib_1)(a_2 + ib_2) = (a_1 a_2 - b_1 b_2) + i(a_1 b_2 + a_2 b_1)$$

$$(r_1 e^{i\theta_1})(r_2 e^{i\theta_2}) = r_1 r_2 e^{i(\theta_1 + \theta_2)} \tag{A1.23}$$

$$(re^{i\theta})^n = r^n e^{in\theta}$$

Division

$$\frac{a_1 + ib_1}{a_2 + ib_2} = \frac{1}{a_2^2 + b_2^2}\left[(a_1 a_2 + b_1 b_2) + i(a_2 b_1 - a_1 b_2)\right]$$

$$\frac{r_1 e^{i\theta_1}}{r_2 e^{i\theta_2}} = \left(\frac{r_1}{r_2}\right) e^{i(\theta_1 - \theta_2)} \tag{A1.24}$$

Complex Conjugate

If $c = a + ib$, $\bar{c} = a - ib$, and $\bar{c}$ is the *conjugate* of c.

$$\overline{c_1 c_2} = \bar{c}_1 \bar{c}_2$$

$$\overline{re^{i\theta}} = re^{-i\theta} \tag{A1.25}$$

If $c = a + ib$, $c\bar{c} = a^2 + b^2$.   If $c = re^{i\theta}$, $c\bar{c} = r^2$.

Trigonometric Functions

$$\sin\theta = \frac{1}{2i}(e^{i\theta} - e^{-i\theta}), \qquad \cos\theta = \tfrac{1}{2}(e^{i\theta} + e^{-i\theta}). \tag{A1.26}$$

Hyperbolic Functions

$$\sinh \theta = \tfrac{1}{2}(e^\theta - e^{-\theta}), \qquad \cosh \theta = \tfrac{1}{2}(e^\theta + e^{-\theta}). \tag{A1.27}$$

Geometry of Conic Sections

$$r = \frac{ep}{1 - e \cos \theta},$$

$e = $ eccentricity
$p = $ distance from focus to directrix

circle	$e = 0$
ellipse	$e < 1$
parabola	$e = 1$
hyperbola	$e > 1$

$$\tag{A1.28}$$

Ellipse

Pericentron (perigee, if the ellipse is about the earth)

Minimum distance from focus to ellipse: $r_1 = \dfrac{pe}{1 + e}$

Apocentron (apogee, if the ellipse is about the earth)

Maximum distance from focus to ellipse: $r_2 = \dfrac{pe}{1 - e}$

Semiminor axis: $a = \dfrac{pe}{1 - e^2}$

Semimajor axis: $b = \dfrac{pe}{\sqrt{1 - e^2}}$

Radius of Curvature

Cartesian Coordinates $\qquad R = \dfrac{\left[1 + \left(\dfrac{dy}{dx}\right)^2\right]^{3/2}}{\left|\left(\dfrac{d^2 y}{dx^2}\right)\right|}$ $\tag{A1.29}$

Polar Coordinates $\qquad R = \dfrac{\left[r^2 + \left(\dfrac{dr}{d\theta}\right)^2\right]^{3/2}}{\left|r^2 + 2\left(\dfrac{dr}{d\theta}\right)^2 - r\dfrac{d^2 r}{d\theta^2}\right|}$

Series Expansion of Functions

Taylor Series for f(x) About x = a

$$f(x) \approx f(a) + f'(a)(x - a) + \frac{1}{2!} f''(a)(x - a)^2 + \cdots$$

$$+ \frac{1}{n!} f^{(n)}(a)(x - a)^n \quad \text{(A1.30a)}$$

Taylor Series for f(x,y) About x = a, y = b

$$f(x,y) \approx f(a,b) + [(x - a)f_x + (y - b)f_y]\Big|_{(a,b)}$$

$$+ \frac{1}{2!} [(x - a)^2 f_{xx} + 2(x - a)(y - b)f_{xy} + (y - b)^2 f_{yy}]\Big|_{(a,b)} + \cdots$$

$$+ \frac{1}{n!} \left\{ \left[(x - a)\frac{\partial}{\partial x} + (y - b)\frac{\partial}{\partial y} \right]^n f \right\}\Big|_{(a,b)} \quad \text{(A1.30b)}$$

Fourier Series for f(x) on the Interval −l < x < l

$$f(x) \approx a_0 + a_1 \cos \frac{\pi x}{l} + a_2 \cos \frac{2\pi x}{l} + \cdots$$

$$+ b_1 \sin \frac{\pi x}{l} + b_2 \sin \frac{2\pi x}{l} + \cdots, \quad \text{(A1.31)}$$

where $\quad a_n = \frac{1}{l} \int_{-l}^{l} f(x) \cos \frac{n\pi x}{l} \, dx, \quad a_0 = \frac{1}{2l} \int_{-l}^{l} f(x) \, dx,$

$$b_n = \frac{1}{l} \int_{-l}^{l} f(x) \sin \frac{n\pi x}{l} \, dx.$$

D. MATRICES

A rectangular array of elements (usually real or complex numbers) is called a *matrix*.

$$\begin{bmatrix} a_{11} & a_{12} & \cdots & a_{1j} & \cdots & a_{1n} \\ a_{21} & a_{22} & \cdots & a_{2j} & \cdots & a_{2n} \\ \cdot & \cdot & \cdots & \cdot & \cdots & \cdot \\ a_{i1} & a_{i2} & \cdots & a_{ij} & \cdots & a_{in} \\ \cdot & \cdot & \cdots & \cdot & \cdots & \cdot \\ a_{m1} & a_{m2} & \cdots & a_{mj} & \cdots & a_{mn} \end{bmatrix}$$

*j*th column

*i*th row

The elements a_{ij} are denoted by two subscripts.

i)   The first (i) is the row position.

ii) The second (j) is the column position.

Thus a_{23} is in the second row and the third column.

A matrix which has m rows and n columns is said to be $m \times n$ dimensional. Matrices are often denoted $[A]$.

Addition

Two matrices of the same dimension can be added:

$[C] = [A] + [B]$.

If $[A]$ and $[B]$ are both $m \times n$, $[C]$ will be $m \times n$. The (ij) element of $[C]$ is

$$c_{ij} = a_{ij} + b_{ij}. \tag{A1.32}$$

Multiplication by a Scalar

If $[B] = k[A]$, and the elements of $[A]$ and $[B]$ are a_{ij} and b_{ij}, we define

$$b_{ij} = ka_{ij}. \tag{A1.33}$$

That is, $k[A]$ means that every element of $[A]$ is multiplied by the same number, k.

Matrix Multiplication

Suppose that $[A]$ is $m \times n$ dimensional and $[B]$ is $n \times p$ dimensional, or in other words, the number of *columns* of $[A]$ is the same as the number of rows of $[B]$; then the matrix product $[A][B]$ is defined as

$[C] = [A][B]$.

where $[C]$ is $m \times p$ dimensional with elements

$$c_{ij} = \sum_{k=1}^{n} a_{ik} b_{kj} \tag{A1.34}$$

This definition is illustrated in Fig. A1.14.

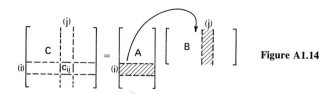

Figure A1.14

That is, the element c_{ij} can be thought of as the dot product of the ith row of $[A]$ with the jth column of $[B]$.

Transpose of a Matrix

If the rows and columns of a matrix $[A]$ are switched, the result is called the *transpose,* and is denoted by $[A]^t$. Thus

$$\begin{bmatrix} a_{11} & a_{12} & \cdots & a_{1n} \\ a_{21} & a_{22} & \cdots & a_{2n} \\ \cdot & \cdot & \cdots & \cdot \\ a_{m1} & a_{m2} & \cdots & a_{mn} \end{bmatrix}^t = \begin{bmatrix} a_{11} & a_{21} & \cdot & a_{m1} \\ a_{12} & a_{22} & \cdot & a_{m2} \\ \cdot & \cdot & \cdot & \cdot \\ \cdot & \cdot & \cdot & \cdot \\ \cdot & \cdot & \cdot & \cdot \\ a_{1n} & a_{2n} & \cdot & a_{mn} \end{bmatrix} \qquad (A1.35)$$

The transpose of a product can be shown to be the product of the transposes taken in reverse order. Thus if $[C] = [A][B]$, then $[C]^t = [B]^t[A]^t$.

Inverse Matrix

Suppose $[A]$ is a *square* matrix, then the *inverse* of $[A]$ is a matrix denoted by $[A]^{-1}$, such that

$$[A][A]^{-1} = [A]^{-1}[A] = [I],$$

where $[I]$ is the *identity matrix:*

$$[I] = \begin{bmatrix} 1 & 0 & 0 & \cdots & 0 & 0 \\ 0 & 1 & 0 & \cdots & 0 & 0 \\ \cdot & \cdot & \cdot & \cdots & \cdot & \cdot \\ 0 & 0 & 0 & \cdots & 0 & 1 \end{bmatrix}$$

The identity matrix takes its name from the fact that

$$[A][I] = [I][A] = [A]$$

for any square matrix $[A]$.

Laplace Expansion of the Inverse Matrix

The Laplace expansion for the inverse of a square matrix $[A]$ proceeds in three steps.

1) Construct the *cofactor* matrix $[M]$. You can obtain the cofactor matrix from $[A]$ by replacing the element a_{ij} by the signed minor associated with the (i,j) position,

$$A_{ij} = (-1)^{i+j} m_{ij},$$

where m_{ij} is the determinant of $[A]$ with the ith row and jth column deleted.

2) The *adjoint* of $[A]$ is defined to be the transpose of the cofactor matrix:

adj $[A] = [M]^t$.

3) The *inverse* of $[A]$ is then

$$[A]^{-1} = \frac{\text{adj } [A]}{|A|}. \tag{A1.36}$$

That is, the inverse of $[A]$ is the adjoint of $[A]$ divided by the determinant of $[A]$.

Example

Let $\quad [A] = \begin{bmatrix} 1 & 0 & 1 \\ 2 & 1 & 2 \\ 0 & 3 & 1 \end{bmatrix}$

The minor associated with position $(1,2)$ is the determinant of $[A]$ with the first row and second columns deleted:

$$m_{12} = \begin{vmatrix} 2 & 2 \\ 0 & 1 \end{vmatrix} = 2.$$

Similarly,

$$m_{11} = -5, \qquad m_{12} = 2, \qquad m_{13} = 6,$$
$$m_{21} = -3, \qquad m_{22} = 1, \qquad m_{23} = 3,$$
$$m_{31} = -1, \qquad m_{32} = 0, \qquad m_{33} = 1.$$

Thus the cofactor matrix is

$$[M] = \begin{bmatrix} -5 & -2 & +6 \\ +3 & +1 & -3 \\ -1 & 0 & +1 \end{bmatrix}.$$

The adjoint matrix is

$$\text{adj } [A] = \begin{bmatrix} -5 & +3 & -1 \\ -2 & +1 & 0 \\ +6 & -3 & +1 \end{bmatrix},$$

and since $|A| = 1$, we get *inverse*

$$[A]^{-1} = \begin{bmatrix} -5 & +3 & -1 \\ -2 & +1 & 0 \\ +6 & -3 & +1 \end{bmatrix}.$$

....

Cramer's Rule

Suppose that $[A]$ is a square matrix. Given that $[A][X] = [C]$ is any matrix equation, we can solve for $[X]$ using the inverse of $[A]$:

$$[X] = [A]^{-1}[C].$$

In the special case in which $[X]$ and $[C]$ are *column* vectors, the matrix equation becomes

$$a_{11}x_1 + a_{12}x_2 + \cdots + a_{1n}x_n = c_1,$$

$$a_{21}x_1 + a_{22}x_2 + \cdots + a_{2n}x_n = c_2,$$

$$\cdot \qquad \cdot \qquad \cdots \qquad \cdot \qquad \cdot$$

$$a_{n1}x_1 + a_{n2}x_2 + \cdots + a_{nn}x_n = c_n.$$

The solution for x_i can be written

$$\xrightarrow{\hspace{3cm}} (i)$$

$$x_i = \frac{\begin{vmatrix} a_{11} & \cdots & c_1 & \cdots & a_{1n} \\ a_{21} & \cdots & c_2 & \cdots & a_{2n} \\ \cdot & \cdots & & \cdots & \cdot \\ a_{n1} & \cdots & c_n & \cdots & a_{nn} \end{vmatrix}}{\begin{vmatrix} a_{11} & a_{12} & \cdots & a_{1n} \\ \cdot & \cdot & \cdots & \cdot \\ a_{n1} & a_{n2} & \cdots & a_{nn} \end{vmatrix}} \qquad (A1.37)$$

That is, the ratio of two determinants:

i)   Numerator: the determinant of $[A]$ with the ith column replaced by the column vector $[C]$.

ii)   Denominator: the determinant of $[A]$.

The above is called *Cramer's rule*.

Appendix 2

MOMENTS OF INERTIA

A. MOMENT OF INERTIA ABOUT A LINE

Consider the rigid body and line l in Fig. A2.1. The moment of inertia of the body about the line l is defined by

$$I_{ll} = \int_V r^2 \, dm, \qquad (A2.1)$$

where r is the perpendicular distance of the element of mass dm from the line l.

This appendix will try to indicate some of the ways in which one might try to evaluate the moment of inertia of a given body. There are essentially three methods which will be considered:

i) Evaluation of (A2.1) directly

ii) Experimental determination of the moment of inertia

iii) Evaluation of the moment of inertia from a Table of Inertias. All three methods have limitations.

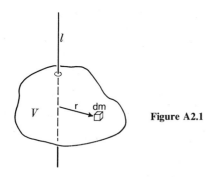

Figure A2.1

i) *Direct Evaluation of the Moment of Inertia*

Consider the uniform plane section in Fig. A2.2. We have placed a grid on the section and divided it into the partial areas $A_1, \ldots, A_{16}$. If the density

ρ is constant, we have the section partitioned into the following elements of mass:

$$dm_1 = \rho A_1,$$

$$dm_2 = \rho A_2, \qquad \rho = \text{mass/area}$$

$$\cdots$$

$$dm_{16} = \rho A_{16}.$$

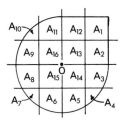

Figure A2.2

For each of these mass elements dm_i, we identify a distance r_i from O to the section A_i. In theory, r_i will *not* be the distance from O to the center of A_i but if A_i is small enough it is not terribly important which point of A_i is used to determine r_i.

The moment of inertia of the section about O is then approximately

$$I^0_{zz} = \sum_{i=1}^{16} (\rho A_i) r_i^2.$$

Clearly the elements which are farthest from O, that is A_1 to A_{12}, contribute the most to the moment of inertia, and the elements which are closest to O, that is A_{13} to A_{16}, contribute the least.

Example A2.1

In Fig. A2.3, a uniform bar of width h and length l is divided into three equal sections, A_1, A_2, A_3. We have

$$A_1 = A_2 = A_3 = \tfrac{1}{3}lh.$$

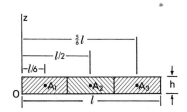

Figure A2.3

We take

$$r_1 = \frac{l}{6}, \qquad r_2 = \frac{l}{2}, \qquad r_3 = \frac{5}{6} l,$$

where r_i is the distance from O to the center of each section (we should anticipate that this is approximate!). Given that ρ is the ratio of mass/area of the bar, $m = \rho l h$ is the total mass of the bar.

Forming (A2.1) with our approximate values, we get

$$I_{zz}^0 \approx \rho A_1 r_1^2 + \rho A_2 r_2^2 + \rho A_3 r_3^2$$
$$= \tfrac{1}{3}\rho l h \left(\frac{l^2}{36} + \frac{l^2}{4} + \frac{25}{36} l^2 \right) = \tfrac{1}{3}\rho l^3 h \left(\tfrac{35}{36}\right)$$

or

$$I_{zz}^0 \approx \tfrac{1}{3} m l^2 \left(\tfrac{35}{36}\right).$$

From the Table of Inertias, we can determine that the exact value of I_{zz}^0 is

$$I_{zz}^0 = \tfrac{1}{3} m l^2 + \tfrac{1}{12} m h^2.$$

If h is very small compared to l, we usually write

$$I_{zz}^0 = \tfrac{1}{3} m l^2.$$

Note that our approximate value is only slightly lower than this value. If, however, $l = h$, the exact value $0.417 m l^2$ is considerably higher than our approximate value $0.324 m l^2$.

It should be obvious why we got a *lower* value for I_{zz}^0. It is a result of our choices of r_1, r_2, and r_3.

....

ii) *Experimental Determination of the Moment of Inertia*

In many circumstances, an experimental determination of the moment of inertia is recommended. Consider the experimental apparatus shown in Fig. A2.4.

A circular platform of radius R_0, which is suspended by three strings, each of length h, is given a slight angular displacement θ, and then released. The platform will oscillate with a certain period, τ_0, which can then be related to the moment of inertia *of the platform* through the equation

$$I_0^C = \frac{m_0 g \tau_0^2}{4\pi^2} \left(\frac{R_0^2}{h} \right), \tag{A2.2}$$

where m_0 is the mass of the platform. (We shall derive this result below.)

Now suppose that a body is placed on the platform, and the experiment is repeated. Let the resulting period of motion be denoted by τ. The total moment of inertia of the *platform and the body is*

$$I_1^C = \frac{(m_0 + m)g\tau^2}{4\pi^2}\left(\frac{R_0^2}{h}\right).$$ (A2.3)

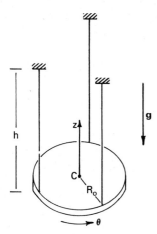

Figure A2.4

Thus the moment of inertia of the body alone is $I^C = I_1^C - I_0^C$ or

$$I^C = \left[\frac{gR_0^2}{4\pi^2 h}\right]\left[m\tau^2 + m_0(\tau^2 - \tau_0^2)\right]$$

or   (A2.4)

$$I^C = \left[\frac{gR_0^2}{4\pi^2 h}\right](m + m_0)g\tau^2 - I_0^C.$$

Equation (A2.4) gives the moment of inertia and I^C of the body about the vertical axis through C, the center of mass.

To derive Eq. (A2.2), consider Fig. A2.5.

If the disk rotates through a small angle θ, each string will have an inclination ϕ to the vertical. And

$$R_0\theta \approx h\phi.$$ (a)

During this rotation, the disk will rise an amount

$$\Delta h \approx h(1 - \cos\phi) \approx \tfrac{1}{2}h\phi^2,$$

and thus, by (a),

$$\Delta h \approx \frac{1}{2}\left(\frac{R_0^2}{h}\right)\theta^2.$$ (b)

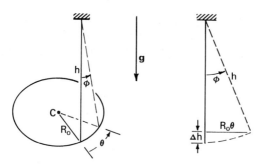

Figure A2.5

The potential energy is then

$$V = m_0 g \, \Delta h \approx \tfrac{1}{2} m_0 \left(\frac{gR_0^2}{h} \right) \theta^2.$$ (c)

The kinetic energy of rotation is

$$T = \tfrac{1}{2} I_0^C \dot{\theta}^2.$$ (d)

(We neglect the kinetic energy due to the vertical motion of the disk.) If

$$\theta = A \sin \omega_n t,$$

and we equate the maximum values of (c) and (d), we obtain

$$\tfrac{1}{2} I_0^C \omega_n^2 A^2 = \tfrac{1}{2} m_0 \left(\frac{gR_0^2}{h} \right) A^2$$

or

$$I_0^C = \frac{m_0 g \tau_0^2}{4\pi^2} \left(\frac{R_0^2}{h} \right),$$

where we have used the fact that $\omega_n \tau_0 = 2\pi$. This establishes (A2.2).

Example A2.2

Suppose the apparatus in Fig. A2.4 is defined by the following parameters: $h = 10.0$ ft, $R_0 = 1.0$ ft, $m_0 g = 8.0$ lb, and $I_0^C = 0.125$ slug-ft^2.

An object which weighs 5.0 lb is placed on the platform. The period τ is then 2.20 sec. From the second form of (A2.4), we have

$$I^C = \left[\frac{(1.0)}{(4)(9.9)(10.0)} \right] (8.0 + 5.0)(4.0) - 0.125 \text{ slug-ft}^2$$

or

$$I^C = 0.034 \text{ slug-ft}^2.$$

Note that if the apparatus were allowed to oscillate without the body on it, the period τ_0 would be $\tau_0 = 2.48$ sec. Thus the apparatus has a longer period unloaded than loaded! This is because the period of the apparatus is determined by (i) the mass and (ii) the inertia of the apparatus, but these have opposite effects, as (A2.2) shows.

....

iii) *Evaluation of the Moment of Inertia from a Table of Inertias*

When a body has a uniform density and a common geometrical shape, the moment of inertia can often be determined from a Table of Inertias. Such a table is included at the end of Appendix A2 for both plane and three-dimensional bodies.

B. MOMENTS AND PRODUCTS OF INERTIA OF THREE-DIMENSIONAL BODIES

i) *Direct Determination of Moments and Products of Inertia*

Consider the rigid body in Fig. A2.6.

The moments and products of inertia about point A with respect to the axes x, y, z are as follows.

$$I_{xx}^A = \int_V (y^2 + z^2)\ dm \qquad I_{xy}^A = I_{yx}^A = \int_V (xy)\ dm$$

$$I_{yy}^A = \int_V (z^2 + x^2)\ dm \qquad I_{yz}^A = I_{zy}^A = \int_V (yz)\ dm \qquad \text{(A2.5)}$$

$$I_{zz}^A = \int_V (x^2 + y^2)\ dm \qquad I_{zx}^A = I_{xz}^A = \int_V (zx)\ dm$$

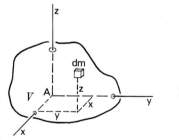

Figure A2.6

If the moments and products of inertia have been determined about some coordinate axes, the moments and products of inertia can be written about any parallel-axis system by means of the *parallel-axis theorem*. (Consider Fig. A2.7.)

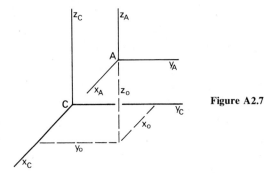

Figure A2.7

Suppose that x_C, y_C, z_C is an axis system through the *center of mass* of a body. Suppose x_A, y_A, z_A is a coordinate system parallel to x_C, y_C, z_C so that

$$x_C = x_A + x_0, \qquad y_C = y_A + y_0, \qquad z_C = z_A + z_0.$$

Then

$$I_{xx}^A = I_{xx}^C + m(y_0^2 + z_0^2)$$

$$I_{yy}^A = I_{yy}^C + m(x_0^2 + z_0^2)$$

$$I_{zz}^A = I_{zz}^C + m(x_0^2 + y_0^2)$$

$$I_{xy}^A = I_{xy}^C + mx_0y_0 \tag{A2.6}$$

$$I_{yz}^A = I_{yz}^C + my_0z_0$$

$$I_{zx}^A = I_{zx}^C + mz_0x_0$$

ii) *Experimental Determination of Moments of Inertia and Principal Axes of a Rigid Body**

If we use the apparatus in Fig. A2.4, we can determine the moment of inertia of a body with respect to the axis through C in the vertical direction. By changing the orientation of the body on the platform, we can determine the moment of inertia about any line through C. The problem is to determine the products of inertia. To accomplish this, we propose the following scheme.

---

* I am indebted to Dr. T. Tielking for pointing out this material.

Consider the axes x, y, z and a, b, c both through the point C in Fig. A2.8.

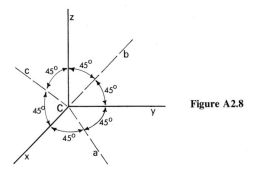

Figure A2.8

We specifically choose a, b, c to be oriented* so that:

a lies in the x, y plane at $45°$ from x,

b lies in the y, z plane at $45°$ from y,

c lies in the z, x plane at $45°$ from z.

The transformation equations for the moments of inertia are

$$I_{aa}^C = \lambda_{ax}^2 I_{xx}^C + \lambda_{ay}^2 I_{yy}^C + \lambda_{az}^2 I_{zz}^C - 2\lambda_{ax}\lambda_{ay} I_{xy}^C - 2\lambda_{ax}\lambda_{az} I_{xz}^C - 2\lambda_{ay}\lambda_{az} I_{yz}^C,$$

and there are two other equations for I_{bb}^C and I_{cc}^C. In these equations, λ_{ax}, . . . , λ_{cz} are the direction cosines. We can evaluate those for our particular case:

$$\lambda_{ax} = \lambda_{ay} = \frac{\sqrt{2}}{2}, \qquad \lambda_{az} = 0,$$

$$\lambda_{by} = \lambda_{bz} = \frac{\sqrt{2}}{2}, \qquad \lambda_{bx} = 0,$$

$$\lambda_{cz} = \lambda_{cx} = \frac{\sqrt{2}}{2}, \qquad \lambda_{cy} = 0.$$

Thus the equations for I_{aa}^C, I_{bb}^C, I_{cc}^C become

$$I_{aa}^C = \tfrac{1}{2}(I_{xx}^C + I_{yy}^C) - I_{xy}^C,$$

$$I_{bb}^C = \tfrac{1}{2}(I_{yy}^C + I_{zz}^C) - I_{yz}^C,$$

$$I_{cc}^C = \tfrac{1}{2}(I_{zz}^C + I_{xx}^C) - I_{zx}^C.$$

---

* The axes a, b, and c do not form an orthogonal set.

Now, if we measure the following quantities,

$$I^C_{xx}, \ I^C_{yy}, \ I^C_{zz}, \qquad I^C_{aa}, \ I^C_{bb}, \ I^C_{cc},$$

then the products of inertia are

$$I^C_{xy} = \tfrac{1}{2}(I^C_{xx} + I^C_{yy}) - I^C_{aa},$$

$$I^C_{yz} = \tfrac{1}{2}(I^C_{yy} + I^C_{zz}) - I^C_{bb}, \qquad\qquad\qquad \text{(A2.7)}$$

$$I^C_{zx} = \tfrac{1}{2}(I^C_{zz} + I^C_{xx}) - I^C_{cc}.$$

From (A2.7) we now know all the moments and products of inertia with respect to one coordinate set x, y, z centered at C. We can use (A2.6) to find the moments and products of inertia in other parallel coordinate frames.

TABLE OF INERTIAS

Body	Centroid	Moment of inertia	Product of inertia
1 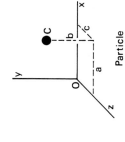 Particle	$x_C = a$ $y_C = b$ $z_C = c$	$I_{xx}^C = 0$ $I_{yy}^C = 0$ $I_{zz}^C = 0$ $I_{xx}^0 = m(b^2 + c^2)$ $I_{yy}^0 = m(a^2 + c^2)$ $I_{zz}^0 = m(a^2 + b^2)$	$I_{xy}^C = 0$ $I_{yz}^C = 0$ $I_{xz}^C = 0$ $I_{xy}^0 = mab$ $I_{yz}^0 = mbc$ $I_{xz}^0 = mac$
2 Thin rod	$x_C = \dfrac{l}{2}$ $y_C = 0$ $z_C = 0$	$I_{xx}^C = 0$ $I_{yy}^C = \frac{1}{12}ml^2$ $I_{zz}^C = \frac{1}{12}ml^2$ $I_{xx}^0 = 0$ $I_{yy}^0 = \frac{1}{3}ml^2$ $I_{zz}^0 = \frac{1}{3}ml^2$	$I_{xy}^C = 0$ $I_{yz}^C = 0$ $I_{xz}^C = 0$ $I_{xy}^0 = 0$ $I_{yz}^0 = 0$ $I_{xz}^0 = 0$

TABLE OF INERTIAS (continued)

Body	Centroid	Moment of inertia	Product of inertia
3 Thin rectangular plate	$x_C = \dfrac{a}{2}$ $y_C = \dfrac{b}{2}$ $z_C = 0$	$I_{xx}^C = \frac{1}{12}mb^2$ $I_{yy}^C = \frac{1}{12}ma^2$ $I_{zz}^C = \frac{1}{12}m(a^2 + b^2)$ $I_{xx}^O = \frac{1}{3}mb^2$ $I_{yy}^O = \frac{1}{3}ma^2$ $I_{zz}^O = \frac{1}{3}m(a^2 + b^2)$	$I_{xy}^C = 0$ $I_{yz}^C = 0$ $I_{xz}^C = 0$ $I_{xy}^O = \frac{1}{4}mab$ $I_{yz}^O = 0$ $I_{xz}^O = 0$
4 Thin arc	$x_C = \dfrac{R\sin\alpha}{\alpha}$ $y_C = 0$ $z_C = 0$	$I_{xx}^C = \dfrac{mR^2}{2}\left[1 - \dfrac{\sin 2\alpha}{2\alpha}\right]$ $I_{yy}^C = \dfrac{mR^2}{2}\left[1 + \dfrac{\sin 2\alpha}{2\alpha} - \dfrac{2\sin^2\alpha}{\alpha^2}\right]$ $I_{zz}^C = mR^2\left[1 - \dfrac{\sin^2\alpha}{\alpha^2}\right]$ $I_{xx}^O = \dfrac{mR^2}{2}\left[1 - \dfrac{\sin 2\alpha}{2\alpha}\right]$ $I_{yy}^O = \dfrac{mR^2}{2}\left[1 + \dfrac{\sin 2\alpha}{2\alpha}\right]$ $I_{zz}^O = mR^2$	$I_{xy}^C = 0$ $I_{yz}^C = 0$ $I_{xz}^C = 0$ $I_{xy}^O = 0$ $I_{yz}^O = 0$ $I_{xz}^O = 0$

5

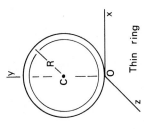

Thin ring

$$I^C_{xx} = \tfrac{1}{2}mR^2 \qquad I^C_{xy} = 0$$
$$I^C_{yy} = \tfrac{1}{2}mR^2 \qquad I^C_{yz} = 0$$
$$I^C_{zz} = mR^2 \qquad I^C_{xz} = 0$$

$$I^O_{xx} = \tfrac{3}{2}mR^2 \qquad I^O_{xy} = 0$$
$$I^O_{yy} = \tfrac{1}{2}mR^2 \qquad I^O_{yz} = 0$$
$$I^O_{zz} = 2mR^2 \qquad I^O_{xz} = 0$$

$$x_C = 0$$
$$y_C = R$$
$$z_C = 0$$

6

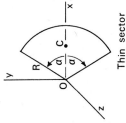

Thin sector

$$I^C_{xx} = \frac{mR^2}{4}\left[1 - \frac{\sin 2\alpha}{2\alpha}\right] \qquad I^C_{xy} = 0$$
$$I^C_{yy} = \frac{mR^2}{4}\left[1 + \frac{\sin 2\alpha}{2\alpha} - \frac{16\sin^2\alpha}{9\alpha^2}\right] \qquad I^C_{yz} = 0$$
$$I^C_{zz} = \frac{mR^2}{2}\left[1 - \frac{8\sin^2\alpha}{9\alpha^2}\right] \qquad I^C_{xz} = 0$$

$$I^O_{xx} = \frac{mR^2}{4}\left[1 - \frac{\sin 2\alpha}{2\alpha}\right] \qquad I^O_{xy} = 0$$
$$I^O_{yy} = \frac{mR^2}{4}\left[1 + \frac{\sin 2\alpha}{2\alpha}\right] \qquad I^O_{yz} = 0$$
$$I^O_{zz} = \frac{mR^2}{2} \qquad I^O_{xz} = 0$$

$$x_C = \frac{2R\sin\alpha}{3\alpha}$$
$$y_C = 0$$
$$z_C = 0$$

TABLE OF INERTIAS (*continued*)

Body	Centroid	Moment of inertia	Product of inertia
7 — Thin disk	$x_C = 0$ $y_C = R$ $z_C = 0$	$I_{xx}^C = \frac{1}{4}mR^2$ $I_{yy}^C = \frac{1}{4}mR^2$ $I_{zz}^C = \frac{1}{2}mR^2$ $I_{xx}^O = \frac{5}{4}mR^2$ $I_{yy}^O = \frac{1}{4}mR^2$ $I_{zz}^O = \frac{3}{2}mR^2$	$I_{xy}^C = 0$ $I_{yz}^C = 0$ $I_{xz}^C = 0$ $I_{xy}^O = 0$ $I_{yz}^O = 0$ $I_{xz}^O = 0$
8 — Thin, hollow disk 	$x_C = 0$ $y_C = R_2$ $z_C = 0$	$I_{xx}^C = \frac{m}{4}\left(R_1^2 + R_2^2\right)$ $I_{yy}^C = \frac{m}{4}\left(R_1^2 + R_2^2\right)$ $I_{zz}^C = \frac{m}{2}\left(R_1^2 + R_2^2\right)$ $I_{xx}^O = \frac{m}{4}\left(R_1^2 + 5R_2^2\right)$ $I_{yy}^O = \frac{m}{4}\left(R_1^2 + R_2^2\right)$ $I_{zz}^O = \frac{m}{2}\left(R_1^2 + 3R_2^2\right)$	$I_{xy}^C = 0$ $I_{yz}^C = 0$ $I_{xz}^C = 0$ $I_{xy}^O = 0$ $I_{yz}^O = 0$ $I_{xz}^O = 0$

9

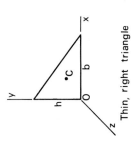

Thin, right triangle

$$x_C = \frac{b}{3}$$

$$y_C = \frac{h}{3}$$

$$z_C = 0$$

$$I_{xx}^C = \frac{m}{18} h^2 \qquad I_{xy}^C = -\frac{m}{36} bh$$

$$I_{yy}^C = \frac{m}{18} b^2 \qquad I_{yz}^C = 0$$

$$I_{zz}^C = \frac{m}{18}(b^2 + h^2) \qquad I_{xz}^C = 0$$

$$I_{xx}^O = \frac{m}{6} h^2 \qquad I_{xy}^O = \frac{m}{12} bh$$

$$I_{yy}^O = \frac{m}{6} b^2 \qquad I_{yz}^O = 0$$

$$I_{zz}^O = \frac{m}{6}(b^2 + h^2) \qquad I_{xz}^O = 0$$

10

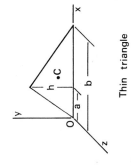

Thin triangle

$$x_C = \frac{a + b}{3}$$

$$y_C = \frac{h}{3}$$

$$z_C = 0$$

$$I_{xx}^C = \frac{m}{18} h^2 \qquad I_{xy}^C = \frac{m}{36} h(2a - b)$$

$$I_{yy}^C = \frac{m}{18}(a^2 - ab + b^2) \qquad I_{yz}^C = 0$$

$$I_{zz}^C = \frac{m}{18}(a^2 - ab + b^2 + h^2) \qquad I_{xz}^C = 0$$

$$I_{xx}^O = \frac{m}{6} h^2 \qquad I_{xy}^O = \frac{m}{12} h(2a + b)$$

$$I_{yy}^O = \frac{m}{6}(a^2 + ab + b^2) \qquad I_{yz}^O = 0$$

$$I_{zz}^O = \frac{m}{6}(a^2 + ab + b^2 + h^2) \qquad I_{zx}^O = 0$$

TABLE OF INERTIAS (continued)

Body	Centroid	Moment of inertia	Product of inertia
11 Rectangular parallelepiped	$x_C = \dfrac{a}{2}$ $y_C = \dfrac{b}{2}$ $z_C = \dfrac{c}{2}$	$I_{xx}^C = \dfrac{m}{12}(b^2 + c^2)$ $I_{yy}^C = \dfrac{m}{12}(a^2 + c^2)$ $I_{zz}^C = \dfrac{m}{12}(a^2 + b^2)$ $I_{xx}^O = \dfrac{m}{3}(b^2 + c^2)$ $I_{yy}^O = \dfrac{m}{3}(a^2 + c^2)$ $I_{zz}^O = \dfrac{m}{3}(a^2 + b^2)$	$I_{xy}^C = 0$ $I_{yz}^C = 0$ $I_{xz}^C = 0$ $I_{xy}^O = \dfrac{m}{4}ab$ $I_{yz}^O = \dfrac{m}{4}bc$ $I_{xz}^O = \dfrac{m}{4}ac$
12 Right, circular cylinder	$x_C = 0$ $y_C = \dfrac{h}{2}$ $z_C = 0$	$I_{xx}^C = \dfrac{m}{12}(3R^2 + h^2)$ $I_{yy}^C = \tfrac{1}{2}mR^2$ $I_{zz}^C = \dfrac{m}{12}(3R^2 + h^2)$ $I_{xx}^O = \dfrac{m}{12}(3R^2 + 4h^2)$ $I_{yy}^O = \tfrac{1}{2}mR^2$ $I_{zz}^O = \dfrac{m}{12}(3R^2 + 4h^2)$	$I_{xy}^C = 0$ $I_{yz}^C = 0$ $I_{xz}^C = 0$ $I_{xy}^O = 0$ $I_{yz}^O = 0$ $I_{xz}^O = 0$

13

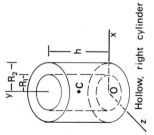

Hollow, right cylinder

$x_C = 0$

$y_C = \dfrac{h}{2}$

$z_C = 0$

$$I_{xx}^C = \frac{m}{12}\left(3R_1^2 + 3R_2^2 + h^2\right) \qquad I_{xy}^C = 0$$

$$I_{yy}^C = \frac{m}{2}\left(R_1^2 + R_2^2\right) \qquad I_{yz}^C = 0$$

$$I_{zz}^C = \frac{m}{12}\left(3R_1^2 + 3R_2^2 + h^2\right) \qquad I_{xz}^C = 0$$

$$I_{xx}^O = \frac{m}{12}\left(3R_1^2 + 3R_2^2 + 4h^2\right) \qquad I_{xy}^O = 0$$

$$I_{yy}^O = \frac{m}{2}\left(R_1^2 + R_2^2\right) \qquad I_{yz}^O = 0$$

$$I_{zz}^O = \frac{m}{12}\left(3R_1^2 + 3R_2^2 + 4h^2\right) \qquad I_{xz}^O = 0$$

14

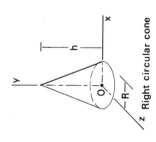

Right circular cone

$x_C = 0$

$y_C = \dfrac{h}{4}$

$z_C = 0$

$$I_{xx}^C = \frac{3m}{80}\left(4R^2 + h^2\right) \qquad I_{xy}^C = 0$$

$$I_{yy}^C = \frac{3m}{10}R^2 \qquad I_{yz}^C = 0$$

$$I_{zz}^C = \frac{3m}{80}\left(4R^2 + h^2\right) \qquad I_{xz}^C = 0$$

$$I_{xx}^O = \frac{m}{20}\left(3R^2 + 2h^2\right) \qquad I_{xy}^O = 0$$

$$I_{yy}^O = \frac{3m}{10}R^2 \qquad I_{yz}^O = 0$$

$$I_{zz}^O = \frac{m}{20}\left(3R^2 + 2h^2\right) \qquad I_{xz}^O = 0$$

TABLE OF INERTIAS (continued)

Body	Centroid	Moment of inertia	Product of inertia
15 Sphere	$x_C = 0$ $y_C = R$ $z_C = 0$	$I_{xx}^C = \frac{2}{5}mR^2$ $I_{yy}^C = \frac{2}{5}mR^2$ $I_{zz}^C = \frac{2}{5}mR^2$ $I_{xx}^O = \frac{7}{5}mR^2$ $I_{yy}^O = \frac{2}{5}mR^2$ $I_{zz}^O = \frac{7}{5}mR^2$	$I_{xy}^C = 0$ $I_{yz}^C = 0$ $I_{xz}^C = 0$ $I_{xy}^O = 0$ $I_{yz}^O = 0$ $I_{xz}^O = 0$
16 Hollow sphere	$x_C = 0$ $y_C = R_2$ $z_C = 0$	$I_{xx}^C = \frac{2m}{5}\frac{R_2^5 - R_1^5}{R_2^3 - R_1^3}$ $I_{yy}^C = \frac{2m}{5}\frac{R_2^5 - R_1^5}{R_2^3 - R_1^3}$ $I_{zz}^C = \frac{2m}{5}\frac{R_2^5 - R_1^5}{R_2^3 - R_1^3}$ $I_{xx}^O = \frac{m}{5}\frac{7R_2^5 - R_1^3(2R_1^2 + 5R_2^2)}{R_2^3 - R_1^3}$ $I_{yy}^O = \frac{2m}{5}\frac{R_2^5 - R_1^5}{R_2^3 - R_1^3}$ $I_{zz}^O = \frac{m}{5}\frac{7R_2^5 - R_1^3(2R_1^2 + 5R_2^2)}{R_2^3 - R_1^3}$	$I_{xy}^C = 0$ $I_{yz}^C = 0$ $I_{xz}^C = 0$ $I_{xy}^O = 0$ $I_{yz}^O = 0$ $I_{xz}^O = 0$

Appendix 3

THE WORK–ENERGY RELATION: RIGID-BODY MOTIONS

The two axioms of rigid-body motion are

i) $\mathbf{F} = \dot{\mathbf{p}}_C$,   (A3.1)

ii) $\mathbf{M}_O = \dot{\mathbf{H}}_O$.   (A3.2)

We have noted in Chapter 8 that (A3.2) can be replaced by an equivalent equation written about the center of mass:

iii) $\mathbf{M}_C = \dot{\mathbf{H}}_C$.   (A3.2′)

Suppose that the velocity of the center of mass is $\mathbf{v}_C$ and the angular velocity is $\boldsymbol{\omega}$. From (A3.1) and (A3.2), we form the following:

$$\mathbf{F} \cdot \mathbf{v}_C = \mathbf{v}_C \cdot \dot{\mathbf{p}}_C \qquad \text{(a)} \qquad\qquad \mathbf{M}_C \cdot \boldsymbol{\omega} = \boldsymbol{\omega} \cdot \dot{\mathbf{H}}_C \qquad \text{(a)}$$

which leads us to define

$$T_C = \int_0^{\mathbf{p}_C} \mathbf{v}_C \cdot d\mathbf{p}_C \qquad \text{(b)} \qquad\qquad T_{R/C} = \int_0^{\mathbf{H}_C} \boldsymbol{\omega} \cdot d\mathbf{H}_C \qquad \text{(b)}$$

the kinetic energy of the center of mass.

the kinetic energy relative to the center of mass.

From the left side of the equation (a), we define

$$P_C = \mathbf{F} \cdot \mathbf{v}_C \qquad \text{(c)} \qquad\qquad P_{R/C} = \mathbf{M}_C \cdot \boldsymbol{\omega} \qquad \text{(c)}$$

the power associated with the center of mass motion.

the power associated with motion relative to the center of mass.

Then from (b) and (c), we have

$$P_C = \frac{dT_C}{dt} \qquad \text{(d)} \qquad \left| \qquad P_{R/C} = \frac{dT_{R/C}}{dt} \right. \qquad \text{(d)}$$

If the total power P is defined as

$$P = P_C + P_{R/C} \tag{e}$$

and the total kinetic energy is defined as

$$T = T_C + T_{R/C}, \tag{f}$$

then from (d), (e), and (f), we have

$$P = \frac{dT}{dt}. \tag{A3.3}$$

The work–energy relation can be obtained from (d) or (A3.3), integrating over time.

$$W_{AB}^C = T_C \Big|_A^B \qquad \text{(g)} \qquad \left| \qquad W_{AB}^{R/C} = T_{R/C} \Big|_A^B \right. \qquad \text{(g)}$$

or

$$W_{AB} = T \Big|_A^B \tag{A3.4}$$

To complete our derivation, we have only to show that P is the power of all the forces acting on the rigid body and that T is the total kinetic energy of the rigid body (i.e., we must show that P and T agree with the expressions we developed for these qualities in Chapter 8.)

First,* consider a rigid body subject to forces $\mathbf{F}_1, \ldots, \mathbf{F}_n$ (any couple acting on the body will be represented by a pair of equal and opposite forces acting on parallel lines). See Fig. A3.1.

If force $\mathbf{F}_i$ is applied to point $\mathbf{r}_i$, the power of $\mathbf{F}_i$ is

$$\mathbf{F}_i \cdot \mathbf{v}_i.$$

The power of all the forces is thus

$$P = \sum_{i=1}^{n} \mathbf{F}_i \cdot \mathbf{v}_i. \tag{a}$$

Let us refer our motion to the center of mass. We have

$$\mathbf{v}_i = \mathbf{v}_C + \boldsymbol{\omega} \times \mathbf{r}_{i/C}. \tag{b}$$

---

* We are repeating here the argument made in Chapter 8, Section 7.

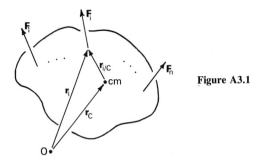

Figure A3.1

Inserting (b) into (a) gives

$$P = \sum_{i=1}^{n} \mathbf{F}_i \cdot \mathbf{v}_C + \sum_{i=1}^{n} [\mathbf{F}_i \cdot (\boldsymbol{\omega} \times \mathbf{r}_{i/C})]. \tag{c}$$

The first term of (c) can be written

$$\left(\sum_{i=1}^{n} \mathbf{F}_i\right) \cdot \mathbf{v}_C = \mathbf{F} \cdot \mathbf{v}_C, \tag{d}$$

where $\mathbf{F}$ is the force resultant acting on the rigid body.

The second term of (c) can be rewritten by (A1.5):

$$\sum_{i=1}^{n} [\boldsymbol{\omega} \cdot (\mathbf{r}_{i/C} \times \mathbf{F}_i)] = \boldsymbol{\omega} \cdot \sum_{i=1}^{n} (\mathbf{r}_{i/C} \times \mathbf{F}_i).$$

But

$$\sum_{i=1}^{n} (\mathbf{r}_{i/C} \times \mathbf{F}_i) = \mathbf{M}_C,$$

the moment resultant about the center of mass. Thus the second term of (c) becomes

$$\mathbf{M}_C \cdot \boldsymbol{\omega}. \tag{e}$$

With (d) and (e), (c) becomes

$$P = \mathbf{F} \cdot \mathbf{v}_C + \mathbf{M}_C \cdot \boldsymbol{\omega}, \tag{A3.5}$$

which is in exact agreement with the power term of (A3.3).

To evaluate the kinetic energy term T, we recall that

$$\mathbf{p}_C = m\mathbf{v}_C \tag{a} \qquad \qquad \mathbf{H}_C = [I^C]\boldsymbol{\omega} \tag{a}$$

where $[I^C]$ is the 3-by-3 matrix representing the inertia tensor.

In order to evaluate

$$T_C = \int_0^{\mathbf{p}_C} \mathbf{v}_C \cdot d\mathbf{p}_C \qquad \qquad T_{R/C} = \int_0^{\mathbf{H}_C} \boldsymbol{\omega} \cdot d\mathbf{H}_C$$

we need to write

$$\mathbf{v}_C = \frac{1}{m}\,\mathbf{p}_C \qquad \qquad \boldsymbol{\omega} = [I^C]^{-1}\mathbf{H}_C$$

where $[I^C]^{-1}$ is the inverse matrix of $[I^C]$.
Then

$$T_C = \int_0 \frac{1}{m}\,(\mathbf{p}_C \cdot d\mathbf{p}_C) \qquad \qquad T_{R/C} = \int_0 [I^C]^{-1}(\mathbf{H}_C \cdot d\mathbf{H}_C)$$

or

$$T_C = \frac{p_C^2}{2m} \qquad \text{(b)} \qquad \qquad T_{R/C} = \tfrac{1}{2}\mathbf{H}_C^t[I^C]^{-1}\mathbf{H}_C \qquad \text{(b)}$$

If we wish to express the kinetic energy in terms of velocity variables, we insert (a) into (b):

$$T_C = \tfrac{1}{2}mv_C^2 \qquad \text{(c)} \qquad \qquad T_{R/C} = \tfrac{1}{2}\boldsymbol{\omega}^t[I^C]\boldsymbol{\omega}. \qquad \text{(c)}$$

Thus the total kinetic energy becomes

$$T = \tfrac{1}{2}mv_C^2 + \tfrac{1}{2}\boldsymbol{\omega}^t[I^C]\boldsymbol{\omega},$$

as expected.

BIBLIOGRAPHY

BIBLIOGRAPHY

The following is a very brief list of some books on mechanics which can be consulted for further reading.

Cabannes, H., *General Mechanics*. Waltham, Mass.: Blaisdell Publishing Co., 1968.

Crandall, S. H., D. C. Karnopp, E. F. Kurtz, and D. C. Pridmore-Brown, *Dynamics of Mechanical and Electromechanical Systems*. New York: McGraw-Hill Book Co., 1968.

Goldstein, H., *Classical Mechanics*. Reading, Mass.: Addison-Wesley Publishing Co., 1959.

Karelitz, G. B., J. Ormondroyd, and J. M. Garrelts, *Problems in Mechanics*. New York: Macmillan Co., 1939.

Lanczos, C., *The Variational Principles of Mechanics*. Toronto: University of Toronto Press, 1970.

Osgood, W. F., *Mechanics*. New York: Dover Publications, Inc., 1965.

Sommerfeld, A., *Mechanics*. New York: Academic Press, Inc., 1952.

Symon, K. R., *Mechanics*. Reading, Mass.: Addison-Wesley Publishing Co., 1960.

Synge, J. L., and B. A. Griffith, *Principles of Mechanics*. New York: McGraw-Hill Book Co., 1959.

ANSWERS
TO
SELECTED
PROBLEMS

ANSWERS TO SELECTED PROBLEMS

CHAPTER 1

1.1. (a) $10\mathbf{i} + 10\mathbf{j}$, $5\mathbf{i} - 5\mathbf{j}$ (b) $5\mathbf{i} + 15\mathbf{j}$ (c) 0 (d) $-2\mathbf{k}$
1.3. (a) 13/15 (b) 13/3 1.5. $8\sqrt{29}$ in^2 = 43.08 in^2
1.11. (a) (i) $T = (ML/F)^{1/2}$, (ii) F/M, (iii) FL
    (b) (i) $L = FT^2/M$, (ii) F/M, (iii) F^2T^2/M
1.13. $[I] = FLT^2 = ML^2$, $[k] = M$
1.15. (i) P, (ii) P/T, (iii) T, (iv) P, (v) $1/T$
1.17. 1065.6 in.-newton, 9760 cm-oz
1.19. (a) 4.515 mi/hr (b) (i) 49.52 sec, (ii) 13.29 min
1.21. 4.453×10^6 gm-cm/sec 1.23. 3.125×10^{-2} oz

CHAPTER 2

2.1. $-18.75\mathbf{i} + 37.5\mathbf{k}$ ft-lb
2.3. (a) $-4\mathbf{i} + 2\mathbf{j} + 5\mathbf{k}$ ft-lb (b) $5\mathbf{i} - 4\mathbf{j} - 4\mathbf{k}$ ft-lb
    (c) $-\mathbf{i} + 2\mathbf{j} - \mathbf{k}$ ft-lb (d) 0
2.5. No 2.7. 5.71°
2.9. (a) 452.5 lb (b) 320 lb (compression)
2.11. (a) $100 \cot \theta$ lb (b) $100 \csc \theta$ lb (c) $14.48° \leqslant \theta \leqslant 90°$
2.13. $7(6/35)^{1/2} = 2.90$ ft
2.15. $T_A = 12mg$, $T_B = 11mg$, $T_C = 9mg$, $T_D = 5mg$
2.17. (a) 125 lb (b) $O_x = 0$, $O_y = 625$ lb
2.19. (a) 62.5 lb (b) $O_x = 0$, $O_y = 62.5$ lb
2.21. (a) 0 (b) 0 (c) 0 2.23. Left pin: $R_y = -3P$
2.25. $R_x = R_y = 0$, $M_O = -4600\mathbf{k}$ ft-lb
2.27. $x_C = 0$, $y_C = \frac{1}{3}$ ft 2.31. $x_C = 0.75$, $y_C = 0.30$

2.35. (a) $\dfrac{L}{3}\left(\dfrac{2\rho_2 + \rho_1}{\rho_2 + \rho_1}\right)$ (b) $\dfrac{L}{2}$
2.37. $x_{cg} = (6.5 \times 10^{-3})a$, $y_{cg} = 0$
2.39. $x_{cg} = y_{cg} = 0$, $z_{cg} = \dfrac{h}{4}\left(\dfrac{6ab - \pi R^2}{3ab - \pi R^2}\right)$
2.41. $-19.28 \leqslant F \leqslant 119.28$ lb 2.43. $-17.32 \leqslant F \leqslant 193.63$ lb

2.51. (a) $\mathbf{F} = -100\mathbf{k}$ and $\mathbf{C} = 0$ at $\mathbf{r}_{OA} = \mathbf{j} + \alpha\mathbf{k}$
   (b) Add at A: $\mathbf{F} = 100\mathbf{k}$, $\mathbf{C} = -400\mathbf{i}$
2.55. (a) 28.9 lb   (b) 33.3 lb   (c) 316.67 lb
2.57. Left end: $N = 0.5W \cot\theta$; right end: $R_x = -0.5W \cot\theta$, $R_y = W$
2.59. Left wall: $\mathbf{N} = 2Q\mathbf{i}$, $\mathbf{f} = 0$; horizontal surface: $\mathbf{N} = W\mathbf{j}$, $\mathbf{f} = -Q\mathbf{i}$
2.61. $\dfrac{\mu(1+\mu)W}{\mu(\mu-1)+(r/R)(1+\mu^2)}$ 2.63. $\mu W\left(\dfrac{R}{r}-1\right)$
2.65. $AB = 141.4$ lb, $BC = 100$ lb, $CD = 0$, $EB = -100$ lb
2.67. (a) 31.25 lb   (b) $A_x = -18.75$ lb, $A_y = 25$ lb, $C_x = 18.75$ lb, $C_y = 25$ lb
2.69. (a) (i) $-250l$ ft-lb, (ii) $-500(l-x)$ ft-lb   (b) (i) 0, (ii) -500 lb
2.71. $\sin^{-1}(4\mu)$
2.75. $F_A = 4\mu mg$ (compression), $F_B = mg(1 + 3.46\mu)$ (tension)
2.76. $R_x = 13.22$ lb, $R_y = -32.20$ lb
2.77. String anchored below and at an angle of 22.32° to the right of vertical. Tension: 34.81 lb.
2.79. $A_x = -13.95$ lb, $A_y = 72.09$ lb
2.83. (a) $A_y = -10^3$ lb, $B_y = 2 \times 10^3$ lb
   (b) $V = 10^3$ lb, $0 \leqslant x < L/2$; $V = -10^3$ lb, $L/2 < x \leqslant L$
   (c) $M = -10^3 x$ ft-lb, $0 \leqslant x \leqslant L/2$; $M = -10^3 (L-x)$ ft-lb, $L/2 \leqslant x \leqslant L$
2.85. $0.596W$
2.87. (a) $0.096W$ (b) 5.19   2.89. (a) $0.77W$ (b) 0.65

CHAPTER 3

3.1. (a) $\mathbf{v} = R\dot\theta(-\sin\theta\mathbf{i} + \cos\theta\mathbf{j})$,
   $\mathbf{a} = R[-(\ddot\theta \sin\theta + \dot\theta^2 \cos\theta)\mathbf{i} + (\ddot\theta \cos\theta - \dot\theta^2 \sin\theta)\mathbf{j}]$
   (b) $\mathbf{v} = R\dot\theta\mathbf{e}_t$, $\mathbf{a} = R\ddot\theta\mathbf{e}_t + R\dot\theta^2\mathbf{e}_n$
3.3. 77.44, 38.72, 25.81 ft/sec²
3.5. $\vec{\mathbf{a}} = \dfrac{2v_0^2}{(1+4x^2)^2}(-2x\mathbf{i} + \mathbf{j})$
3.7. (a) $R = \sec x$ (b) Curve is flat
3.9. (a) 6.25   (b) 8.33   (c) 10.0   (d) 16.67 in/sec²
3.11. (a) $\mathbf{v} = \left[10 + 20 \sin\left(\dfrac{\pi}{6}t\right)\right]\mathbf{i}$, $\mathbf{a} = \left[\dfrac{10\pi}{3}\cos\left(\dfrac{\pi}{6}t\right)\right]\mathbf{i}$
   (b) $t = 7, 11$ sec; $a = -\frac{5\pi}{3}\sqrt{3}, +\frac{5\pi}{3}\sqrt{3}$ ft/sec²   (c) 167.44 ft
3.13. (a) $t = 0, (2v_0/g)$ (b) $y = x - \frac{1}{2}(g/v_0^2)x^2$
   (c) $x = (v_0^2/g)$ at $t = (v_0/g)$
3.15. $8\sqrt{h}$ ft/sec
3.17. (a) $x = R(\theta - \sin\theta)$, $y = R(1 - \cos\theta)$
   (b) (i) $\theta = 0$, (ii) $\theta = \pi$ rad   (c) $R\omega_0^2\mathbf{j}$
3.19. 7.17 ft
3.21. (a) $\mathbf{v} = 88(\mathbf{i} + 176t\mathbf{j})$ ft/sec, $\mathbf{a} = 1.549 \times 10^4\mathbf{j}$ ft/sec²
   (b) $\mathbf{v} = 88(\mathbf{i} + 2x\mathbf{j})$ ft/sec, $\mathbf{a} = 1.549 \times 10^4\mathbf{j}$ ft/sec²
3.23. $\mathbf{v} = \omega_0[\sinh(\omega_0 t)\mathbf{e}_r + \cosh(\omega_0 t)\mathbf{e}_\theta]$, $\mathbf{a} = 2\omega_0^2 \sinh(\omega_0 t)\mathbf{e}_\theta$
3.25. (a) $\mathbf{r} = d[(l^2 + m^2)^{1/2}\mathbf{e}_r + (1 - l^2 - m^2)^{1/2}\mathbf{k}]$ at $\theta = \tan^{-1}(m/l)$
   (b) $\mathbf{v} = v_0[(l^2 + m^2)^{1/2}\mathbf{e}_r + (1 - l^2 - m^2)\mathbf{k}]$

3.27. (a) $r = R(10 - \theta)$ (b) $\mathbf{a} = R\alpha_0(2\theta^2 - 20\theta - 1)\mathbf{e}_r + 5R\alpha_0(2 - \theta)\mathbf{e}_\theta$
3.29. $\dot{\mathbf{a}} = (\dddot{r} - 3\dot{r}\dot{\theta}^2 - 3r\dot{\theta}\ddot{\theta})\mathbf{e}_r + (3\ddot{r}\dot{\theta} + 3\dot{r}\ddot{\theta} + r\dddot{\theta} - r\dot{\theta}^3)\mathbf{e}_\theta + \dddot{z}\mathbf{k}$
3.33. 30.55 ft

CHAPTER 4

4.1. (a) $\mathbf{v} = (1 - \cos t)\mathbf{i} + \sin t\mathbf{j} + 10\mathbf{k}$ ft/sec
   (b) $F_z = 0$ (c) $\mathbf{r} = (t - \sin t)\mathbf{i} + (1 - \cos t)\mathbf{j} + 10t\mathbf{k}$ ft
4.5. 15.96 ft/sec   4.9. $x = (2e^t/\sqrt{4 - e^{2t}})$ ft
4.13. $x = 10 + \frac{1}{2}(\sqrt{1 + 8t} - 1)$ ft
4.15. (a) 12.8 ft/sec^2 (b) 6294 lb
4.17. 31.39°
4.19. (b) $\ddot{\theta} + (g/l) \sin \theta - (a_0/l) \cos \theta = 0$ (c) $\tan^{-1}(a_0/g)$
4.21. (a) $\omega = [\omega_0^2 - (g/l)\theta^2]^{1/2}$ (b) $\theta = \omega_0(l/g)^{1/2}$
4.23. (a) $\cos^{-1}[1 - v_0^2/2gl]$ (b) $\cos^{-1}[1 - v_0^2/2g(l - a)]$
4.25. (a) $H_{Ox} = H_{Oy} = 0$, $H_{Oz} = m(x\dot{y} - y\dot{x})$
   (b) $M_{Ox} = M_{Oy} = 0$, $M_{Oz} = xF_y - yF_x$
   (c) $xF_y - yF_x = m(x\ddot{y} - yx)$
4.27. (a) $H_{Or} = -mrz\dot{\theta}$, $H_{O\theta} = m(z\dot{r} - r\dot{z})$, $H_{Oz} = mr^2\dot{\theta}$
   (b) $M_{Or} = -zF_\theta$, $M_{O\theta} = zF_r - rF_z$, $M_{Oz} = rF_\theta$
   (c) $zF_\theta = mz(2\dot{r}\dot{\theta} + r\ddot{\theta})$, $zF_r - rF_z = m(z\ddot{r} - r\ddot{z} - rz\dot{\theta}^2)$, $rF_\theta = mr(2\dot{r}\dot{\theta} + r\ddot{\theta})$
4.29. Either $\mathbf{r}_{AB} \| \mathbf{a}_A$ or $\mathbf{a}_A = 0$
4.31. (a) $\dot{\theta}^2 = [g/(l^2 - R^2)^{1/2}]$ (b) $\tau^2 = (4\pi^2/g)(l^2 - R^2)^{1/2}$
4.33. (a) 866.03 ft-lb   (b) 984.8 ft-lb
4.35. (a) 2.48×10^5 ft-lb   (b) Same as in (a)
4.37. (1) $5 \ln (17/2) = 10.7$, (2) $5 \ln (32/17) = 3.163$, (3) $10 \ln (4) = 13.862$ ft-lb
4.39. $F_0 R\pi$
4.41. (b) (i) $V(x)$ . . . a minimum, (ii) Bounds: $V(x) = E$
4.45. (a) $F_x = -100(2x + y)$ lb, $F_y = -100(x + 2)$ lb
   (b) $F_x = -400$ lb, $F_y = -300$ lb
   (c) $F = 500$ lb   (d) 216.87° to the $+x$ axis
4.47. (a) $F_x = -(2/x^2y)$, $F_y = -(2/xy^2)$ lb   (b) $F = 2[(x^2 + y^2)^{1/2}/(xy)^2]$ lb
   (c) Inclination of 243.43° to the $+x$ axis
4.51. $v_0 = 25.3$ ft/sec, $v_B = 50.6$ ft/sec
4.53. $\dot{z} = [2gh/(1 + R^2/k^2)]^{1/2}$
4.55. (a) $P = (2/\pi)mg$ (b) 39.54°
4.57. (a) $P = mg \sin \theta$ (b) $P(\theta) = (50 + 16 \sin \theta)$ lb
4.59. (a) $P = mg \left[\dfrac{y + (y^2 + d^2)^{1/2}}{2d - y}\right]\left[\dfrac{d^2 + (2d - y)^2}{d^2 + y^2}\right]^{1/2}$ (b) 22.19 ft/sec
4.60. (a) $v = 8(1 - x^2)^{1/2}$ ft/sec   (b) $\dot{x} = \left(\dfrac{1 - x^2}{1 + 4x^2}\right)^{1/2}$ ft/sec
4.61. (a) $5mg$ (b) $\frac{5}{8}mg$
4.63. (a) 6.08×10^{24} kg   (b) 4.17×10^{23} slug
4.65. $[2\gamma m_e h/(R^2 + Rh)]^{1/2}$
4.67. (a) $[(k/m)(3x - x^2)]^{1/2}$ ft/sec   (b) $x = \frac{3}{2}$ft, $v = \frac{3}{2}\sqrt{k/m}$ ft/sec

4.69. (a) $f = -2A^2 p_0 V_0 \left(\dfrac{x}{V_0^2 - A^2 x^2} \right)$ (b) $V = -p_0 V_0 \ln (V_0^2 - A^2 x^2)$

4.71. (a) $f = p_0 V_0 A \left(\dfrac{1}{V_0 + Ax} \right)$ (b) $V = -p_0 V_0 \ln (V_0 + Ax)$

(c) $\left[\dfrac{2 p_0 V_0}{m} \ln (1 + Al/V_0) \right]^{1/2}$

4.73. 9.26 ft/sec

4.75. (a) $\ddot{\theta} + \mu \dot{\theta}^2 + (g/R)(\sin \theta + \mu \cos \theta) = 0$ (b) $\ddot{\theta} - \mu \dot{\theta}^2 + (g/R)(\sin \theta - \mu \cos \theta) = 0$

4.77. (a) $V = \frac{1}{4} \alpha (x^2 + y^2)^2$ (b) 4.47 ft/sec   (c) 4.47 ft/sec

4.83. (a) $V = \frac{1}{2} k [(r^2 + z^2)^{1/2} - l_0]^2$

(b) $F_r = -k [r - l_0 r/(r^2 + z^2)^{1/2}]$, $F_\theta = 0$, $F_z = -k [z - l_0 z/(r^2 + z^2)^{1/2}]$

4.85. (a) $mg(1 + 2h)$ (b) $mg(1 + 4h)$ (greatest)   (c) mg (least)

4.87. (a) $v = \dfrac{mg}{c_0} (1 - e^{-c_0 t/m})$ (b) $v = \dfrac{mg}{c_0}$

4.89. (a) $\dot{x} = -\left[80 \ln \left(\dfrac{20}{10 + x} \right) \right]^{1/2}$ ft/sec   (b) -7.45 ft/sec

4.91. $\eta = \left(\dfrac{g}{2 v_0^2 \cos^2 \theta} \right) \xi^2$

4.93. (a) $\frac{1}{2}(v_0^2/g) \sin^2 \theta$ (b) $(v_0^2/g) \sin \theta \cos \theta$

4.95. (a) 10.95 ft/sec   (b) 9.03 ft/sec

4.97. (a) $y = -0.213 x^2 + 0.577 x$ ft   (b) 2.71 ft

4.99. (a) 2.6 m/sec   (b) 14.24 m/sec   (c) 97.43°

4.101. $\dfrac{dy}{dx} = \left(\dfrac{1}{v_0 \cos \theta} \right) (v_0^2 \sin^2 \theta - 2gy)^{1/2}$, $y = -\left(\dfrac{g}{v_0^2} \right) x^2 + x$

4.103. (a) $\theta = \tan^{-1} [50.63(64 - 15 x^2)^{-3/2}]$

(b) $\theta_{min} = 5.65°$ at $x = 0$, $\theta_{max} = 81.02°$ at $x = \pm 2$ mi

4.105. 48.19°   4.107. $N = mg(3 \cos \theta - 2)$

4.109. (a) $N = -3 mg(1 - \cos \theta)$ (b) $F = \frac{5}{2} mg (R/x_0)$ (c) $F = \dfrac{m}{t_0} (5 gR)^{1/2}$

4.113. (a) $N_A = 2 mg(a/b)$ (b) $N_B = mg[1 + 4(b/a)^2]$

4.115. $v_A = 5.66$ ft/sec, $v_B = 4.0$ ft/sec, $v_C \to \infty$

4.117. (a) $f = \dfrac{p_0 V_0 A}{V_0 + Ax}$ (b) $V = -p_0 V_0 \ln (V_0 + Ax)$

(c) $p_0 = \dfrac{5 mgR}{2 V_0} \left[\dfrac{1}{\ln (1 + Al/V_0)} \right]$

4.119. $H = R \left(\dfrac{2.37 mg + 3.66 C}{mg - 2C} \right)$

4.121. (a) $\dot{r} = \left[v_0^2 - \left(\dfrac{k}{m} - \omega_0^2 \right) r^2 \right]^{1/2}$ (b) $N = 2 m \omega_0 \left[v_0^2 - \left(\dfrac{k}{m} - \omega_0^2 \right) r^2 \right]^{1/2}$

4.123. (a) $C_1 = \frac{1}{2}(r_0 + v_0/\omega_0)$, $C_2 = \frac{1}{2}(r_0 - v_0/\omega_0)$ (b) $v_0 = -r_0 \omega_0$

CHAPTER 5

5.1. 500 ft²/sec   5.3. (a) 62.83 sec   (b) 15.71 sec

5.5. (b) $r = f[\theta(t)]$ or $r = (C/\dot{\theta})^{1/2}$

5.7. (a) $t(10 + 2.5t)$ rad   (b) 6.47 rev   (c) 127.32 rev

5.9. (a) $\frac{1}{2}\pi\left(1 + \frac{18}{\pi} t\right)^{1/3}$ rad   (b) $\pi\left(1 + \frac{18}{\pi} t\right)^{1/3}$ ft

5.12. (a) $250m$ ft-lb   (b) $1300m$ ft-lb   (c) $1050m$ ft-lb

5.13. (a) -5×10^3 $(1/r^2)$ ft-lb   (b) 1050 ft-lb

5.15. $y^2 - 1.20xy + 480.36x^2 = 100$ in^2

5.17. (a) No   (b) $\mathbf{H}_O = m[-rz\dot{\theta}\mathbf{e}_r + (\dot{r}z - r\dot{z})\mathbf{e}_\theta + r^2\dot{\theta}\mathbf{k}]$, $\mathbf{M}_O = mgr\mathbf{e}_\theta$
(d) No

5.19. E.g., O and m connected by a dashpot

5.21. (a) $\ddot{r} + \left(\frac{k}{m}\right)r - R^4\omega^2\left(\frac{1}{r^3}\right) = 0$ (b) $\dot{r}^2 = \left(\omega^2 R^2 - \frac{k}{m} r^2\right)\left[1 - \left(\frac{R}{r}\right)^2\right]$

5.23. (a) $\ddot{r} - 800\left(\frac{1}{r^3}\right) + 16 = 0$ ft/sec^2 (b) $\dot{r}^2 = 264 - 32r - 800\left(\frac{1}{r^2}\right)$ (ft/sec)^2

5.25. (a) $E = -1.14 \times 10^8$ (mi/hr)^2, $C = 5.66 \times 10^7$ mi^2/hr
(b) $e = 0.733$, $pe = 2550$ mi

5.27. (a) $\ddot{r} + \left(\frac{c_1}{m}\right)\dot{r} - C^2\left(\frac{1}{r^3}\right) = 0$ (b) $\frac{d}{d\theta}\left(\frac{1}{r^2}\frac{dr}{d\theta}\right) + \frac{c_1}{Cm}\frac{dr}{d\theta} - \frac{1}{r} = 0$

5.29. (a) $\frac{1}{2}p$,   (b) p, p

5.31. $v_0 = 2.474 \times 10^4$ mi/hr, $r = [(8.238 \times 10^3)/(1 - \cos\theta)]$ mi

5.35. (a) 1.75×10^4 mi/hr   (b) 1.47 hr

5.37. (a) $e \approx 1$, $ep = 1.02$ mi   (b) Elliptical orbit which is very nearly parabolic

5.39. (a) $r = \dfrac{(r_0v_0)^2/(\gamma m_e)}{1 - [1 - r_0v_0^2/\gamma m_e]\cos\theta}$ (b) $r_1 = r_0$, $r_2 = \dfrac{(r_0v_0)^2}{2\gamma m_e - r_0v_0^2}$
(c) $v_0 = (\gamma m_e/r_0)^{1/2}$

5.41. (a) $E = 0$, $C = (2\gamma m_e R)^{1/2}\sin\alpha$ (b) $e = 1$, $p = 2R\sin^2\alpha$

5.43. (a) 6.92×10^3 mi/hr   (b) 1.67×10^4 mi/hr

5.45. (a) 3.62×10^3 mi/hr   (b) 1.50×10^3 mi/hr   (c) 3.62×10^3 mi/hr

5.46. 3.17×10^{25} mi^3/yr^2 5.47. 0.611 year

5.51. (a) 10 mi^2/min   (b) $\theta = \dfrac{10}{10 - t}$ rad, $r = (10 - t)$ mi, t . . . min

5.53. (a) 10 ft^2/sec   (b) $t = 2.5(1 - e^{-4\theta})$ sec
(c) $\theta = -0.25\ln(1 - 0.4t)$ rad, $r = 10(1 - 0.4t)^{1/2}$ ft

CHAPTER 6

6.1. $\ddot{x} + \dfrac{1}{m}(k_1 + k_2)x = 0$ 6.2. $\ddot{x} + \dfrac{1}{m}\left(\dfrac{k_1k_2}{k_1 + k_2}\right)x = 0$

6.3. $\ddot{x} + \left(\dfrac{c}{m}\right)\dot{x} + \left(\dfrac{k}{m}\right)x = 0$ 6.5. $\ddot{x} + \dfrac{1}{m}\left(\dfrac{c_1c_2}{c_1 + c_2}\right)\dot{x} + \left(\dfrac{k}{m}\right)x = 0$

6.7. $\ddot{\theta} + \dfrac{c}{m}\left(\dfrac{a}{l - a}\right)^2\dot{\theta} + \left[\dfrac{k}{m}\left(\dfrac{a}{l - a}\right)^2 + \dfrac{g}{l - a}\right]\theta = 0$

6.9. $\ddot{x} + \dfrac{1}{m}(c + c_0)\dot{x} + \dfrac{k}{m}x = \dfrac{1}{m}c_0A_0\omega\sin\omega t$

6.11. $\ddot{y} + (\gamma A/m)y = 0$ 6.12. $\ddot{x} + (2p_0A^2/mV_0)x = 0$

6.13. $\ddot{x} + \dfrac{p_0 A V_0^{\gamma}}{m}\left[\dfrac{1}{(V_0 - Ax)^{\gamma}} - \dfrac{1}{(V_0 + Ax)^{\gamma}}\right] = 0$, $\ddot{x} + \left(\dfrac{2\gamma A^2 p_0}{mV_0}\right)x = 0$

6.15. $\ddot{x} + \left(\dfrac{c}{m + m_0}\right)\dot{x} + \left(\dfrac{k}{m + m_0}\right)x = \left(\dfrac{m_0}{m + m_0}\right)\omega^2 A_0 \sin \omega t$

6.16. (a) $\ddot{r} + \left(\dfrac{k}{m} - \omega_0^2\right)r = -g \sin \omega_0 t$

(b) $\ddot{r} + \left(\dfrac{k}{m} - \omega_0^2\right)r = \dfrac{kR}{m} - g \sin \omega_0 t$

6.17. $\ddot{r} + \left(\dfrac{k}{m} - \omega_0^2\right)r = \dfrac{kR}{m} + g \cos \omega_0 t - a_0 \sin \omega_0 t$

6.19. $\ddot{x} + (k/m)x = \mu g$ 6.20. $\ddot{x} + (2F/mh)x = 0$

6.21. (a) $\ddot{x} + \dfrac{2k}{m}\left(1 - \dfrac{a}{h}\right)x = 0$ (b) $a = 0$; $\ddot{x} + \dfrac{2k}{m}x = 0$

6.23. $\left[\dfrac{1}{m}\left(\dfrac{k_1 k_2}{k_1 + k_2}\right)\right]^{1/2}$ rad/sec   6.25. $A\left(\dfrac{2p_0}{mV_0}\right)^{1/2}$ rad/sec

6.27. $\left(\dfrac{2F}{mh}\right)^{1/2}$ rad/sec   6.29. $\dfrac{1}{2}\left(\dfrac{k}{2m}\right)^{1/2}$ rad/sec

6.31. (b) $\frac{1}{2}(v_0/\sqrt{gl}\,)$ rad   (c) $\theta = \frac{1}{2}(v_0/\sqrt{gl}\,)\sin\sqrt{g/l}\,t$

6.35. (a) $\ddot{r} - \omega_0^2 r = 0$ (b) $r = Re^{-\omega_0 t}$

6.37. $f(t) = 2.236 \sin (t + 0.46)$

6.39. $f(t) = -2.236 \sin (\pi t - 0.46)$

6.41. $\omega_n = \left(\dfrac{k}{m} + \dfrac{g}{l}\right)^{1/2}$ rad/sec   6.43. $\omega_n = \dfrac{1}{2}\left(\dfrac{k}{m}\right)^{1/2}$ rad/sec

6.45. $f = 0.5$ cycle/min $= 8.33 \times 10^{-3}$ Hz, $\tau = 2.0$ min $= 120$ sec

6.47. $f = 3.183$ Hz, $\tau = 0.314$ sec

6.49. $\omega = 35$ rad/sec, $f = 5.57$ Hz, $\tau = 0.18$ sec

6.51. (a) $f = 0.159$ Hz, $\tau = 6.28$ sec   (b) $f = 0.477$ Hz, $\tau = 2.09$ sec
(c) $f = 0.159$ Hz, $\tau = 6.28$ sec   (d) $f = 1.59$ Hz, $\tau = 0.628$ sec
(e) $f = 0.477$ Hz, $\tau = 2.09$ sec   (f) Not periodic

6.53. (a) $l = 7.3$ ft   (b) Need the pendulum amplitude   (c) $\omega_n = 2.09$ rad/sec

6.55. (a) $l = (m_s g/k)$, m_p arbitrary

(b) $(\omega_n)_p = \left[\dfrac{k}{m_s}\left(1 + \dfrac{a_0}{g}\right)\right]^{1/2}$, $(\omega_n)_s = \left(\dfrac{k}{m_s}\right)^{1/2}$

6.57. $\omega_n = \left(\dfrac{k}{m} - \omega_0^2\right)^{1/2}$, $F_0 = (a_0^2 + g^2)^{1/2}$, $\omega = \omega_0$

6.59. (a) $r = \dfrac{Rk}{k - m\omega_0^2} - mg\left(\dfrac{1}{k - 2m\omega_0^2}\right)\sin \omega_0 t$ (b) About $r = \dfrac{Rk}{k - m\omega_0^2}$

6.61. (a) $\omega_n = \left(\dfrac{k}{m} - \omega_0^2\right)^{1/2}$, $\omega = \omega_0$, $F_0 = -(a_0^2 + g^2)^{1/2}$

(b) $r = \dfrac{Rk}{k - m\omega_0^2}$ (c) m, k, ω_0 (d) F_0

6.63. $x = (\mu mg/k) + (x_0 - \mu mg/k) \cos \sqrt{k/m}\,t$ (b) $x = (2\mu mg/k) - x_0$

6.65. (a) (i) 1.33, (ii) 2.29, (iii) 1.78   (b) (i) 0.417 ft, (ii) 0.714 ft, (iii) 0.556 ft

6.69. $x = -1.6 \sin 3t + 2.4 \sin 2t$

6.73. (a) $\ddot{y} + (k/m)y = (k/m)y_0 + g$ (b) $\ddot{\eta} + (k/m)\eta = 0$
  (c) $y = y_0 + (mg/k) + (m/k)^{1/2}v_0 \sin \sqrt{k/m}\ t$, $\eta = (m/k)^{1/2}v_0 \sin \sqrt{k/m}\ t$

6.75. $\dfrac{c_1 c_2}{2\sqrt{mk}(c_1 + c_2)}$ 6.77. $\dfrac{c}{2m} \left| \dfrac{\left(\dfrac{a}{1-a}\right)^2}{\left[\dfrac{k}{m}\left(\dfrac{a}{1-a}\right)^2 + \left(\dfrac{g}{1-a}\right)\right]^{1/2}} \right|$

6.79. (a) $a < 2$ (b) $a = 2$ (c) $a > 2$ . . . all lb-sec/ft
6.83. Light damping so that $(\Delta x/x)$ is small compared to 1
6.85. $\zeta = 0.199$, $\omega_n = 1.69$ rad/sec
6.87. $\delta = 0.04$, $\zeta = 6.5 \times 10^{-3}$, $\omega_n = 7.06$ rad/sec
6.91. $x = 3e^{-1.5t} \cos 4.77t$ in.
6.93. $x = 2e^{-10.5t} \cosh 7.83t$ in.
6.95. $x = e^{-4.8t}(0.277 \sin 3.6t + 0.208 \cos 3.6t)$ in.
6.97. $x = (0.5 + 7.5t)e^{-15t}$ ft

6.101. $\omega_n = \left(\dfrac{k}{m}\right)^{1/2}$, $\zeta = \dfrac{c + c_0}{2\sqrt{mk}}$, $F_0 = \dfrac{c_0 A_0 \omega}{m}$

6.103. $\omega_n = \left(\dfrac{k}{m + m_0}\right)^{1/2}$, $\zeta = \dfrac{c}{2\sqrt{k(m + m_0)}}$, $F_0 = \left(\dfrac{m_0}{m + m_0}\right)\omega^2 A_0$

6.105. Amp. = 0.089 ft, phase = $126.96° = 2.22$ rad
6.107. Amp. = 0.08 ft, phase = $90° = 1.57$ rad
6.109. Amp. = 1.66 ft   6.111. Amp. = 0.147 ft
6.113. (a) 0.22   (b) 0.167   (c) 1 at $\omega = 0$
6.115. (a) 0.783   (b) 0.625   (c) 1 at $\omega = 0$
6.117. $x = 0.089 \sin (6t - 2.22) + e^{-3t}(0.071 + 0.534t)$
6.119. $x = 0.042 \sin (3t - 1.47) + e^{-12t}(0.309 \sinh 11.31t + 0.041 \cosh 11.31t)$ in.
6.121. (a) $c = 2$ lb-sec/ft, $\omega = 4$ rad/sec   (b) $x = 0.25[\sin (4t - 1.57) + e^{-4t}]$ ft
6.123. $W = (F^2\pi/\omega_n c)$
6.127. (a) $h(t - t^*) = t - t^*$ (b) $x = (F_0/\omega^2)(\omega t - \sin \omega t)$
6.129. $x = 0.078(t - 0.125 \sin 8t)$

6.131. $x = \dfrac{F_0}{a^2 + \omega_n^2}\left[e^{at} - \left(\dfrac{a}{\omega_n}\right) \sin \omega_n t - \cos \omega_n t\right]$

6.135. $m_0 = (\Delta mr/A_0)$, $k_0 = (\Delta mr\omega^2/A_0)$
6.137. (a) $k_0 = 200$ lb/ft   (b) $a = 2$ ft

CHAPTER 7

7.1. (a) $0.67\mathbf{i} - 0.67\mathbf{j} + 2.50\mathbf{k}$ ft
  (b) $\mathbf{v}_C = 0.67\mathbf{i} + 0.17\mathbf{j} + \mathbf{k}$ ft/sec, $\mathbf{p}_C = 2.0\mathbf{i} + 0.5\mathbf{j} + 3\mathbf{k}$ lb-sec
7.2. (a) $\mathbf{r}_C = 0$ (b) $\mathbf{v}_C = 4\mathbf{i}$ ft/sec
7.3. $\mathbf{r}_C = 4t\mathbf{i}$
7.5. (a) $(\mathbf{i} + \mathbf{j})$ ft/sec   (b) $(t\mathbf{i} + t\mathbf{j} - 16t^2\mathbf{k})$ ft
7.7. (a) $L/2$ (b) $(H^2/L)[\sqrt{1 + (L/H)^2} - 1]$

7.8. $\mathbf{v} = 2v_0(\mathbf{i} - 5\mathbf{j})$ 7.9. $x = \dfrac{mA}{m + M}(\omega t - \sin \omega t)$

7.11. (a) $y = -(mg/2k)$ (b) $y = \dfrac{mg}{2k}(\cos \sqrt{k/m}\, t - 1) - \left(\dfrac{mgh}{2k}\right)^{1/2} \sin \sqrt{k/m}\, t$

7.13. $16\mathbf{k}$ slug-ft^2/sec

7.15. (a) $4\mathbf{i} + 5\mathbf{j}$ lb   (b) (i) $8\mathbf{k}$ ft-lb, (ii) $4\mathbf{k}$ ft-lb

7.17. (a) $-48\mathbf{j}$ lb   (b) (i) $-72\mathbf{k}$ ft-lb, (ii) 0

7.19. (a) $\frac{1}{4}v_0\mathbf{i}$ (b) $v_0/4R$

7.21. (b) $\ddot{\theta} + (g/l)\sin\theta = 0$

7.23. (a) 32 ft-lb   (b) 8 ft-lb

7.27. $T_{\text{before}} = \frac{1}{2}mv_0^2$, $T_{\text{after}} = \frac{1}{2}m(102v_0^2)$

7.31. (a) $2v_0(\mathbf{i} - 5\mathbf{j})$ (b) $101mv_0^2$

7.33. (b) $\dot{x}^2 = \dfrac{(1 + b^2)v_0^2 - 2(kb^2/m)\sin^2 x}{1 + b^2 \cos^2 x}$

　　　(c) $v_0^2 = \dfrac{2kb^2}{m(1 + b^2)}$ (d) No

7.37. (a) 10 rad/sec   (b) $10(2\mathbf{i} + \mathbf{j})$ ft/sec

7.38. (a) $\ddot{y} + (k/m)y = 0$ (b) $\omega_n^2 = (k/m)$

7.39. (a) $\ddot{\theta} + \dfrac{3kl^2}{m(3l^2 + 2a^2)}\theta = 0$ (b) $\omega_n^2 = \dfrac{3kl^2}{m(3l^2 + 2a^2)}$

7.41. (a) $\ddot{y} + \left(\dfrac{2k}{3m}\right)y = 0$ (b) $\omega_n^2 = \dfrac{2k}{3m}$

7.43. (a) $[(2g/l)(1 - \cos\theta)]^{1/2}$ (b) $(2gl)^{1/2}$

7.49. (a) 8.33 ft/sec   (b) 41.67 ft-lb

7.52. $v_1 = v_2 = \frac{1}{2}u$

7.53. $v_1 = 0.147u$, $v_2 = 0.853u$

7.54. (a) $\theta_1 = (\frac{2}{3})^{1/2}\theta_0$ (b) $\theta_2 = (\frac{2}{3})\theta_0$, $\theta_n = (\frac{2}{3})^{(1/2)m}\theta_0$

7.55. $\frac{1}{2}|\theta_1 - \theta_2|$

7.57. (a) $(m/k)^{1/2}u_0$ (b) $[(0.9)(m/k)]^{1/2}u_0$,

　　　(c) $(0.9)(m/k)^{1/2}u_0$, $(0.9)^{(1/2)m}(m/k)^{1/2}u_0$

7.59. $m_2 = 2m$, $v_2 = \frac{1}{2}u_0$

7.61. $e = 0$ 7.63. $e = 0$ 7.65. $e = 0.816$

7.67. (a) $x = v_0 e^{-\omega_n \zeta t}\left(\dfrac{1}{\omega_n \sqrt{1 - \zeta^2}} \sin \omega_n \sqrt{1 - \zeta^2}\, t\right)$

　　　(b) $f = m\omega_n v_0 e^{-\omega_n \zeta t}\left(\dfrac{1 - 2\zeta^2}{\sqrt{1 - \zeta^2}} \sin \omega_n \sqrt{1 - \zeta^2}\, t + 2\zeta \cos \omega_n \sqrt{1 - \zeta^2}\, t\right)$

　　　(c) $v_1 = -v_0 e^{-(2\sqrt{3}\pi/9)} = -(0.298)v_0$

　　　(d) $e = 0.298$ (coefficient of restitution)

7.69. $(9.9)l\,\theta_0$ 7.71. 3.48 ft

7.73. $\mathbf{v}_1 = 5.0\mathbf{i} - 1.44\mathbf{j}$ ft/sec, $\mathbf{v}_2 = 5.05\mathbf{j}$ ft/sec

7.75. 10 lb: $\mathbf{v} = 75\mathbf{i} - 25\mathbf{j}$ ft/sec; 20 lb: $\mathbf{v} = 62.5\mathbf{i} + 37.5\mathbf{j}$ ft/sec

7.77. (a) $c_0 = 0.032$ sec^{-1} (b) $y = \dfrac{v_e}{c_0}(1 - c_0 t)\ln(1 - c_0 t) + v_e t - \frac{1}{2}gt^2$

7.79. (a) $\ddot{y} - (g/l)y = 0$ (b) $y = y_0 \cosh \sqrt{g/l}\, t$

7.81. (a) $v_0 e^{-(cl/v_0 m_0)}$ (b) $m_0[e^{(cl/v_0 m_0)} - 1]$

CHAPTER 8

8.1. (a) $\mathbf{v}_P = 15.20\mathbf{i} + 3.0\mathbf{j}$ ft/sec   (b) $\mathbf{a}_P = -8.40\mathbf{i} + 11.89\mathbf{j}$ ft/sec^2

8.3. (a) $\dot{\theta} = 4$ rad/sec, $\ddot{\theta} = 14$ rad/sec^2 (b) $\mathbf{v}_P = 0$, $\mathbf{a}_P = -80(\mathbf{i} + \mathbf{j})$ ft/sec^2

8.5. (a) $\theta = 1.0$ rad/sec
   (b) $\mathbf{v} = 7.07\mathbf{e}_t = 5(\mathbf{i} + \mathbf{j})$ ft/sec, ($\mathbf{e}_t$ = unit vector along bar).

8.7. (a) $\mathbf{v}_P = \mathbf{i}[v_0(1 - \sin\theta) + (l - v_0 t)\theta\cos\theta] + \mathbf{j}[v_0\cos\theta + (l - v_0 t)\dot{\theta}\sin\theta]$
   (b) $\bar{\mathbf{a}}_p = \bar{\mathbf{i}}\{(l - v_0 t)\ddot{\theta}\cos\theta - 2v_0\dot{\theta}\cos\theta - (l - v_0 t)\dot{\theta}^2\sin\theta\}$
   $+ \bar{\mathbf{j}}\{(l - v_0 t)\ddot{\theta}\sin\theta - 2v_0\dot{\theta}\sin\theta + (l - v_0 t)\dot{\theta}^2\cos\theta\}$.

8.9. (a) $\mathbf{v}_{A/O} = R\dot{\theta}(\cos\theta\,\mathbf{i} + \sin\theta\,\mathbf{j})$, $\mathbf{v}_{P/A} = l\dot{\phi}(\cos\phi\,\mathbf{i} + \sin\phi\,\mathbf{j})$,
   $\mathbf{v}_{P/O} = \mathbf{i}(R\dot{\theta}\cos\theta + l\dot{\phi}\cos\phi) + \mathbf{j}(R\dot{\theta}\sin\theta + l\dot{\phi}\sin\phi)$
   (b) $\mathbf{a}_{A/O} = R[\mathbf{i}(\ddot{\theta}\cos\theta - \dot{\theta}^2\sin\theta) + \mathbf{j}(\ddot{\theta}\sin\theta + \dot{\theta}^2\cos\theta)]$
   $\mathbf{a}_{P/A} = l[\mathbf{i}(\ddot{\phi}\cos\phi - \dot{\phi}^2\sin\phi) + \mathbf{j}(\ddot{\phi}\sin\phi + \dot{\phi}^2\cos\phi)]$
   $\mathbf{a}_{P/O} = \mathbf{i}[R(\ddot{\theta}\cos\theta - \dot{\theta}^2\sin\theta) + l(\ddot{\phi}\cos\phi - \dot{\phi}^2\sin\phi)]$
   $+ \mathbf{j}[R(\ddot{\theta}\sin\theta + \dot{\theta}^2\cos\theta) + l(\ddot{\phi}\sin\phi + \dot{\phi}^2\cos\phi)]$

8.13. (a) $\omega_1 = 2$, $\omega_2 = 0$, $\omega_3 = 4$ rad/sec   (b) 20 ft/sec

8.15. (a) 2.0 rad/sec   (b) 1.15 rad/sec   (c) 4.23 ft/sec

8.17. IC is $\frac{1}{2}l$ above the pin joint at the instant indicated.

8.19. (a) (i) Point A, (ii) 10 ft above lowest point of circular path
   (b) (i) $\omega = 1$ rad/sec, (ii) $\omega = 1$ rad/sec

8.21. (a) 5 ft below the point of contact with cart
   (b) Geometric center of cylinder

8.23. (a) $\mathbf{v}_B = -l\omega_1\mathbf{i}$, $\mathbf{a}_B = -(l\alpha_1\mathbf{i} + l\omega_1^2\mathbf{j})$
   (b) $\mathbf{v}_A = -l(\omega_1 + \omega_2)\mathbf{i}$, $\mathbf{a}_A = -l(\alpha_1 + \alpha_2)\mathbf{i} - l(\omega_1^2 + \omega_2^2)\mathbf{j}$
   (c) $\omega_1 = \omega_2 = 0$, $\alpha_1 = -\alpha_2$

8.25. (b) $\mathbf{a}_{IC} = -\left(\dfrac{R}{r}\right)(R - r)\dot{\phi}^2\mathbf{e}_1$

8.31. (b) $\tan^{-1}(\ddot{\theta}/\dot{\theta}^2)$ 8.37. $t = (v_0/3\mu g)$

8.39. $x = -4.8t^2 + 10.0t$, $t < 1.15$ sec

8.41. (a) $\dfrac{2g\sin\theta}{2 + \left(\dfrac{R}{r}\right)^2}$ (b) $f = \dfrac{mg\sin\theta}{1 + 2\left(\dfrac{r}{R}\right)^2}$

8.43. (a) $\theta \leqslant \tan^{-1}(2\mu)$ (b) $\dfrac{2g}{3}(\sin\theta - 2\mu\cos\theta)$

8.47. $\frac{1}{12}m(a^2 + b^2)$ 8.49. $\dfrac{m[ab(a^2 + b^2) - 18\pi R^4]}{12(ab - \pi R^2)}$

8.51. $\frac{2}{5}mR^2$ 8.53. $m\left(\dfrac{b^2}{20} + \dfrac{3a^2}{7}\right)$ 8.55. 33.33 ft-lb

8.57. 16.67 ft-lb   8.59. 500 ft-lb   8.61. 1160.84 ft-lb

8.63. $\dfrac{43mv_0^2}{6} = 7.17mv_0^2$ 8.67. $\dot{\theta} = \left(\omega_0^2 - \dfrac{3k}{m}\theta^2\right)^{1/2}$ 8.69. $-2\Delta\mu mg\cos\theta$

8.71. (a) $(R - r)[P\sin\phi + Q(1 - \cos\phi)]$ (b) $\phi = \cos^{-1}\left[\dfrac{(mg - Q)^2 - P^2}{(mg - Q)^2 + P^2}\right]$

8.73. $0.637mg$

8.77. (a) $\frac{1}{3}\tan\theta$ (b) (i) mgH, (ii) $mgH(1 - \mu\cot\theta)$

8.78. (a) $t_1 = \dfrac{R\omega_0 + v_0}{3\mu g}$ (b) $v_0 = \frac{1}{2} R\omega_0$

8.79. (a) $t_1 = \dfrac{R\omega_0 - v_0}{3\mu g}$ (b) $T_0 = \frac{1}{2} m(v_0^2 + \frac{1}{2} R^2 \omega_0^2)$, $T_1 = \frac{1}{12} m(2v_0 + R\omega_0)^2$

8.80. (a) $-P\left(\dfrac{rR}{mR^2 + I^C}\right)$ (b) $-P\left(\dfrac{mrR}{mR^2 + I^C}\right)$

8.81. (a) $P_{\min} = \dfrac{\mu mg}{\mu + \left(\dfrac{mrR}{mR^2 + I^C}\right)}$

   (b) $\ddot{x} = -\mu(g - P/m)$

$$\left.\begin{array}{l} \\ \ddot{\theta} = -\dfrac{1}{I^C}\left[P(r + \mu R) - \mu mgR\right] \end{array}\right\} \quad P_{\min} < P < mg$$

8.82. (a) $\ddot{x} = P\left[\dfrac{R(R-r)}{I^C + mR^2}\right]$, $\ddot{x}_1 = P\left[\dfrac{(R-r)^2}{I^C + mR^2}\right]$

8.83. (a) $f = P\left[1 - \dfrac{mR(R-r)}{I^C + mR^2}\right]$ (b) $P_{\min} = \dfrac{\mu mg(I^C + mR^2)}{I^C + mrR}$

   (c) $\ddot{x} = P/m - \mu g$, $\ddot{\theta} = \dfrac{1}{I^C}(\mu mgR - Pr)$ where $P > P_{\min}$

8.85. (a) $A_x = 0$, $A_y = \frac{1}{2} mg$ (b) $A_x = 0$, $A_y = \frac{1}{4} mg$

8.87. (a) $T_A = \frac{1}{2} mg$ (b) $T_A = \frac{1}{4} mg$

8.89. (a) $\dot{\theta}^2 = 3\left[\dfrac{g}{l}(1 - \cos\theta) - \dfrac{k}{m}\sin^2\theta\right]$

   (b) $\theta = 0$ (unstable), $\theta = \cos^{-1}(mg/2kl)$ (stable)

8.91. (a) $v = [g(R - r)]^{1/2}$ (b) Same as in part (a)

8.92. $v = [\frac{11}{3} g(R - r)]^{1/2}$

8.93. Unless $\mu = \infty$, the analysis is not strictly valid.

8.95. $\left[\dfrac{g}{b}\left(a^2 + \dfrac{8}{3} b^2\right)\right]^{1/2}$

8.97. (a) $\dot{\theta} = \left(\dfrac{I^0 + ma^2}{I^0 + mr^2}\right)\omega_0$ (b) $\dot{r}^2 = \left(\dfrac{I^0 + ma^2}{I^0 + mr^2}\right)(r^2 - a^2)\omega_0^2$

8.99. $\dfrac{2}{R}\left(\dfrac{m_0 v_0 \cos\alpha}{m + 2m_0}\right)$

8.101. (a) $\dot{x} = 30.98 \cos\theta\left[\dfrac{1 - \cos\theta}{1 + 12\cos^2\theta}\right]^{1/2}$ ft/sec

   (b) $F = 43.82 \sin\theta\left(\dfrac{1 + 12\cos^2\theta}{1 - \cos\theta} + 24\dfrac{\sqrt{1 - \cos\theta}}{(1 + 12\cos^2\theta)^{3/2}}\cos\theta\right)$ 16.

8.103. $l = I^0/mL$ 8.107. $h = \frac{7}{5} R$

8.109. $\dot{\theta}^2 = \dfrac{3g}{2l}(3\sqrt{2} - 4\cos\theta - 2\sin\theta)$

8.111. (a) $\omega_n = \left(\dfrac{3g}{2l}\right)^{1/2}$ (b) $l_1 = \frac{2}{3} l$

8.113. $\omega_n = \left(\dfrac{g}{2R}\right)^{1/2}$ 8.115. $\omega_n = \left(\dfrac{8k}{3m}\right)^{1/2}$

8.121. $\omega_n = \left[\dfrac{3g}{l} \left(\dfrac{m_1 + 2m_2}{2m_1 + 9m_2} \right) \right]^{1/2}$

8.123. (a) $x = x_0 + \dfrac{2A_0}{3} (\omega t - \sin \omega t)$ (b) $\omega \geqslant \left(\dfrac{3\mu g}{A_0} \right)^{1/2}$

8.125. (a) $\ddot{\theta}(\tfrac{7}{2} - 2 \cos \theta) + \dot{\theta}^2 \sin \theta + (g/R) \sin \theta = 0$ (b) $\omega_n = \left(\dfrac{2g}{3R} \right)^{1/2}$

INDEX

INDEX

Acceleration, 5, 11, 93, 95, 100, 108, 113, 442
  angular, 6, 11
  of gravity, 150
Apocentron, 238, 239
Apogee, 238, 239

Brahe, Tycho, 221

Cauchy, 317
Center of gravity, 41, 354
Center of mass, 126, 353, 354, 360
Center of oscillation, 518, 519
Center of percussion, 514, 519
Central force motion, 177, 220
Centroid, 42
Coefficient of sliding friction, 46, 152
Coefficient of static friction, 46
Coefficient of restitution, 386
Cohn-Vossen, 236
Collision problems, 351, 380
  central, 383, 393
  direct, 382, 383
  elastic, 387, 393
  oblique, 383, 393
  plastic, 387, 392
Complex numbers, 535
Composite bodies, 464
Conic sections, 220, 233, 236, 237, 241, 536
Convolution, 313

Coordinate system, 2
  Cartesian, 99, 113, 527, 530, 532, 534
  cylindrical, 106, 113, 530, 532, 534
  natural, 93, 113
  spherical, 111, 113, 531, 532, 534
Couple, 32
Cramer's rule, 541
Crandall, S. H., 275, 279
Curl, 534
Cycle, 280

D'Alembert's principle, 179
Damping, critically damped, 292, 296
  overdamped, 292, 296
  underdamped, 292, 296
Damping ratio, 291, 302
Dashpot, 153
Deformation, 20
Differential geometry, 97
Directional derivative, 529
Directrix, 233, 236
Dimensions, 4
  absolute, 5, 6
  gravitational, 5, 6
Dimensional analysis, 4

Eccentricity, 233, 236, 237, 238
Electrical transformer, 31
Energy, 6, 11
  kinetic, 134, 136, 368, 369, 371, 466, 559

potential, 141, 150, 152
   total mechanical, 143
Equilibrium, particle, 20
   rigid-body, 20, 25, 34
Euler–Savary equation, 445
Euler's theorem, 270, 535

First integral, 144
Focus, 233, 236
Force, 2, 5, 11, 18, 43, 148
   conservative, 139, 141, 153, 533
   constraint, 43, 46, 148
   external, 352
   friction, sliding, 46, 152
   friction, static, 45
   gravity, 43, 144, 149
   internal, 352
   nonconservative, 143
   restraint, built-in, 44, 45
   restraint, pin-joint, 43, 45
   restraint, socket, 44
   resultant, 21, 23
   static, 43
Force system, equivalent, 35
Fourier series, 537
Free-body diagram, 18, 48, 177
Frenet–Serret formulas, 97
Function, harmonic, 279, 280
   hyperbolic, 536
   impulse (Dirac delta), 309
   periodic, 279, 293
Frequency, 6, 11, 280
   natural, 269, 273, 274, 283, 291

Gradient, 530
Gravitational field (*see also* Force,
   gravity), 231

Hilbert, D., 236
Hornung, P., 164
Hydraulic transformer, 31

Impulse, 6, 11, 513
   angular, 6, 11, 513
   Dirac delta, 309
   response, 309, 310, 311

Instant center, acceleration, 443
   velocity, 106, 436, 467

Kepler, J., 221, 243
Kimball, R., 210
Kinematics, Chap. 3, 431
   plane motion, 430
Kinetics, 92–114, 351, 445

Laplace transform, 313
Lighthill, M. J., 309
Linearity, 268, 284
Line integral, 531
Logarithmic decrement, 294

Magnification factor, 284, 302
Mass, 2, 5, 11
Matrix, 537
   inverse, 539
   transpose, 539
McLachlan, N. W., 263
Moment, 6, 11, 21, 125
   resultant, 23
   transformation rule, 27, 35, 450
Moment of inertia, 6, 11, 447, 454
      457, 542, 547, 551
Moment of momentum, 6, 11, 125,
      129, 359, 365, 446
   transformation rule, 365, 450
Moment–moment of momentum
      relation, 125, 129, 358, 362,
      446, 450
Momentum, 3, 6, 11, 126, 354

Newton's laws, 2, 125, 221

Parallel axis theorem, 460, 462, 548
Pars, 279
Particle, 2
Pericentron, 238, 239
Perigee, 238, 239
Period, 6, 11, 244, 280
Phase angle, 272, 284, 303
Plane motion, 430
Position, 93
Power, 133, 559

Power-preserving elements, 30
Product of inertia, 447, 454, 547
Ptolemy, 3

Radius of curvature, 96, 536
Rayleigh's method, 275
Reference frame, 2
 inertial, 2, 20, 92
Relative motion, 431
Relativity, 126
Resonance, 302
Rigid body, 25, 351, 366, 367, 375, 430

Scalar, 523
Semimajor axis, 238, 239
Semiminor axis, 238, 239
Speed, terminal, 161
Springs, 150
Static deflection, 284, 301, 337
Statics, 18
String, 47

Taylor series, 537
Tielking, T., 548
Translation, 438
Trigonometry, 534

Unit system, 4, 7, 11
 British absolute, 10
 British gravitational, 8
 metric absolute, 8
 metric gravitational, 9
Units, conversion table, 10
Universal gravitational constant, 149

Variable mass, 127, 351, 396
Vector, 523
 calculus, 529
 cross product, 525, 528
 dot product, 525, 528
 scalar triple product, 526, 529
 vector triple product, 526
Velocity, 5, 11, 93, 94, 100, 107, 113
 angular, 6, 11, 435
 areal, 6, 11, 220, 222, 244
 escape, 241, 242
Vibration, 263, 477
 absorber, 314, 316, 348
 isolation, 313
Vibratory motion, complete solution,
 320
 free motion–damped system, 264,
 291, 319
 free motion–undamped system, 264,
 268, 284, 319
 forced motion–arbitrary force, 264,
 309, 321
 forced motion–damped system,
 264, 299, 320
 forced motion–undamped system,
 283, 284, 319

Weightless bar, 47
Work, 6, 11, 133, 134, 369, 370, 377,
 467
Work–energy relation, 133, 135, 366,
 376, 465, 468, 559
Wrench, 36, 38